The Philosophy of Information Quality

SYNTHESE LIBRARY

STUDIES IN EPISTEMOLOGY, LOGIC, METHODOLOGY, AND PHILOSOPHY OF SCIENCE

VOLUME 358

For further volumes:
http://www.springer.com/series/6607

Luciano Floridi • Phyllis Illari
Editors

The Philosophy of Information Quality

Springer

Editors
Luciano Floridi
Oxford Internet Institute
University of Oxford
Oxford, Oxfordshire, UK

Phyllis Illari
Department of Science and Technology Studies
University College London
London, UK

ISBN 978-3-319-07120-6 ISBN 978-3-319-07121-3 (eBook)
DOI 10.1007/978-3-319-07121-3
Springer Cham Heidelberg New York Dordrecht London

Library of Congress Control Number: 2014946731

Printed on acid-free paper

Springer is part of Springer Science+Business Media (www.springer.com)

Acknowledgements

This volume is the outcome of a project funded by the Arts and Humanities Research Council, under the title "Information Quality Standards and their Challenges", during the academic years 2011–2013. We are extremely grateful to the AHRC for its generous funding, without which our research would have been impossible.

Many people played a significant role throughout the project that led to this volume. Naomi Gummer, Sarah Hunter, and Marco Pancini, all at Google, were pivotal in shaping the earlier stage of the project. Their support was essential and one of us (Luciano) imposed on their kind availability more than once. He knows his moral debt cannot be repaid. During the development of the project, we relied on the extensive help and support of many colleagues, in particular Carlo Batini, Matteo Palmonari and Gianluigi Viscusi, Università di Milano Bicocca; of Andrew Bass, Christian Benninkmeijer, Ian Dunlop, Suzanne Embury, Matthew Gamble, Carole Goble, Pedro Mendes and Sandra Sampaio, Manchester University; of Meredith Nahm, Duke University; and of Monica Scannapieco, Italian national institute of statistics. Particular thanks are due to Suzanne Embury who set up the meetings for one of us (Phyllis) in Manchester and, along with Sandra Sampaio, answered many questions. Throughout the project, Penny Driscoll provided exceptional managerial support. We are also grateful to speakers and participants in the symposium on information quality, organised as part of the AISB/IACAP World Congress 2012, July 2–6, Birmingham; and in the workshop on information quality, organised at the University of Hertfordshire, December 14 2012. A final thanks goes to all the authors of the chapters in this volume, for their time and scholarship, and for their commitment to a rather demanding research agenda.

Contents

Chapter 1
Introduction

Luciano Floridi and Phyllis Illari

Abstract This chapter describes the project that generated this book.

The most developed post-industrial societies live by information, and Information and Communication Technologies (ICTs) keep them oxygenated. So the better the quality of the information exchanged the more likely it is that such societies will prosper. But what is information quality (IQ) exactly? This is a crucial and pressing question but, so far, our answers have been less than satisfactory. Here are two examples.

In the United States, the *Information Quality Act*, also known as the Data Quality Act, enacted in 2000 (http://www.whitehouse.gov/omb/fedreg_reproducible), left undefined virtually every key concept in the text. So it required the Office of Management and Budget "to promulgate guidance to agencies ensuring the quality, objectivity, utility, and integrity of information (including statistical information) disseminated by Federal agencies" (Congressional Report Service 2004). Unsurprisingly, the guidelines have received much criticism and have been under review ever since (United States Government Accountability Office 2006).

In the UK, some of the most sustained efforts in dealing with IQ issues have concerned the health care system. Already in 2001, the *Kennedy Report* (http://goo.gl/uhFRgk) acknowledged that: "The assessment of the performance of clinicians and information for the benefit of patients depend on the collection, analysis and dissemination of data". However, in 2004, the NHS Information Quality Assurance

L. Floridi
Oxford Internet Institute, University of Oxford,
1 St Giles, Oxford OX1 3JS, UK
e-mail: luciano.floridi@oii.ox.ac.uk

P. Illari (✉)
Department of Science and Technology Studies,
University College London, London, UK
e-mail: phyllis.illari@ucl.ac.uk

L. Floridi and P. Illari (eds.), *The Philosophy of Information Quality*,
Synthese Library 358, DOI 10.1007/978-3-319-07121-3_1,

Consultation (http://tinyurl.com/mm6qbxh) still stressed that "Consideration of information and data quality are made more complex by the general agreement that there are a number of different aspects to information/data quality but no clear agreement as to what these are".

Lacking a clear and precise understanding of IQ standards (such as accessibility, accuracy, availability, completeness, currency, integrity, redundancy, reliability, timeliness, trustworthiness, usability, and so forth) causes costly errors, confusion, impasse and missed opportunities. Part of the difficulty lies in putting together the right conceptual and technical framework necessary to analyse and evaluate IQ. Some steps have been taken to rectify the situation. The first *International Conference on Information Quality* (http://mitiq.mit.edu/ICIQ/2013/) was organised in 1996. In 2006, the Association of Computing Machinery launched the new *Journal of Data and Information Quality* (http://jdiq.acm.org/) The *Data Quality Summit* (http://www.dataqualitysummit.com/) now provides an international forum for the study of information quality strategies. Pioneering investigations (including Wang and Kon (1992), Tozer (1994), Redman (1996), Huang et al. (1999), Gasser (2004), Wang et al. (2005), Al-Hakim (2007), Lee et al. (2006)) and research programs (see the Information Quality Program at MIT, http://mitiq.mit.edu/) have addressed applied issues, plausible scenarios and the codification of best practices. So there is a wealth of available results that could make a difference. However, such results have had limited impact because research concerning IQ has failed to combine and cross-fertilise theory and practice. Furthermore, insufficient work has been done to promote the value-adding synthesis of academic findings and technological know-how. Within this context, the research project "Understanding Information Quality Standards and their Challenges" – funded by the Arts and Humanities Research Council in the UK during the academic years 2011–12 and 2012–13 – sought to bridge the gap between theoretically sound and technologically feasible studies. A primary goal was to apply current investigations in the philosophy of information to the analysis and evaluation of information quality, by combining them with the technical expertise in information management offered by Google UK, a partner in the project.

Initially, the aim was to deliver a white paper on IQ standards that would synthesise conceptual and technological expertise on the challenges posed by the analysis of IQ, address the critical issues that affect the life-cycle (from creation to use) of high IQ, and enable us to share the conceptual and technical understanding of the new challenges posed by ICTs with respect to the identification and evaluation of high IQ resources, in view of creating a conducive environment for exchanges of results on IQ standards. The outcome exceeded our initial expectations. The project succeeded in bridging different communities and cultures that, so far, had failed to interact successfully, leading to a sharing of knowledge that may have a significant and lasting impact on IQ standards. Despite the potential challenges represented by different methodologies, technical vocabularies and working cultures, it soon became clear that more could be achieved, thanks to the fruitful collaboration of many experts, who generously contributed their knowledge, time, and research. A series of successful and fruitful meetings led to a more ambitious outcome, namely this book.

The contents of the specific contributions are discussed in the first chapter, where we provide an overview of the field of IQ studies and of how the remaining chapters fit within it.

Here, we would like to highlight the fact that we hope the completion of this project will contribute to enhancing Google's capacity to deliver information that measures up to the high standards expected by its users. The impact of the project should foster both good practices in IT management and better design principles favouring IQ among internet service providers. The project should also benefit policy-makers seeking to improve IQ standards and the delivery and effectiveness of public information services. We shall be delighted if the project raises users' awareness on the issues affecting the value and reliability of online information, and encourage a more responsible production, usage, and sharing of digital resources. We need to increase the number of people using and trusting information online, and hence enjoying its cultural and social benefits. For this purpose, IQ can deliver huge benefits to the quality of life and health, insofar as they both depend on improvements in the management of IQ issues. Thus the project may help to empower users, by offering them the framework necessary to evaluate the quality of the information they access. Finally, we offer the results of this project to all those researchers working on availability, accessibility and accountability of information; on the philosophy of information; on information management; and on data quality auditing systems. We hope that the delivered analysis will further improve our understanding of what features contribute to improve the quality, and hence usability, of information.

References

Al-Hakim, L. (2007). *Information quality management: Theory and applications*. Hershey: Idea Group Pub.

Congressional Report Service. (2004). *Report for congress, the information quality act: Omb's guidance and initial implementation*. http://tinyurl.Com/nbbtj47

Gasser, U. (Ed.). (2004). *Information quality regulation: Foundations, perspectives, and applications*. Baden-Baden/Zèurich: Nomos/Schulthess.

Huang, K.-T., Lee, Y. W., & Wang, Y. R. (1999). *Quality information and knowledge*. London: Prentice-Hall.

Lee, Y. W., Pipino, L. L., Funk, J. D., & Wang, R. Y. (2006). *Journey to data quality*. Cambridge, MA: MIT Press.

Redman, T. C. (1996). *Data quality for the information age*. Boston: Artech House.

Tozer, G. V. (1994). *Information quality management*. Oxford: Blackwell.

United States Government Accountability Office. (2006). *Information quality act: Expanded oversight and clearer guidance by the office of management and budget could improve agencies' implementation of act: Report to congressional requesters*. Washington, DC: U.S. Govt. Accountability Office.

Wang, Y. R., & Kon, H. B. (1992). *Toward quality data: An attributes-based approach to data quality*. Cambridge, MA: MIT Press.

Wang, R. Y., Pierce, E. M., Madnik, S. E., Zwass, V., & Fisher, C. W. (Eds.). (2005). *Information quality*. Armonk/London: M.E. Sharpe.

Chapter 2
Information Quality, Data and Philosophy

Phyllis Illari and Luciano Floridi

Abstract This chapter introduces the volume.

2.1 Understanding Information Quality

In this opening chapter, we review the literature on information quality (IQ). Our major aim is to introduce the issues, and trace some of the history of the debates, with a view to situating the chapters in this volume – whose authors come from different disciplines – to help make them accessible to readers with different backgrounds and expertise. We begin in this section by tracing some influential analyses of IQ in computer science. This is a useful basis for examining some examples of developing work on IQ in Sect. 2.2. We look at some cases for applying IQ in Sect. 2.3, and conclude with some discussion points in Sect. 2.4.

2.1.1 The MIT Group

The issue of IQ came to prominence in computer science, where a research group at MIT launched and defined the field of IQ in the 1990s. The MIT group was committed to the idea that considerably more could be done about IQ problems. Members

P. Illari (✉)
Department of Science and Technology Studies,
University College London, London, UK
e-mail: phyllis.illari@ucl.ac.uk

L. Floridi
Oxford Internet Institute, University of Oxford,
1 St Giles, Oxford OX1 3JS, UK
e-mail: luciano.floridi@oii.ox.ac.uk

L. Floridi and P. Illari (eds.), *The Philosophy of Information Quality*,
Synthese Library 358, DOI 10.1007/978-3-319-07121-3_2,

Table 2.1 Wang's categorisation (Source: Wang 1998)

IQ category	IQ dimensions
Intrinsic IQ	Accuracy, objectivity, believability, reputation
Accessibility IQ	Access, security
Contextual IQ	Relevancy, value-added, timeliness, completeness, amount of data
Representational IQ	Interpretability, ease of understanding, concise representation, consistent representation

of the group were enormously influential, and generated a large and thriving community: the 18th annual IQ conference was held in 2013 in Arkansas, USA.

The consistent primary message of the MIT group is that quality information is information that is fit for purpose, going far beyond mere accuracy of information. This message bears repeating, as mere accuracy measures are still sometimes informally described as IQ measures. Since the MIT group conceives of IQ projects initially as data management for business, it presses this message as a recommendation to consider 'information consumers': constantly ask what it is that consumers of information need from it, treating data as a valuable and important product, even if the consumers of that product are internal to the organization.

The idea that IQ is a *multidimensional* concept, with accuracy as only one dimension, is now embedded. Much of the MIT group's early work aimed to identify and categorise the dimensions of IQ. This work falls into two different methodological approaches, also identified by Batini and Scannapieco (2006, p. 38).

The first methodological approach is called 'empirical' by Batini and Scannapieco, and by members of the MIT group. Broadly speaking, it consists in surveying IQ professionals, both academics and practitioners, on what they rate as important IQ dimensions, and how they classify them. In the past, some work also examined published papers on IQ, and surveyed professional practitioners at conferences, to identify important IQ skills. The empirical approach is based on initial work by Wand and Wang in 1996, making a citation count, actually given in Wang (1998). In line with the focus on information users, data consumers were also interviewed (Batini and Scannapieco 2006, p. 38).

The categorisation of Wang (1998) is one of the earliest and still most influential categorisations of IQ dimensions. Table 2.1 is a reconstruction of the table given in the paper (Wang 1998, p. 60).

Another important paper is Lee et al. (2002), who give us two comparison tables of classifications of IQ dimensions, one for academics reconstructed in Table 2.2 (Lee et al. 2002, p. 134), laid out according to the Wang (1998) categories, and one for practitioners (Lee et al. 2002, p. 136).

The main difference is that academic approaches try to cover all aspects of IQ, where practitioners focus on particular problems of their context. This separation between academic approaches and practice is interesting, because the MIT group are academics, yet they run the practice-oriented Total Data Quality Management program, which we will discuss shortly.

Note that the aforementioned papers, and others in the tradition, generally do not *define* IQ dimensions, such as objectivity, timeliness, and so on. They primarily

Table 2.2 Classification for academics (Source: Lee et al. 2002, p. 134)

	Intrinsic IQ	Contextual IQ	Representational IQ	Accessibility IQ
Wang and Strong	Accuracy, believability, reputation, objectivity	Value-added, relevance, completeness, timeliness, appropriate amount	Understandability, interpretability, concise representation, consistent representation	Accessibility, ease of operations, security
Zmud	Accurate, factual	Quantity, reliable/ timely	Arrangement, readable, reasonable	
Jarke and Vassiliou	Believability, accuracy, credibility, consistency, completeness	Relevance, usage, timeliness, source currency, data warehouse currency, non-volatility	Interpretability, syntax, version control, semantics, aliases, origin	Accessibility, system availability, transaction availability, privileges
Delone and McLean	Accuracy, precision, reliability, freedom from bias	Importance, relevance, usefulness, informativeness, content, sufficiency, completeness, currency, timeliness	Understandability, readability, clarity, format, appearance, conciseness, uniqueness, comparability	Usableness, quantitativeness, convenience of access[a]
Goodhue	Accuracy, reliability	Currency, level of detail	Compatibility, meaning, presentation, lack of confusion	Accessibility, assistance, ease of use (of h/w, s/w), locatability
Ballou and Pazer	Accuracy, consistency	Completeness, timeliness		
Wand and Wang	Correctness, unambiguous	Completeness	Meaningfulness	

[a]Classified as system quality rather than information quality by Delone and McLean.

categorise them. In referring back to the Wang (1998) paper, members of the MIT group talk of having 'empirically derived' quality dimensions. However, note that they generally aim merely to ask practitioners, academics, and information consumers what they take good quality information to be. These surveys certainly make the initial point: information consumers need more than merely accurate information. Yet this point has been made effectively now, and further surveys might best be used to examine more novel aspects of IQ practice. A natural question arises as to what methodology should be used next to help understand IQ in general.

A second methodological approach is adopted in Wand and Wang (1996). The 1996 paper is referred to, but less than other early papers such as Wang (1998). In

Table 2.3 The 'ontological' approach to IQ (Source: Wand and Wang 1996, p. 94)

DQ dimension	Mapping problem	Observed data problem
Complete	Certain real world (RW) states cannot be represented	Loss of information about the application domain
Unambiguous	A certain information system (IS) state can be mapped back into several RW states	Insufficient information: the data can be interpreted in more than one way
Meaningful	It is not possible to map the IS state back to a meaningful RW state	It is not possible to interpret the data in a meaningful way
Correct	The IS state may be mapped back into a meaningful state, but the wrong one	The data derived from the IS do not conform to those used to create these data

the paper itself, Wand and Wang refer to it as an 'ontological' approach. Batini and Scannapieco (2006) call it a 'theoretical' approach. We adopt the earlier terminology.

There are various summaries in the paper, but our point is best illustrated by Table 2.3, reconstructed from Wand and Wang (1996, p. 94).

Wand and Wang are attempting to understand how IQ errors can be generated. They may also be interested in relations *between* dimensions that surveys may miss. Batini and Scannapieco comment on this paper:

> All deviations from proper representations generate deficiencies. They distinguish between design deficiencies and operation deficiencies. Design deficiencies are of three types: incomplete representation, ambiguous representation, and meaningless states. … Only one type of operation deficiency is identified, in which a state in RW might be mapped to a wrong state in an IS; this is referred to as garbling. (Batini and Scannapieco 2006, p. 36)

Ultimately,

> A set of data quality dimensions are defined by making references to described deficiencies. (Batini and Scannapieco 2006, p. 37)

These dimensions are: complete, unambiguous, meaningful and correct.

Methodologically, the paper is laid out analogously to a mathematical proof, with conclusions apparently derived from axioms or assumptions. In the end, of course, such material can only be analogous to a mathematical proof, and the source of assumptions and the derivations from them are not always clear. Nevertheless, the conclusions are interesting, and it is perhaps better to interpret them as the suggestion of highly experienced academics, who have been thinking about IQ and practising IQ improvement for some time. Then, the test of such conclusions would seem to be whether or not they enhance IQ practice.

Overall, the IQ literature is still seeking a settled method for advancing theoretical understanding of IQ, while even today the field has not fully assimilated the implications of the purpose-dependence of IQ.

2.1.2 IQ Improvement Programmes

There have been huge improvements in IQ practice. The MIT group runs what they call a 'Total Data Quality Management' program (TDQM), helping organizations improve their IQ in practice. A further important context for current work has been the development of tools for this programme.

Wang et al. (2003, p. 2) summarize the idea of TDQM thus:

> Central to our approach is to manage information as a product with four principles [...]:
>
> 1. Understand data consumers' needs,
> 2. Manage information as the product of a well-defined information production process,
> 3. Manage the life cycle of the information product, and
> 4. Appoint information product managers.

Since the 1990s, the focus of TDQM has been to get organizations to ask themselves the right questions, and give them the tools to solve their own IQ problems. The right questions involve understanding the entire process of information in the organization, where it goes and what happens to it, and understanding all the different people who try to use the information, and what they need from it. Then, and only then, can organizations really improve the quality of their information. So the first theme of TDQM is to get information producers to understand, map and control their entire information production process. This is an ongoing task, and TDQM recommends the appointment of information executives on the board of directors of companies, with specific responsibility for managing the company's information flows.

Interwoven with this first theme, the second theme is to allow information consumers to assess for themselves the quality of the information before them, interpret the data semantics more accurately, and resolve data conflicts. This is largely approached using metadata, that is, data about data. Data items are tagged with metadata that allow information users to assess their quality. Such metadata now range widely from an overall IQ score, to something much simpler, such as a source of the data, or an update date and time. This tagging procedure was discussed by Wang et al. (1993a, p. 1):

> In this paper we: (1) establish a set of premises, terms, and definitions for data quality management, and (2) develop a step-by-step methodology for defining and documenting data quality parameters important to users. These quality parameters are used to determine quality indicators, to be tagged to data items, about the data manufacturing process such as data source, creation time, and collection method. Given such tags, and the ability to query over them, users can filter out data having undesirable characteristics.

Here, they are beginning to build the step-procedure that would become central to TDQM. Wang et al. (1993b, p. 2) write:

> It is not possible to manage data such that they meet the quality requirements of all their consumers. Data quality must be calibrated in a manner that enable consumers to use their own yardsticks to measure the quality of data.

They try to show how to do this for some key dimensions: interpretability, currency, volatility, timeliness, accuracy, completeness and credibility. Wang et al. (1995, p. 349) are explicit:

> Because users have different criteria for determining the quality of data, we propose tagging data at the cell level with quality indicators, which are objective characteristics of the data and its manufacturing process. Based on these indicators, the user may assess the data's quality for the intended application.

There are some formal problems with this kind of tagging. The most obvious is that of computational power. If you are already struggling to maintain and control a lot of data, tagging every data item with one or more tags quickly multiplies that problem. Further, one cannot always tag at the cell level – the level of the basic unit of manipulation – as one would prefer. Nevertheless, the idea of the importance of information consumers is being strongly supported in the IQ improvement practice by the use of tagging by metadata aimed at enabling consumers to make their own assessment of information quality.

To achieve this, it is essential to know what an organization does with its information, and what it needs from its information. In a paper where the language of TDQM appears early on, Kovac et al. (1997, p. 63) write:

> Two key steps are (1) to clearly define what an organization means by data quality and (2) to develop metrics that measure data quality dimensions and that are linked to the organization's goals and objectives.

The whole system must be properly understood to provide real quality information, instead of improving only on a department-by-department, stop-gap basis.

This leads to the development of information product maps (IP-MAP) as an improvement of the earlier 'polygen model' (Wang and Madnick 1990). Wang (1998) starts using the term 'information product' (IP), and is clearly building the idea of mapping information:

> The characteristics of an IP are defined at two levels. At the higher level, the IP is conceptualized in terms of its functionalities for information consumers. As in defining what constitutes an automobile, it is useful to first focus on the basic functionalities and leave out advanced capabilities (for example, optional features for an automobile such as air conditioning, radio equipment, and cruise control). … Their perceptions of what constitute important IQ dimensions need to be captured and reconciled. (Wang 1998, p. 61)

He continues:

> 'At a lower level, one would identify the IP's basic units and components and their relationships. Defining what constitutes a basic unit for an IP is critical as it dictates the way the IP is produced, utilized and managed. In the client account database, a basic unit would be an ungrouped client account.' (Wang 1998, p. 63) In summary: 'The IP definition phase produces two key results: (1) a quality entity-relationship model that defines the IP and its IQ requirements, and (2) an information manufacturing system that describes how the IP is produced, and the interactions among information suppliers (vendors), manufacturers, consumers, and IP managers.' (Wang 1998, p. 63)

The IP-MAP is developed in more detail, the basic elements of such a map are defined, and the purpose explained:

> Using the IP-MAP, the IP manager can trace the source of a data quality problem in an IP to one or more preceding steps in its manufacture. We define the property of traceability as the ability to identify (trace) a sequence of one or more steps that precede the stage at which a quality problem is detected. Also, given two arbitrary stages in the IP-MAP, we must be able to trace the set of one or more stages, in progressive order, between the two. Using the metadata, the individual/role/department that is responsible for that task(s) can be identified and quality-at-source implemented. (Shankaranarayanan et al. 2000, p. 15)

The MIT group have already achieved a great deal in expanding understanding of IQ and IQ practice far beyond simple accuracy measures. This has impacted on all current work. Although they structure their thinking in terms of a business model, we will shortly look at IQ applications in science, and in government, the law and other societal institutions.

2.1.3 The 'Italian School'

Batini and Scannapieco (2006) is an excellent overview of work on IQ, a presentation of their own work, and a guide to where new work is needed. Batini and Scannapieco are both academics who also practise, and much more of their work – at least the work from which they draw their examples – is work on government-held data, such as address data. They work broadly along the lines of the TDQM programme, but offer extensions to the IP-MAP better to represent the differences between operational processes, using elementary data, and decisional processes using aggregated data, and to track information flows better. They offer a way to compute and represent quality profiles for these. They also offer 'Complete Data Quality Management' (CDQM) which is their improved version of TDQM to take into account the extra resources they have provided. The particular details are not important to this introductory review, and are thoroughly described in Batini and Scannapieco (2006).

Methodologically, Batini and Scannapieco seem to favour what they call the 'intuitive' approach to developing a theoretical understanding of IQ. They write:

> There are three main approaches adopted for proposing comprehensive sets of the dimension definitions, namely, theoretical, empirical, and intuitive. The theoretical approach adopts a formal model in order to define or justify the dimensions. The empirical approach constructs the set of dimensions starting from experiments, interviews, and questionnaires. The intuitive approach simply defines dimensions according to common sense and practical experience. (Batini and Scannapieco 2006, p. 36)

In line with the intuitive approach, Batini and Scannapieco focus firmly on understanding IQ in practice, by allying discussion of dimensions of IQ and their categories with discussion of examples of metrics used to measure those dimensions. They also categorise IQ *activities*. The idea is to examine common things that are done in the process of improving IQ, and understand what the tools and common methods and problems are. They categorise many activities (Batini and Scannapieco 2006, pp. 70–71), but their aim can be illustrated by looking briefly at the four activities they examine in detail in Chaps. 4, 5, and 6.

One very common activity they call 'object identification'. (It is also sometimes called 'record linking', 'record matching', or 'entity resolution'.) This is when you have two or more sets of data, such as the address data of two different government departments, and you have to identify the records that match the same real-world object – in this case the real house. Data integration is the activity of presenting a unified view of data from multiple, often heterogeneous, sources, such as two sets of address data. Quality composition defines an algebra for composing data quality dimension values. For example, if you have already worked out an IQ score for the completeness of A, and of B, then you need to compute the completeness of the union of A and B. Finally, error localization and correction is the activity performed when the rules on data are known, and you search to find tuples and tables in your data that don't respect the rules, and correct values so that they do. This focus on common activities is a useful practice-oriented way of approaching understanding IQ.

Batini and Scannapieco emphasize that a great deal of work along the lines they have begun is still needed. They write:

> a comprehensive set of metrics allowing an objective assessment of the quality of a database should be defined. Metrics should be related to a given data model or format (e.g., relational, XML, or spreadsheets), to a given dimension (typically a single one), and to different degrees of data granularity. (Batini and Scannapieco 2006, p. 222)

No such comprehensive set is available to date. A great deal has been achieved in IQ, and some very good practice has been developed, but much remains to do. Batini and Scannapieco summarise in their preface:

> On the practical side, many data quality software tools are advertised and used in various data-driven applications, such as data warehousing, and to improve the quality of business processes. Frequently, their scope is limited and domain dependent, and it is not clear how to coordinate and finalize their use in data quality processes.
>
> On the research side, the gap, still present between the need for techniques, methodologies, and tools, and the limited maturity of the area, has led so far to the presence of fragmented and sparse results in the literature, and the absence of a systematic view of the area. (Batini and Scannapieco 2006, p. IX)

Thus IQ research has achieved a great deal both in academia and in practice, but still faces significant challenges. The IQ field is vibrant, still finding out what is possible, and facing many challenges with enthusiasm.

2.2 Developing Work

IQ literature and practice is now so sprawling that we cannot hope to offer anything approaching a comprehensive survey of current work. Instead, as a guide, we offer a look at some of the main areas of development, to illustrate the excitement of current work on IQ. Naturally, we focus on issues relevant to the papers in the rest of the book, and we are guided by the conversations we have been privileged enough to have during the course of our project. This makes for an eclectic tour, which illustrates the fascinating diversity of work on IQ.

Data has been growing, but also diversifying. Single databases with well-defined data schemas are no longer the primary problem. Instead, the challenge is to understand and manage different kinds of systems. Peer-to-peer systems do not have a global schema, as peers donating data determine their own schemas, and schema mappings are needed to allow queries across data. On the web, data can be put up in multiple formats, often with no information about provenance. The most important developments are in extending what has already been well understood, in the safer and easier domain of structured data, to the far messier but more exciting domain of unstructured or partially structured data, and to under-examined forms of data, such as visual data.

In this section, we will examine: how work on provenance and trust is applied to assess quality of unstructured data; attempts to build a mid-level understanding to mediate between theory and practice; the extension of well-understood IQ activities, such as object identification, to unstructured data; work on visual data and data visualization; and understanding IQ by understanding error.

The first major area of developing research is IQ in unstructured data, particularly on trust, provenance and reputation. The core idea is very simple: where do the data come from (provenance), are they any good (trust) and is their source any good (reputation)? The approach develops further the idea of the polygen model, which dealt for the first time with the problem of multiple heterogeneous sources. Provenance is generally offered to the user by tagging data with where it comes from, and what has happened to it before it gets to the user. But much more work is needed on how to model and measure the trustworthiness of data and the reputation of particular sources.

An example of work in progress is early research on metrics for trust in scientific data by Matthew Gamble at the University of Manchester.[1] Gamble is working on how scientists trust information from other scientists. This is an interesting correlate of the problem of crowdsourced data: there is equally a problem of the quality of expert-sourced data. The gold standard for most scientists is to be able to reproduce the data – or at least a sample of the data – themselves. But this is often impossible, for reasons of cost, complexity, or simply because of lack of access to necessary technologies. Cost and risk are important, in Gamble's work, as cost and risk frame judgements of good enough quality. If many people are reporting similar results, meaning that they are not very risky, while the results would be costly to reproduce, then further reducing the risk is not worth the high cost. The published results are likely to be trusted (Gamble and Goble 2011). In this context, Gamble is using provenance traces of data to estimate likely quality of a piece of data. Part of the provenance given is the experimental technique used to generate the data, although frequently there is information missing, such as average rate of false positives. Trust measures indicators of likely quality, such as the number of citations of a paper. Gamble is borrowing available metrics, and using Bayesian probabilistic networks to represent these metrics in order to calculate likely quality, based on provenance,

[1] We are very grateful to Matthew Gamble for meeting with Phyllis Illari to explain the overview of his project.

trust, and so on, currently applied to the likelihood of the correctness of chemical structure. Representing metrics as Bayesian net fragments enables one to join them together, and also to compare them more formally.

In general, the suite of metrics Gamble is developing all have to be adapted to particular situations, but in theory the fragments could be put together with provenance to yield a 'Situation Specific Bayesian Net' to compute an overall quality score of data. In theory, scientists could use it to dump data, or to weight their own Bayesian net according to the quality score of the data. However, this is unlikely in practice. At this stage the work is more likely to yield a benchmark for metrics so that they can be understood and compared in a common way. It also helps to push forward the idea of being able to move from provenance to metrics.

The second area we will look at also explores the connections between theory and domain-specific metrics. Embury and Missier (Chap. 3, this volume) explain that work on identifying and categorising dimensions of IQ is no longer proving useful to their practice, and an alternative approach is needed. They developed what they call a 'Quality View' pattern, which is a way of guiding the search for IQ requirements and the information needed for practitioners to create executable IQ measurement components. They survey how they applied this approach in projects involving identifying proteins, in transcriptomics and genomics, and in handling crime data for Greater Manchester Police. The idea is that Quality View patterns guide the application of decision procedures to data. Although they are mid-level between theory and practice, they guide the development of domain-specific metrics appropriate to the particular data in each case. In this way, Embury and Missier, like Gamble, are exploring the space between work on what IQ is, and the development of highly domain-specific metrics.

The third example of developing work is in extending those things we can do well for structured data, in order to figure out how to perform the same tasks for unstructured data. For example, Monica Scannapieco is working on how to extend one of the common IQ activities for structured data – entity matching or record linkage – to unstructured data. Scannapieco calls this 'object matching'. This is the problem of putting together two or more sets of data, when one faces the task of identifying which data in each set refers to the same worldly object. For example, there are many thousands of web pages containing information about cities. How do we decide which pages are all about London, which are about Paris, and so on?

Scannapieco (Chap. 6, this volume) examines the problem with respect to two different kinds of relatively unstructured data: linked open data and deep web data. Linked open data are data made available on the web, but linked to related data, most obviously, data about the same real world object – such as data about Paris. For example, DBPedia makes the content of the infoboxes on Wikipedia (the structured part of Wikipedia pages) available in Resource Description Framework (RDF) format, which gives the relationship between items, how they are linked, along with both ends of that link. This is in contrast with what is known as deep web data, which is not directly accessible by search engines, because it consists of web pages dynamically generated in response to particular searches, such as the web page an airline site generates in response to a query about flights on a particular day to a particular destination. Object matching is an issue for both cases, as is the size of the data sets. Scannapieco surveys the issues for addressing object matching, and more general quality issues, in such data. A particular concern is settling on a characterization of identity of two objects.

The fourth example of developing work is work on visualization and visual data. The vast majority of the work on data quality to date has been on the quality of numbers or texts such as names stored in databases, yet presenting data visually is now quite standard. For example, in O'Hara (Chap. 11, this volume), maps are used to present crime data to citizens via a website. In Chen, Floridi and Borgo (Chap. 5, this volume) the practice of visualisation of data is examined, and the standard story that the purpose of visualisation is to gain insight is questioned. Chen et al. argue, by looking at various examples, that the more fundamental purpose of visualization is to save time. Notably, time can be saved on *multiple* tasks that the data are used for, which may of course include gaining insight. In addition to allowing there to be multiple purposes for visualisation, this approach also removes any requirement that it be impossible to perform such tasks without using data visualisation. With these arguments in place, Chen et al. argue that the most important metric for measuring the quality of a visualization process or a visual representation is whether it can save the time required for a user or users to accomplish a data handling task.

Batini, Palmonari and Viscusi (Chap. 4, this volume) aim to move beyond the much-studied information quality paradigm case of the traditional database, to examine information quality 'in the wild'. They re-examine traditional concepts of information quality in this new realm. In this, they share a great deal with Scannapieco's work, arguing that traditional dimensions, and approaches such as in the ISO standard issued in 2008 (ISO/IEC 25012:2008), still need extensive rethinking. Batini et al. study schemaless data by examining the quality of visual data, such as photographs, which are ignored by the ISO standard. They suggest that we can define the quality of an image as the lack of distortion or artefacts that reduce the accessibility of its information contents. Common artefacts are blurriness, graininess, blockiness, lack of contrast and lack of saturation. They note that there are going to be ongoing problems with data quality of, for example, diagrams, as even the most objective-seeming accessibility or readability guidelines for creating diagrams show cultural specificity. They offer the example that most diagrammers try to have straight lines, with as few crossing lines as possible, but Chinese professors prefer diagrams with crossing and diagonal lines.

The fifth developing area concerns understanding information quality by examining failures in that quality – by better understanding error. This is much as Batini et al. do in categorising good images as ones that avoid known classes of problems. This approach has been in the literature at least since Wand and Wang (1996), but it is still being pursued. It is adopted by Primiero (Chap. 7, this volume), who sets out to 'define an algorithmic check procedure to identify where a given dimension fails and what kind of errors cause the failure.' (p. 107) Primiero proceeds by applying a broad categorization of errors, in accordance with three main kinds of requirements that can fail when there is error: validity requirements, which are set by the logical and semantic structure of the process; correctness requirements, which are the syntactic conditions for the same process; and physical requirements, which are the contextual conditions in which the information processing occurs. This cross-cuts with three modes of error: conceptual, which relates to configuration and design of the information process; material, or aspects of implementation of the process; and executive, or relating to successful execution of the process. This finally yields four main cases of error (as not all combinations are possible). Primiero uses these to re-examine traditional IQ dimensions such as consistency, accuracy, completeness and accessibility, and assess how failures occur.

Fallis (Chap. 8, this volume) uses a similar approach but in a different way. He analyses IQ by classifying various kinds of disinformation – which he takes to be deliberate misinformation. He writes:

> But disinformation is particularly dangerous because it is no accident that people are misled. Disinformation comes from someone who is actively engaged in an attempt to mislead. Thus, developing strategies for dealing with this threat to information quality is particularly pressing. (p. 136)

Fallis points out that disinformation, unlike a lie, does not have to be a statement but could, instead, be something like a misleading photograph, and disinformation could be true but still designed to mislead by omission. Fallis examines the many different types of disinformation, in an extended attempt to characterize disinformation. He illustrates the variety of kinds of disinformation.

Finally, Stegenga (Chap. 9, this volume) illustrates how various approaches to evaluating information quality in medical evidence are attempts to avoid kinds of error. In sum, the attempt to understand error is clearly yielding interesting work, although it may well not yield a unitary approach to information quality, as might have been hoped. This is not surprising if the purpose-dependence of IQ is taken seriously. Just as particular virtues of information are more important for different purposes, so are particular errors. For some users, late but accurate information is better than speedy but inaccurate information, but not for others.

IQ practice is diversifying, and constantly pushing the boundaries of what is possible. In particular, it is applying existing abilities to unstructured data, such as in understanding the uses and limitations of crowdsourcing, and how to apply techniques that have been developed for structured data in databases to other forms of data such as visual data.

2.3 Applying IQ

Alongside the deepening theoretical understanding of IQ there have been some extraordinary developments in IQ practice, as information has come to pervade almost all of human activity. For example, the increasing availability of data and its use by multiple people and groups in science means that databases are increasingly crucial infrastructure for science. We refer philosophers in particular to the work of Sabina Leonelli (Leonelli 2012, 2013; Leonelli and Ankeny 2012). For data sharing to be effective, data has to be maintained in a form understandable from multiple disciplinary backgrounds, and frequently integrated from multiple sources. So there are extensive applications of the original home of IQ, databases, and newer approaches, such as trust and provenance, in science. The importance of quality information to the well-functioning of society as well is also now hard to underestimate. Frequently, the accessibility of that data to the relevant people is a serious problem, and some data must now be available to all citizens. The two issues of data in science and in society often come together. For example, the absence of longitudinal funding for many scientific databases is a serious impediment in some sciences, and directly impacts society with the handling of medical data (Baker 2012).

Again, we cannot hope to be comprehensive. We will illustrate the issues of applying IQ by looking at examples of applications to medical data, and to social data.

2.3.1 *Medical Data and Evidence*

There has been a buzz about medical data in recent years, so much so that everyone knows there is a potential problem. But what is interesting on investigation is that there are so many facets of IQ problems in medicine, as it arises in medical discovery, treatment, and maintaining patient records so that patients can be treated appropriately over a lifetime.

One of the core challenges of managing records in healthcare systems is the sheer number of people trying to use the data, and their multiple purposes. Patient records have to be maintained, to be usable by many people with widely varying expertise, including at least family doctors, consultants, nurses, and administrators, and they have to be kept confidential. What is wanted is an efficient, accessible, easy to update system that can be used without ambiguity by multiple people for multiple purposes. Patient records therefore nicely illustrate how far IQ problems outstrip mere accuracy.[2]

At the moment, such databases are constrained using integrity constraints on what data can be input, which force consistency. First, there are constraints on what data *have* to be input for each patient, such as name, address, sex, age and so on; and text is not usually entered free-text, but from a list of constrained choices. For example, diagnoses of illnesses are coded, and entered by code. Second, there may be constraints across these choices, to weed out errors at the data input stage. For example, a patient cannot be 3 years old and pregnant; or completely healthy and in the intensive care ward.

Such coding systems can be incredibly frustrating for thousands of busy people whose job is not to maintain data, but to care for patients. There is often a separation between the people who use the data, and those who gather it. Those forced to

[2]We thank Andy Bass, Computer Science, Manchester, who works on patient record systems, for personal conversation about these issues.

gather it may not be as medically informed as those using it, and so struggle to make nuanced choices in difficult to classify cases. Errors are frequent. Further, different *users* of data will maintain data differently. People and institutions are better at maintaining the data that determine what they are paid, and regulated items, such as prescriptions.

Quality assessment is also an issue in the evaluation of medical evidence. First, evidence is assessed for quality when making decisions about effective treatments, and also licensing them to be used, which is done by bodies such as the Food and Drug Administration agency in the US, and the National Institute for Care Excellence in the UK. This issue is addressed by Stegenga (Chap. 9, this volume). A great deal of work has been done to articulate and standardise methods of assessment of evidence in medicine, particularly by international bodies such as the Cochrane Collaboration (http://www.cochrane.org/). The general idea is to articulate best practice. However, the upshot is often to generate a one-size-fits-all assessment of quality based solely on the method by which the evidence was gathered, without reference to its purpose. Almost all approaches to medical data prioritise evidence produced by Randomised Controlled Trials over other forms of studies. Stegenga examines various Quality Assessment Tools that have been designed and used to assess the quality of evidence reported in particular scientific papers, in an attempt to aggregate evidence and make a decision about the effectiveness of a treatment – and ultimately decide whether it should be licensed. A serious problem with these tools is that different tools often do not agree about the quality of a particular study, and different users of the same tool will often not agree about the quality of a particular study. There are many serious problems of assessing the quality of medical evidence (Clarke et al. 2014; Osimani 2014).

Information about diseases and effective treatments is available on the web, and patients access it. Further, medical professionals need some way to keep up their expertise once they have finished their formal training, and they also turn to web information. Ghezzi, Chumbers and Brabazon (Chap. 10, this volume) describe a variety of measures available to help assess internet sources of medical information. They also describe a course they have designed to train medical students to assess medical evidence on the web, to allow them to update their own expertise, and to talk with patients who may have been misled by what they have read online. Even relatively simple measures, such as checking whether the information comes from a source that is attempting to *sell* the treatment, and searching for references to scientific papers, have proven very effective at weeding out bad information.

IQ is also a problem in the general move to repurpose medical data. Given the expense of data gathering, the ongoing need for more data, and the idea that there are rich resources in data that often go unmined, it is not particularly surprising that there are various moves afoot to make data gathered in one study available for further use. The Food and Drug Administration agency (FDA) in the USA is encouraging this, as is Health Level 7 in Europe. There are significant challenges, as illustrated by the project involving Meredith Nahm, in bioinformatics at Duke,[3] which defined

[3] We thank Meredith Nahm for discussions.

the data elements for schizophrenia that the FDA intends to require to be released to the central database before the FDA will license treatments (Nahm 2012; Nahm et al. 2012). Even with FDA backing for these kinds of projects, trying to get support from experts and funding bodies proved quite a challenge. Ultimately, the project used the DSM-IV, which is the diagnostics manual for psychiatry, and the paperwork generated by clinical professionals, to extract a set of suggested data elements, before engaging in consultation exercises with experts to finalise data elements. However, just before the publication of the updated DSM-V, the NMIH, a major funder of research in psychiatry, announced that it will preferentially fund projects that ignore the DSM categories in favour of their own system. The challenge is that categories of disease and relevant data elements are not settled in psychiatry, or in medicine, and will have to be updated continuously. Projects of this kind will be ongoing.

2.3.2 *Social Data*

Data have now become a huge concern of society. Again we illustrate the diversity of the impact of information quality on society by examining three cases. First, we look at the quality of personal digital archives. Then we examine the increasingly pressing issue of how to admit only quality information into law courts, given the impossibility of jurors making an informed assessment of such information. Thirdly, we examine the increasing drive to making government data open. This continues the theme of the law, as we will look at crime data, which clearly comes full circle to impact on the private lives of citizens.

First, personal digital archives, such as Facebook timelines, personal and professional files, or family photographs and albums, have become important to people in managing and enjoying their lives. How we disseminate such information, manage its quality, and protect it, is of deep personal and professional concern.

John (Chap. 13, this volume) uses the expertise of a professional who manages digital archives for the British Library, to examine the quality of digital archives as they are managed by private individuals.

John lays out seven aspects of quality for digital archives, as a background. But he argues that we should also pay attention to the quality of digital archives 'in the wild' – not only when they enter a specialised repository. This is partly to assist in the job of repositories, but also because the role of personal archives means that their quality affects people's lives. John argues that thinking from an evolutionary perspective – examining how information varies, and is replicated and selected – can help us ask the right questions about quality of personal digital information, and understand better how such information grows, inheriting characteristics of previous archives, such as the growth of a family's archive. This perspective should help, as natural selection has proven good at creating adaptability in the face of uncertainty, which is just what such personal digital archives need. A crucial question for investigation, then, is: are there predictable ways in which digital archives grow in the wild, predictable 'selection pressures' that we can come to understand better, and so better control and compensate for?

The second area we will examine is the quality of expert evidence in the law, specifically in law courts. There is variation across countries, of course, but judges are often asked to ensure that only good quality evidence gets presented in court, and there have been some notable failures. There are currently proposed new rules on expert evidence in the UK. In practice, up until now relatively simple proxy indicators of quality have been favoured, such as the professional qualifications of the expert, membership of professional societies, and peer review and citations of scientific work referenced. Schafer (Chap. 12, this volume) discusses how digital media can change this, with particular reference to how digital media can change peer review, which is currently a favoured quality mechanism.

One crucial problem of forensic information being presented in court is the availability of a sensible reference database. The need for such a database to allow estimations of relevant probabilities came to the fore with DNA, and the situation is much worse for many other kinds of evidence. For example, if you lack a reference database for, say, earprints, then how alike earprints are cannot be estimated accurately, and evidence as to how similar the earprint recovered from the scene is to that of the accused cannot really be given. Schaffer argues that the digital revolution will help with this problem in the future, by allowing access to non-regulated, informal datasets that can allow forensic practitioners to estimate base rates and standards in an unprecedented way.

Schafer also argues that the digital revolution can help with a second problem: the possibility of lawyers and judges assessing whether abstract scientific theories used by experts are 'generally accepted in the scientific community'. Peer review itself cannot indicate whether an idea has come to general acceptance. But digital media are supporting new forms of engagement with science, and allowing access to ongoing discussion of already published papers, including information about post-publication *withdrawal* of papers. Schafer envisages that, in the future, such venues might be routinely data-mined to allow more quantitative assessment of whether research is generally accepted, and suggests that IQ research can help with this task.

The third area we will consider is open data, which O'Hara (Chap. 11, this volume) discusses with respect to government data. Open data is made available to anyone who might wish to use it, so it is explicitly presented with no specific user or purpose in mind. This raises similar problems as the repurposing of data in medicine. O'Hara looks at heuristics and institutional approaches to quality in open data, and at how the semantic web might support mechanisms to enhance quality. One idea associated with open data is that increased scrutiny will improve the quality of data, by detecting errors, leading to the idea of crowdsourced data improvement.

O'Hara discusses a particular initiative to make local crime data available to citizens in the UK, to allow them to take it into account in decisions such as where to live, and routes to travel. The project met problems integrating data from 43 different police forces in the UK, lacking any national geodata coding standard. Further, burglaries and assaults have a definite location that can be mapped, but this is not true of all crimes, such as identity theft. It was also difficult to maintain anonymity. If a burglary is shown as taking place at your address, then you are identified as the victim of that crime, perhaps against your wishes. Reasonable data accuracy was reconciled with the need for some anonymity by making the data available on location vaguer, giving number of crimes by small geographical area, rather than a precise location for each one. O'Hara suggests that data producers designing such a system should interact with likely users to make the data accessible. The decision was to compensate for problems in the data by making users as aware as possible of the possible limits of the data they were given, using metadata. So note that in the end getting such open data systems to work is difficult without *some* attention to possible users of the information.

Ultimately, then, these three cases illustrate how pervasive information quality issues are, and how they impact on the daily lives of everyone in society.

2.4 Conclusion: Theoretical Challenges

The concluding two papers of the book finish where we started, as Illari and Floridi examine the theoretical problem of purpose-dependence of IQ, as pressed by the MIT group. Illari (Chap. 14, this volume) takes up purpose-dependence alongside the practical problem that successful metrics for measuring IQ are highly domain-specific and cannot be transferred easily. She argues that both theoretical and practical approaches to IQ need to be framed in terms of an understanding of these deep problems. She supports a categorisation of IQ dimensions and metrics that highlights, rather than obscures, these problems.

Floridi (Chap. 15, this volume) examines purpose-dependence alongside the argument that the problem of big data is often not the amount of data, but the difficulty of the detection of small patterns in that data. IQ concerns the possibility of detecting these patterns. Floridi argues for a 'bi-categorical' approach to IQ that allows it to be linked explicitly to purpose.

These issues play out in many of the papers in the volume. Purpose-dependence inhibits the possibility of inter-level theorising about IQ, creating understanding that lies between what IQ is, dimension categorisations, and domain-specific metrics. This is addressed by the Embury and Missier paper (Chap. 3, this volume) and in the work by Gamble that we have discussed, and shows the importance of this work.

This background also illuminates the attempt to address IQ comprehensively by categorising error, shared in this volume by Primiero, Fallis and in some ways by Batini et al. and Stegenga. This approach is undeniably valuable, but a comprehensive assessment may be too much to hope for. It is likely that different kinds of error are more or less important for different purposes.

In medical evidence, discussed by Stegenga (Chap. 9, this volume), we see the impact of pursuing an ideal of a purpose-independent estimation of quality of evidence. The way traditional evidence assessments proceed, quality of evidence is ideally independent of *everything* except the method used to generate the evidence.

Against the background of this IQ literature, the deep difficulties with such an approach are clear.

The scale of data now available in medical research also underlines the small patterns problem. Increasingly, our ability to process data – to find the small patterns we seek – is the critical problem. Purpose rules here, too. More data is no good if it merely obscures the pattern you are looking for in your dataset. There needs to be more attention explicitly to discriminating amongst purposes in assessing fitness for purpose, allowing us better to recognise which data is worth holding on to.

This is an interesting backdrop to the moves to assess information quality in the wild, which we find here in both Batini et al., and John. Learning to deal with information in its natural form, and extract what we need from it there, should help address this problem. This is aligned, then, with work on dealing with unstructured data, such as examining object matching (Scannapieco, Chap. 6, this volume), and making data open partly to allow increased scrutiny (O-Hara, Chap. 11, this volume).

In short, IQ is a challenging and exciting area of research, already bearing fruit, and certain to reward further research.

References

Baker, M. (2012). Databases fight funding cuts. *Nature, 489*(19). doi: 10.1038/489019a

Batini, C., & Scannapieco, M. (2006). *Data quality: Concepts, methodologies and techniques.* Berlin/New York: Springer.

Clarke, B., Gillies, D., Illari, P., Russo, F., & Williamson, J. (2014). Mechanisms and the evidence hierarchy. *Topoi.* doi:10.1007/s11245-013-9220-9.

Gamble, M., & Goble, C. (2011, June 14–17). *Quality trust and utility of scientific data on the web: Towards a joint model.* Paper presented at the WebSci'11, Koblenz.

Kovac, R., Lee, Y. W., & Pipino, L. L. (1997). *Total data quality management: The case of IRI.* Paper presented at the conference on information quality, Cambridge, MA.

Lee, Y. W., Strong, D. M., Kahn, B. K., & Wang, R. Y. (2002). AIMQ: A methodology for information quality assessment. *Information & Management, 40*(2), 133–146. doi:10.1016/s0378-7206(02)00043-5.

Leonelli, S. (2012). Classificatory theory in data-intensive science: The case of open biomedical ontologies. *International Studies in the Philosophy of Science, 26*(1), 47–65.

Leonelli, S. (2013). Integrating data to acquire new knowledge: Three modes of integration in plant science. *Studies in the History and the Philosophy of the Biological and Biomedical Sciences: Part C, 4*(4), 503–514.

Leonelli, S., & Ankeny, R. (2012). Re-thinking organisms: The epistemic impact of databases on model organism biology. *Studies in the History and Philosophy of the Biological and Biomedical Sciences, 43*, 29–36.

Nahm, M. (2012). *Knowledge acquisition from and semantic variability in schizophrenia clinical trial data.* Paper presented at the ICIQ 2012, Paris.

Nahm, M., Bonner, J., Reed, P. L., & Howard, K. (2012). *Determinants of accuracy in the context of clinical study data.* Paper presented at the ICIQ 2012, Paris.

Osimani, B. (2014). Hunting side effects and explaining them: Should we reverse evidence hierarchies upside down? *Topoi.* doi:10.1007/s11245-013-9194-7.

Shankaranarayanan, G., Wang, R. Y., & Ziad, M. (2000). *IP-Map: Representing the manufacture of an information product.* Paper presented at the 2000 Conference on Information Quality, MIT, Cambridge, MA.

Wand, Y., & Wang, R. Y. (1996). Anchoring data quality dimensions in ontological foundations. *Communications of the ACM, 39*(11), 86–95. doi:10.1145/240455.240479.

Wang, R. Y. (1998). A product perspective on total data quality management. *Communications of the ACM, 41*(2), 58–65. doi:10.1145/269012.269022.

Wang, R. Y., & Madnick, S. E. (1990). *A polygen model for heterogeneous database-systems: The source tagging perspective.* In D. McLeod, D. R. Sacks, & H. Schek (eds.) *The 16th international conference on very large data bases* (pp. 519–538). Las Altos: Morgan Kaufman.

Wang, R. Y., Kon, H. B., & Madnick, S. E. (1993a). *Data quality requirements analysis and modelling.* Paper presented at the ninth international conference of data engineering, Vienna.

Wang, R. Y., Reddy, M. P., & Gupta, A. (1993b). *An object-oriented implementation of quality data products.* Paper presented at the WITS-'93, Orlando.

Wang, R. Y., Reddy, M. P., & Kon, H. B. (1995). Toward quality data: An attribute-based approach. *Decision Support Systems, 13*(3–4), 349–372. doi:http://dx.doi.org/10.1016/0167-9236(93)E0050-N.

Wang, R. Y., Allen, R., Harris, W., & Madnick, S. E. (2003). *An information product approach for total information awareness.* Paper presented at the IEEE aerospace conference, Big Sky, pp. 1–17.

Chapter 3
Forget Dimensions: Define Your Information Quality Using Quality View Patterns

Suzanne M. Embury and Paolo Missier

Abstract When creating software components that aim to alleviate information quality problems, it is necessary to elicit the requirements that the problem holders have, as well as the details of the existing technical infrastructure that will form the basis of the solution. In the literature, standard sets of IQ dimensions have been proposed as a means of initiating and structuring the information gathering and design processes involved.

Over the past decade, we have been involved in several projects to develop IQ assessment components. In the earlier projects, we tried hard to make use of the standard IQ dimensions in this way, but found that we derived little benefit from this approach. In some cases, the IQ problem we were focussed on could not be assigned cleanly to one dimension or another. In others, the dimension was clear, but we found that that knowledge saved us very little of the work we had to do when the dimension was not identified up front.

However, IQ problems are typically very challenging, and some sort of guiding principles are needed. In this paper, we propose our earlier notion of the Quality View (QV) as an alternative (or additional) technique to IQ dimensions for developing IQ management components. We reflect on our experiences in using QVs in three quite different IQ-related projects, and show how our initial basic pattern turned out to be a good starting point for the information gathering and design tasks involved, replacing IQ dimensions in the role originally envisaged for them.

S.M. Embury (✉)
School of Computer Science, University of Manchester,
Oxford Road, Manchester M13 9PL, UK
e-mail: suzanne.embury@manchester.ac.uk

P. Missier
School of Computing Science, Newcastle University,
Claremont Tower 9.08, Newcastle upon Tyne NE1 7RU, UK
e-mail: paolo.missier@ncl.ac.uk

L. Floridi and P. Illari (eds.), *The Philosophy of Information Quality*,
Synthese Library 358, DOI 10.1007/978-3-319-07121-3_3,

3.1 Introduction

When attempting to come up with a crisp and useful definition for the concept of "information quality" (IQ), the starting point for many IQ researchers and practitioners is the standard definition of *quality* as referring to the degree to which its subject is *fit for purpose*. Thus, we see definitions in the IQ literature such as the following from Larry English, which states that information is high quality if it supports "consistently meeting or exceeding all Knowledge Workers and end-Customer expectations" (English 2009, p. 32).

Definitions like this one are very useful as far as they go, since they remind us that there is no absolute definition of good IQ. IQ can only be assessed relative to some specific goal or problem. Such definitions also encourage us to broaden our notion of what IQ means, beyond the more obvious concerns relating simply to accuracy of data. Data can be accurate, but still not be fit for purpose, if it is in a format we cannot decipher, uses terms with which we are not familiar, or is structured in a way that does not support the forms of query we need to ask of it. These definitions also remind us that IQ is a multi-faceted concept. What they do not do, however, is provide us with much help in discovering what the needs of information consumers are or how close our information is to meeting them at any one time. There is clearly a huge gap between a definition such as "fit for purpose" and the construction of a useful, concrete software system for managing information quality levels in real systems.

In an attempt to bridge this gap, researchers and practitioners set about mapping the range of different concerns that fall under the heading of IQ, leading to proposals for sets of IQ *dimensions*. Perhaps the most influential of these is the study by Wang and Strong (1996), in which a survey of information consumers about the kinds of IQ issue they felt were important is described, leading to the identification of 15 major IQ dimensions grouped into 4 categories. The dimensions proposed ranged from accuracy and completeness, to reputation, interpretability and accessibility. Many other proposals for taxonomies of IQ dimensions followed, with some becoming a key element of IQ management methodologies (such as AIMQ [Lee et al. 2002] and CDQ [Batini and Scannapieco 2006]). These methodologies typically advise their users to begin by eliciting and documenting the IQ dimensions (from some given list) that correspond to their requirements. The next step is for users to derive specific, implementable measurement procedures from these identified dimensions. In the case of AIMQ, for example, it is suggested that surveys for information stakeholders are designed, in order to gather information about perceived IQ in relation to the identified dimensions. In other methodologies, it is expected that a piece of software will be written, to compute the metric automatically from the data present in the organisation's information systems. Or else, a manual procedure will be instigated, involving spot checks of data against the real world state (such as checking correctness of customer addresses when they call with customer service issues).

Over the course of a number of years, we have attempted to use IQ dimensions in this way, as a starting point for the development of software components that measure IQ, in a variety of domains, but most particularly in e-Science. We began with the expectation that the standard dimensions would indeed be a helpful starting point for the design of IQ metrics, but our experiences across several different projects strongly suggested otherwise. We found that domain experts preferred to use their own domain-specific terms rather than the generic dimensions we were interested in, and often proposed measures that did not fall clearly under the heading of any one dimension, but instead combined aspects of several. Even where a clear dimension could be identified up-front (as in the case of our experiments with completeness measures for a class of functional genomics data), that knowledge did not help us in any meaningful way to design the measurement procedures.

The difficulties we encountered in working with the standard IQ dimensions led us to propose an alternative conceptual model for thinking about and operationalising IQ measures: namely, the Quality View (QV) pattern (Missier et al. 2006). We originally defined this pattern as a means of facilitating the specification of reusable and sharable software components for assessing IQ in highly-specific domains. In this paper, we reflect on our experiences from previous projects where we applied the QV pattern to a diverse set of real problem domains. Our aim is to assess its suitability as a possible alternative to the standard IQ dimensions as a means of guiding the elicitation of IQ requirements and the information needed to create executable IQ measurement components. We also report on the limitations of the standard IQ dimensions that we encountered in each of the projects, as well as showing how our simple basic QV pattern has shown itself capable of defining measurement and transformation options for a range of different IQ types.

We begin by presenting the motivation for the development of the QV pattern, followed by an overview of the pattern itself. We then show how the QV pattern has been employed in a range of different projects, to measure the quality of proteomics data, the completeness of SNP data, and to address data duplication problems in scientific and governmental data. Finally, we conclude with some discussion points for the community as a whole.

3.2 From Dimensions to Measurements?

Many proposals for sets of IQ dimensions have appeared in the literature (e.g., English 1999; Eppler and Muenzenmayer 2002; Loshin 2004; Batini and Scannapieco 2006). While there are major areas of overlap between these proposals (most include the common IQ forms, such as accuracy, completeness and currency of information, for example), there is no general agreement on the complete and correct set of dimensions, nor on the most appropriate organisation of those dimensions into higher-level categories (Illari and Floridi 2012). Even the dimensions themselves are often only vaguely and imprecisely "defined", with conflicting terminology and meanings. More frustrating still, the literature does not at present provide much guidance on how to convert dimensions of interest into detailed quality measurement procedures. The best we have found to date are the lists of specific measurement techniques that are applicable to particular dimensions provided by McGilvray (2008). But, though this sort of guidance is extremely helpful, there is still a major hurdle that IQ practitioners have to jump in eliciting and implementing measurement procedures for specific dimensions that have been identified as being of interest.

So, while the breakdown of IQ into multiple dimensions is undoubtedly helpful, it does not in itself move us towards a position where we can create precise *operational* definitions of IQ. Nor does it seem worthwhile for the community to engage in a process of Dimension Debates in an attempt to find one agreed set of dimensions, terms and definitions, with one agreed hierarchical organisation, until we are able to associate much more precise meanings and boundaries to proposals for dimensions than is currently the case.

Part of the reason for the difficulty of translating dimensions into measures is that IQ measurement procedures tend to be highly domain- and application-specific, being based on the particular business rules and data semantics around which the information system is constructed. For example, we saw this point emerge strongly in our work on the Qurator project (Missier et al. 2006), which looked at IQ in the context of scientific data. In this project, we helped proteomics scientists to design an accuracy-based quality measure (called the PMF score [Preece et al. 2006a]) for use in their work, identifying which proteins are present in organisms under specific disease or other conditions. However, there are two major technological approaches to proteomics: protein mass fingerprinting, and tandem MS. Our quality metric is only applicable to the first of these. If we wished to design a similar accuracy metric for tandem MS data, we would need to start from scratch, and look for specific features of that data that could be used to identify false positive matches.

We also looked at quality issues in the (not so very different) fields of transcriptomics and functional genomics (specifically, single nucleotide polymorphism data) and again found ourselves designing metrics from scratch, based around the particular semantics of the applications we were aiming to support (Missier et al. 2007). Nor is our experience in this respect limited to scientific data. In more recent work, we have been helping the Greater Manchester Police Authority with data duplication issues (Hedeler et al. 2013), and some of the rules we formulated are unlikely to be directly applicable to other data sets, even containing similar types of data. This is because they are so tightly focussed on the specific needs of the GMP context. The same phenomenon can be seen in (for example) the papers published in conferences such as ICIQ every year, discussing the specific IQ features of specific application domains.

The highly domain-specific nature of IQ measures means that the conceptual gap between the abstract dimensions and the measures is substantial. At the very least, some intermediate structure is needed to help IQ stakeholders to bridge the gap, and avoid continual retreading of the same design ground. Or, alternatively, some mechanism for arriving at a dimensional classification from the bottom-up (i.e., from the collection of measures that proved themselves to be useful) might be the way forward.

However, even supposing some useful alternative design approach can be found, there is still the question of what concrete benefits the IQ dimensions bring when designing IQ measures. As we have said, our domain experts were not very interested in the dimensions themselves, with the exception of the basic dimensions that corresponded to fairly common English words, such as consistency and completeness. They preferred to use their own domain-specific terms and concepts to describe the problems they were experiencing with IQ. Even when we could establish that a dimension was of interest early on, it gave us little help in eliciting a suitable measurement procedure. In some cases, the domain experts already had clear ideas about what needed to be measured, and we were left with the task of retro-fitting the measurement procedures to the dimensions; although it was far from clear what the value of this kind of post-hoc classification relative to the standard dimensions was. We postulated that the dimensions might help problem holders to find and reuse suitable quality views, if used as a high-level index into a repository of QVs, and created a formal ontology of IQ dimensions to support this, but as yet have no evidence that the dimensions are sufficient in themselves to allow discovery of IQ measures for reuse.

Our experience suggests, therefore, that rather than needing some intermediate point between dimensions and measures, we may in fact need a completely new starting point, to either replace or complement the standard IQ dimensions. In the next section, we present an overview of the QV pattern, and discuss its suitability as a candidate for the role of such a starting point.

3.3 The Quality View Pattern

Although it is common in the IQ literature to talk of "measuring", "evaluating" or "assessing" the quality of information, in practice the best we can hope for is to compute a close *estimate* of quality. Consider, for example, the common case of customer address data, which must be assessed for accuracy/correctness. However rich and detailed the data model, there is no attribute (or collection of attributes) stored within standard address data that can tell us, by itself, whether the person named is indeed currently resident at the given address. Instead, the best we can manage is to estimate the accuracy of each record, by making a complex sequence of reasonability checks using the data set and other relevant sources of information to try to bridge the gap between the raw information stored and the real world semantics it reflects. For example, trusted reference data sets might be used to determine the validity of postcodes or zip codes appearing in the addresses, and to check that the given street and town name is consistent with the code. Other sources containing related information, such as database of records of bill payments, might be used to cross-check against the address data, since inconsistencies could indicate that the person has changed their address. At the end of all this, the best we can achieve is to combine the results from the various checks to make a defensible guess at the quality of the data, rather than a definitive, absolute measure of its quality.

Our experience of designing and implementing IQ measurement components suggests that, in practice, the process of estimating information quality involves the application of one or more *decision procedures* to the data set under study, based on a set of identified features of the data. These decision procedures have the task of assigning each item in the data set to some specific quality class or rank. For example, data might be classified by applying a set of thresholds to a computed numerical "score" (as in the case of the PMF score mentioned earlier for proteomics data [Stead et al. 2006]), or a more complex process might be required, involving the application of a decision tree, a clustering algorithm or a set of learnt association rules (as in the quality rules proposed by Burgoon et al. [2005]).

In our previous work, we exploited this observed common pattern to propose the notion of a *quality view* (Missier et al. 2006). A quality view (QV for short) is an instantiation of the generic quality assessment pattern that has been specialised for use with a particular type of data and a particular definition of IQ. The user wishing to measure a particular type of IQ plugs domain-specific components into a quality view specification. The specification is then submitted to a model-driven compiler, to generate an executable software component that implements the desired quality management behaviour (Missier et al. 2006).

Viewed from the outside, a QV is a software component that performs a transformation on an input data set (i.e., the data that is to be the subject of the quality measurement) to produce (one or more) output data sets. The QV assesses the quality of each item in the input data set (using the domain-specific definition of IQ), transforms it accordingly, and adds it to (one or more of) the output data sets. Several standard forms of quality manipulation can be specified through this model. For example, one obvious example is an IQ-based filter. In this case, the transformation applied to the data set involves removing all items that do not meet some user-specified quality threshold. In other situations, however, it is more useful for the user to see the quality classification of each item, rather than hiding the poor data from view. In this case, the transformation step performed by the quality view would be to augment each item with an additional attribute containing its quality classification. Alternatively, a quality view might act as a data cleaning mechanism, by identifying low quality data items and acting on them to correct or improve them before reincorporating them in the output data set.

The internals of the QV pattern are illustrated in Fig. 3.1. As the figure shows, a QV consists of a layered configuration of three kinds of component. The role of the top layer of components is to gather the raw evidence which will be used as the basis on which to make decisions about the quality of the items in the input data set (labelled *DS* in the figure). As we have already discussed, it is rarely the case that the input data set contains sufficient information in itself to make a convincing estimate of IQ. More commonly, additional data sources must be queried and further quality metadata must be computed, in order to provide the details needed by the quality decision procedures. It is the responsibility of this layer of *quality evidence functions* (QEFs) to gather this additional information, and to make it available for later processing in the QV. Each QEF web service is responsible for gathering a different collection of evidence types, describing a different aspect of the input data set.

The task of actually assessing the quality of elements in the input data set is the responsibility of the middle layer of components. These are called *quality assertions*

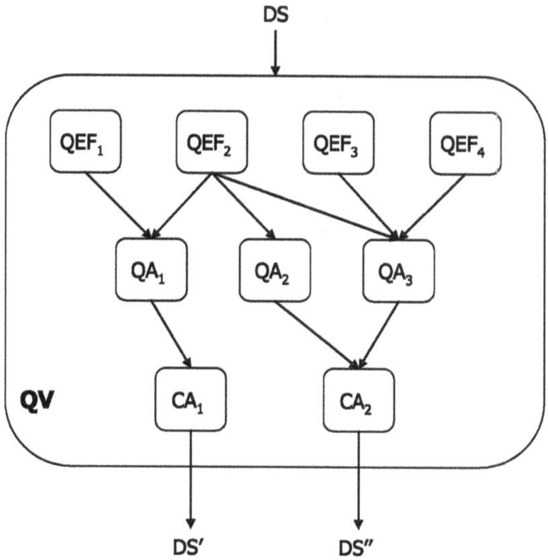

Fig. 3.1 The Qurator quality view pattern

(QAs). Each QA is a (web service) implementation of a decision procedure that maps the elements of the input data set *DS* onto a specific quality classification scheme. A quality classification scheme is a partially ordered set of labels (for example, *high* >> *medium* >> *low*), with one label for each level of quality that is relevant to the form of IQ assessed by the QA. To make its decision, each QA may consume as many or as few items of quality evidence as are produced by the layer of QEFs. Each QA contributes one quality classification label per item in the data set to the state of the QV.

The final layer of components implement the quality-oriented transformation of the data set, based on the evidence and the quality classifications produced by the QEFs and the QAs. The transformations are implemented as condition-action rules (CAs), with the conditions being stated declaratively within the QV specification and the actions being (in our current framework) either a call to a transforming web service or the application of an XSLT expression. It is possible to chain CAs to produce a single output stream, or to apply them in parallel to produce multiple output streams (as illustrated in the figure). This latter option is particularly useful for routing data of different quality levels to different processing components after exit from the QV. For example, high quality data could be passed to a browser for inspection by the user, while low quality data is diverted to an error queue for off-line examination and possible correction.

All quality views implement a common interface, so that they are (potentially) reusable and exchangeable. The input data sets, for example, must be a collection of XML elements, each of which has an identifier attribute. The output(s) of the QV should also be in XML, but are otherwise unconstrained. However, for convenience, we also export an additional data set from each QV, containing the quality classifications created by the operation of the QV. This is supplied in the form of a matrix

Table 3.1 Example of the quality classification data structure

Identifier	QA_1	QA_2	...	QA_n
UP764532.2	Low	5.4	...	Low
UP723943.5	Average	5.2	...	Good
UP829222.4	High	6.9	...	High
⋮	⋮	⋮	⋮	⋮

of values, as illustrated in Table 3.1. This matrix contains one row for each element in the input data set. The first column contains the identifiers of the elements, and the following columns give the classification as produced by each of the quality assertions in the QV. By exporting this internal data as standard from the QV web service (organised according to an XML Schema defined within the Qurator project), we can provide easy programmatic access to quality decisions without having to constrain the form of the transformation implemented by each QV.

Interfaces are also defined for the internal QEF and QA components, so that they can be reused in different quality views. Following the semantic Web service approach (Medjahed et al. 2003), each QV, QEF and QA Web service is annotated with information about its semantics, and its role in the overall quality assessment process, relative to a specially designed IQ ontology (Preece et al. 2006b). This facilitates discovery and consistent reuse of QVs and their components, but it also allows us to create compilers for QVs that generate IQ management software components for use in the information environments preferred by users. To date, we have created a compiler for QVs that creates quality assessment sub-workflows that can be embedded within larger workflows for the Taverna workflow engine (Hull et al. 2006), and also provide a mechanism for invoking QVs as standalone Web services through a semantic registry (Preece et al. 2006b).

As an example, in Fig. 3.2 we show the QV instantiation for the PMF Score we devised for use with PMF proteomics data, mentioned earlier. The QV has a single QEF, which actually computes three kinds of evidence for each of the input protein identifications. It extracts all three from the "Pedro" data file (which contains a full description of the identifications, indexed by identification identifier). The single QA uses the evidence, plus some pre-defined thresholds, to classify each protein identification as being good, okay or poor. Finally, the CA-rule filters out undesirable identifications, so that the user is presented only with good quality ones. The idea is that a scientist might apply this QV to her data if she has found that interpretation of it is too difficult because of the level of noise in the identifications. She can then look at just the good identifications, and lower the threshold of the QV to include okay identifications if she does not find what she is looking for. Further details can be found elsewhere (Missier et al. 2006).

It should be noted that the elicitation of the form of this quality metric was *not* assisted by any *a priori* selection of an IQ dimension as a starting point by the

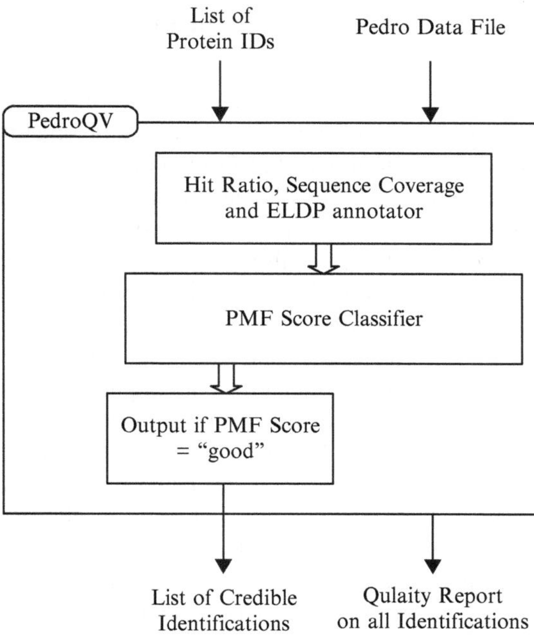

Fig. 3.2 A quality view based on the PMF score quality measure

domain experts (a proteomics academic and a proteomics research fellow). The domain experts worked instead from their deep understanding of the data, to identify characteristics that would lead them to have lowered confidence in an identification. We then collaborated to turn these into an IQ measure. We noted, post-hoc, that there was no easy way to map the measures thus created to the standard dimensions: they had elements of accuracy, completeness, believability and several others, but no single dimension encompassed the measure alone. It then emerged that such a mapping was largely irrelevant to our project partners, and we gave up trying to fit their ideas into the straight-jacket of the standard dimensions.

In contrast, the QV pattern (in its first serious outing) gave us a very helpful framework against which to coordinate our knowledge elicitation activities and our design/implementation work. This experience suggested that we might use the QV pattern as a starting point for the elicitation of IQ requirements, by asking problem owners questions that allowed us to identify not the broad dimensions of interest, but the objective forms of evidence that could be readily computed using the information and computational resources at hand, and the kind of decision procedure that would be of value (was it a classification task or a ranking task, for example). We could also ask questions that would elicit the kinds of data transformations that would be needed to deliver a useful solution (filtering, annotating or cleaning, for example). Then, the achievable connections between these elements could be explored, and the computational elements required to bridge any gaps not supported by the existing infrastructure could be identified and created.

3.4 Measuring Completeness of Information

In the Qurator project, we were able to demonstrate the value of the QV pattern by devising several different quality views for use in a variety of scientific domains. However, we expected that the pattern would probably not be suitable for expressing IQ measures across the full range of IQ dimensions. We therefore set out to test it by applying it to a wider range of applications and measures.

Our first attempt at stretching the capabilities of the basic QV pattern was to build a completeness measure for a functional genomics application using data on Single Nucleotide Polymorphisms (SNPs). Without going into detail, these analyses look for single nucleotide base changes in genes that may indicate a role for the gene in some biological function or state (such as being implicated in the onset of a particular disease). This is done by comparing portions of the genome of an individual of interest with the set of observed base changes for the individual's species. However, at present, for some species, no single database exists that records all the known SNPs. Running an analysis using an incomplete set of observed SNPs can mean that good candidate genes are missed, and the experimental stage of the work which follows the analysis may be fruitless. Therefore, it is valuable for scientists to check completeness of the data set they plan to use as their set of SNPs against the full population of observed SNPs for the species of interest.

Having determined that we would be working under the IQ dimension of completeness, the next step was to work out exactly which kind of completeness our application required, since several different kinds of completeness have been identified in the literature (see, for example, proposals by Fox et al. [1994], Motro and Rakov [1998] and Pipino et al. [2002]). However, in this application we needed a very specific form of completeness that was not defined in the literature in any detail: we needed a form of tuple-oriented completeness that determines the degree to which the individuals within the data set under study covers the full population of entities the user is interested in (Emran et al. 2008).[1] We were therefore left with the task of defining the concept in sufficient detail to allow it to be implemented as a meaningful computational component (Emran et al. 2013). We gave it a name (population-based completeness) to distinguish it from the other forms of completeness mentioned in the literature (such as column-based completeness), and then set about using the QV pattern to produce a useful operational version of the concept specialised to our domain (Emran 2011).

We expected this form of completeness to present a challenge to our QV pattern, since it operates at a different granularity to the measures we had studied so far. In the previous measures we had tackled, we were scoring the quality of individual records in the input set. In population-based completeness, the completeness of the entire data set is scored, relative to a defined reference population. Contrary to our expectations, however, our QV model proved to be quite resilient to this change. All we had to do was to supply the entire input data set, as the single record being

[1] The concept is mentioned, but not defined, by Pipino et al. (2002).

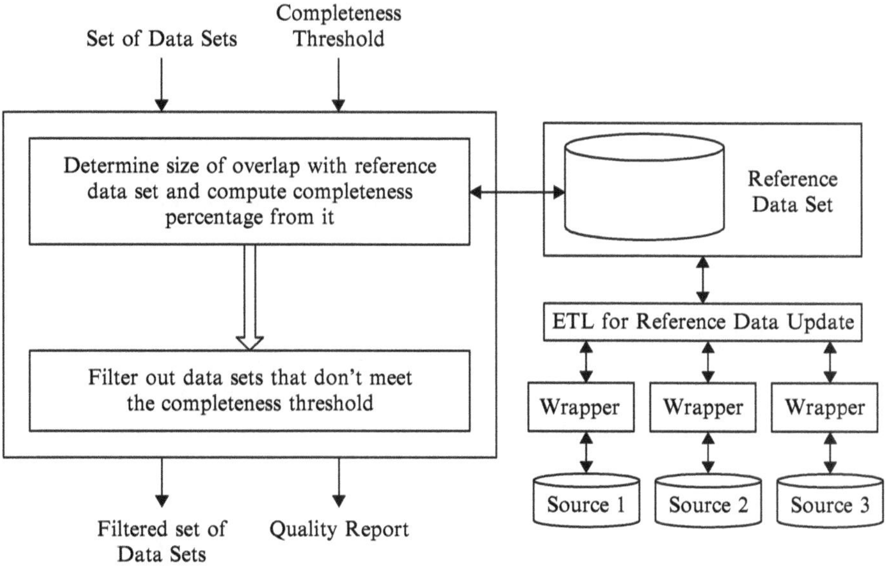

Fig. 3.3 Architecture of the QV (on the *left*) and supporting infrastructure (on the *right*) for population-based completeness QV

passed in on the main input channel. If the data set was small, this could be done by nesting the complete data collection in the input file, so that it appeared as a single item at the top level. Or, where the data set is too large for this to be feasible, the input to the QV can consist of a list of references to the data sets in question, giving URIs from which the required data can be queried by the QEF components when the QV is evaluated. The QAs can then produce a single completeness score for each data set in the input list.

After this realisation, the QV pattern for this kind of completeness turned out to be very simple. The inputs were the data sets to be measured, plus an additional parameter giving details of the reference set to be used in the completeness measure. One QEF was needed to execute the query to determine the percentage of items in the input data set that are also in the reference source. We did not need any Quality Assertion components to perform any complex decision procedure on this score, and used a simple filter CA to remove data sets from the input list that did not meet the threshold set by the person configuring the QV.

However, the QV by itself was not sufficient, since we also needed a mechanism for creating and maintaining the reference data sets on which the completeness measure depends. In the SNP domain, no single source existed that could take on the role of a reference data set, and therefore we needed an ETL pipeline to create the reference data set we needed by integration from several other sources. Figure 3.3 illustrates the architecture of the system.

By contrast with the development of the PMF Score QV, in this project we knew from the beginning very clearly which dimension we were working with. And if we had been dealing with one of the forms of completeness already well documented in the literature, that may have been helpful to us in moving more quickly towards a solution. As it was, we were forced to design our QV and supporting infrastructure from scratch. The QV pattern did provide some help in structuring our ideas, and we were pleased to find that it could be used unchanged for this kind of measurement. However, it was so simple that it did not present a very taxing test of the abilities of the pattern to cope with a variety of situations. Moreover, although we did not need to extend the QV pattern itself, we did need to create the additional components needed to create and maintain the reference source. We can conclude from this that the QV pattern may not be sufficient in itself, but may need to be combined with a range of other patterns, describing the supporting information architecture for IQ measurement that may be required for any given measure. These patterns again could be mapped onto the existing infrastructure, so that the gaps between the desired provision and the existing components can be identified and created.

3.5 Assessing and Improving Data Duplication Issues

Later projects have given us the opportunity to test the applicability of the QV pattern in a solution to the widespread and costly data duplication problem (sometimes also referred to as the entity resolution problem, amongst other names), in which an attempt is made to determine which records within a data set correspond to which real world individuals in the set represented by the data set, and to merge them so that each individual is represented by a single record. In some cases, it can be far from obvious whether two records refer to one individual or two. Suppose for example that we have two records for an "F. Smith" living at "93 Acacia Avenue, Toptown". Is this a case of data duplication, or do two people with the same first initial and surname live at this address? The data duplication problem occurs across many domains, and leads to wasted resources, missed opportunities and sometimes even fatal errors (Elmagarmid et al. 2007). In our work, we have examined the data duplication problem for biological data, as well as for address-based data in a separate project for the Greater Manchester Police Authority.

A typical approach to data de-duplication uses some form of clustering to identify records that are similar and that may refer to the same real world individual. It proved to be easy to fit this kind of processing into our QV pattern. Figure 3.4 illustrates the basic components needed: a collection of QEFs that compute various similarity scores between the individuals in the data set; a QA that uses the similarity scores to cluster similar records together; and finally some CA rules to determine which clusters have sufficiently high similarity that the records within them can be merged (and to carry out the merge), and which have not. Other CAs can route problematic cases to queues for manual handling.

However, there is a significant difference between this simplistic pattern and the complexities of a real instance identification solution for some specific domain. For example, data quality errors (such as typos in manually entered data) can make

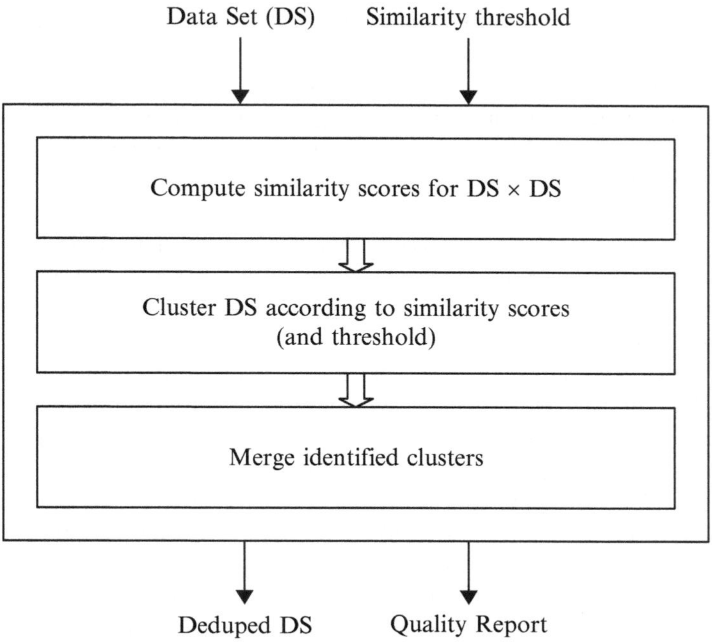

Fig. 3.4 QV for typical data duplication approach

the computation of similarity scores difficult (especially those based on domain semantics, rather than textual similarity). To deal with this, we can chain this data de-duplication QV with other QVs designed to detect and (where possible) resolve these contributing data quality problems before the deduping QV gets to work on the data. Although the technical environment we were working with made implementation as pure QVs impracticable, we used the QV model as a design pattern to structure our thinking. Once again, it proved useful to focus on:

- The outputs the users wanted, and the quality-oriented actions that were needed to produce them (CA components); for example, de-duped records, references to master records, and problematic cases.
- The available resources, and the QEFs that they suggested would be implementable; for example, the availability of reference sets (such as lists of valid postcodes) suggests the feasibility of QEFs to determine fitness of records for certain kinds of textual or domain-specific similarity methods.
- The assertions needed to link the available evidence to the desired transformations; for example, comparison with profiles of various error cases to produce a decision as to whether a record is suitable for a particular de-duplication approach.

As in the case of the PMF Score, it was not possible to identify a single clear dimension that the data de-duplication problems we looked at belonged to. Data de-duplication is a complex problem that does not have one specific cause: as we have said, typically a whole host of different IQ problems in the data lead to difficulties in identifying when records refer to the same real world individual. For example, syntactic and semantic inaccuracies in (for example) postcodes/zipcodes, representational issues (such as might occur when address data from multiple countries is combined), currency problems (such as changes in postcode boundaries that are not reflected in historical data) are all dimensions that are present in the problem being tackled. However, there is a difference here with the situation we encountered in the case of the PMF Score. There, a single measure had the characteristics of several dimensions (and of no one dimension alone). Here, we have multiple problems coinciding, each of which may be counted as an example of a different dimension, or of a mix of dimensions.

This leads us to another of the weaknesses of the "dimensions" approach for tackling IQ problems. Having identified that we have several problems corresponding to several dimensions at work, we have no way of making a composite of the dimensions to explain how they inter-relate in our problem setting. While we can easily and quickly describe how to chain a QV to fix a specific IQ problem with a QV that performs de-duplication, resulting in a quite precise description of how the different solution components will work together, we cannot achieve a similar precision of description using dimensions alone. It is not clear, for example, what we mean when we say we want to pipeline the accuracy dimension with the currency dimension. We could add words to explain our meaning, but the dimension concepts themselves don't help us to express such composites of IQ problems, or to compare their differences. This suggests another requirement for our "starting point" (or "intermediate point" for those contexts where the IQ dimensions have value) for IQ problem/solution modelling: we need a conceptual language that allows us to describe IQ problems that are composites of smaller problems, that helps us to tease apart the various issues into sub-problems that can be solved independently, and that allows us to compare competing solutions. The QV pattern as it stands is clearly not the last word on such a conceptual language. But it may point the way towards the creation of one.

3.6 Conclusions

In this paper, we have reported on our efforts to define forms of information quality not in terms of abstract dimensions, but in terms of concrete, realisable measurement and improvement patterns. The patterns are built from components that gather objective evidence, perform more complex decision procedures over that evidence, and apply rules to determine what improvement or correction activities to take, based on the decision outcomes. The QV pattern provides us with a framework for embedding domain-specific components into a generic quality management framework. The domain-specific components conform to common interfaces, so that they too can be shared and reused, in different QV configurations.

Our original three-layer QV pattern has proven to be quite resilient, and can express IQ measures from a range of IQ dimensions as well as IQ measures that are not easily classified by any current dimension. It therefore provides a way of defining IQ measurements and improvement steps that is orthogonal to traditional IQ dimensions, and which may even replace them, for certain application contexts.

Our work with QVs has also thrown into focus some of the limitations with the use of the standard IQ dimensions as a means of kick-starting the elicitation of IQ requirements, and of guiding us towards a concrete, operational solution. One can envisage IQ dimensions working well in domains where the IQ problems fall clearly into one of the existing dimensions, or a clearly delineated subset of them, and where detailed measurement patterns for the dimension have been documented. However, in cases where the IQ problem does not fall neatly into one dimension, where it is a composite of many other IQ problems, or where a new form of measurement pattern is needed for a standard dimension, then another modelling approach may be more beneficial.

In these contexts, the standard IQ dimensions are not at the right level of abstraction for the work we want them to do. The challenge for the information quality community, then, is to find a way to bridge the gap between the abstract characterisations of information quality that attempt to define and direct the field, and the highly domain-specific implementations of quality solutions that seem to be necessary for success in practice. In other fields where this kind of dichotomy exists (for example, software quality), patterns have provided a useful bridging concept. Perhaps the same could be true for information quality?

While the QV pattern we have been working with so far is almost certainly not the final answer to this challenge, it does perhaps point us in a useful direction for future research. Perhaps, as a community, we should be working to identify a library of useful patterns for describing IQ problems and solutions, and for linking the two? Our experience in working with QV patterns suggests that such patterns will have value if they allow us not only to elicit IQ requirements, but also to characterise the IQ actions that are practicable with the information architecture at hand, and to guide us in bridging the two.

We also need to find ways of communicating experience and expertise gained in the solution of IQ problems in practice in ways that can lead to the creation of a body of more generally applicable IQ wisdom. The documentation of individual point solutions is useful up to a point, but only has lasting value if we can abstract some transferable knowledge from them. Again, in other fields, patterns have proven to have a useful role in this respect, providing a vocabulary for the discussion and comparison of proposed solutions. Development of an IQ pattern language to facilitate this conversation could have value for the future development of this challenging and important field.

Acknowledgements The work reported in this paper on Quality Views was supported by a grant from the EPSRC. The opinions of the authors have been greatly improved by discussions with their colleagues on the Qurator project team, in the Information Management Group at Manchester and the Informatics research group at the University of Cardiff.

References

Batini, C., & Scannapieco, M. (2006). *Data quality: Concepts, methodologies and techniques*. Berlin: Springer.

Burgoon, L., Eckel-Passow, J., Gennings, C., Boverhof, D., Burt, J., Fong, C., & Zacharewski, T. (2005). Protocols for the assurance of microarray data quality and process control. *Nucleic Acids Research, 33*(19), e172.

Elmagarmid, A., Ipeirotis, P., & Verykios, V. (2007). Duplicate record detection: A survey. *IEEE Transactions on Knowledge and Data Engineering, 19*(1), 1–16.

Emran, N. (2011). *Definition and analysis of population-based data completeness measurement*. Ph.D. thesis, The University of Manchester.

Emran, N., Embury, S., & Missier, P. (2008). Model-driven component generation for families of completeness. In P. Missier, X. Lin, A. de Keijzer, & M. van Keulen (Eds.), *Proceedings of the international workshop on quality in databases and management of uncertain data*, Auckland, New Zealand, pp. 123–132.

Emran, N., Embury, S., Missier, P., Mat Isa, M., & Kamilah Muda, A. (2013). Measuring data completeness for a microbial genomics database. In A. Selamat et al. (Eds.), *Proceedings of 5th Asian Conference on Intelligent information and database systems (ACIIDS'13)* (LNAI 7802, pp. 186–195). Kuala Lumpur: Springer.

English, L. (1999). *Improving data warehouse and business information quality*. New York: Wiley.

English, L. (2009). *Information quality applied: Best practices for improving business information, processes and systems*. Indianapolis: Wiley.

Eppler, M., & Muenzenmayer, P. (2002). Measuring information quality in the web context: A survey of state-of-the-art instruments and an application methodology. In C. Fisher & B. Davidson (Eds.), *Proceedings of 7th international conference on information quality (IQ 2002)* (pp. 187–196). Cambridge, MA, USA: MIT.

Fox, C., Levitin, A., & Redman, T. (1994). The notion of data and its quality dimensions. *Information Processing and Management, 30*(1), 9–19.

Hedeler, C., Embury, S. M., & Paton, N. W. (2013). *The role of reference data sets in data deduplication*. In preparation.

Hull, D., Wolstencroft, K., Stevens, R., Goble, C., Pocock, M., Li, P., & Oinn, T. (2006). Taverna: A tool for building and running workflows of services. *Nucleic Acids Research, 34*(Web Server issue), W729–W732.

Illari, P., & Floridi, L. (2012). IQ: Purpose and dimensions. In *17th International Conference on Information quality (ICIQ 2012)*, Paris.

Lee, Y., Strong, D., Kahn, B., & Wang, R. (2002). AIMQ: A methodology for information quality assessment. *Information & Management, 40*(2), 133–146.

Loshin, D. (2004). *Enterprise knowledge management – The data quality approach* (Series in Data management). Morgan Kaufmann.

McGilvray, D. (2008). *Executing data quality projects: Ten steps to quality data and trusted information*. Amsterdam/Boston: Morgan Kaufmann.

Medjahed, B., Bouguettaya, A., & Elmagarmid, A. (2003). Composing web services on the semantic web. *VLDB Journal, 12*(4), 333–351.

Missier, P., Embury, S., Greenwood, R., Preece, A., & Jin, B. (2006). Quality views: Capturing and exploiting the user perspective on data quality. In U. Dayal et al. (Eds.), *Proceedings of the 32nd International Conference on Very large data bases (VLDB'06)* (pp. 977–988). Seoul: ACM Press.

Missier, P., Embury, S., Hedeler, C., Greenwood, M., Pennock, J., & Brass, A. (2007). Accelerating disease gene identification through integrated SNP data analysis. In *Proceedings of the 4th International Workshop on Data integration in the life sciences* (pp. 215–230). Springer.

Motro, A., & Rakov, I. (1998). Estimating the quality of databases. In *Proceedings of the Third International Conference on Flexible query answering systems (FQAS'98)* (pp. 298–307). Springer-Verlag.

Pipino, L. L., Lee, Y. W., & Wang, R. Y. (2002). Data quality assessment. *Communications of the ACM, 45*(4), 211–218.

Preece, A., Jin, B., Missier, P., Embury, S., Stead, D., & Brown, A. (2006a). Towards the management of information quality in proteomics. In *Proceedings of 19th IEEE International Symposium on Computer-based medical systems (CBMS'06)* (pp. 936–940). Salt Lake City: IEEE Computer Society Press.

Preece, A., Jin, B., Pignotti, E., Missier, P., & Embury, S. (2006b). Managing information quality in e-science using semantic web technology. In *Proceedings of 3rd European Semantic web conference (ESWC06)* (LNCS 4011, pp. 472–486). Springer.

Stead, D., Preece, A., & Brown, A. (2006). Universal metrics for quality assessment of protein identifications by mass spectrometry. *Molecular and Cell Proteomics, 5*(7), 1205–1211.

Wang, R., & Strong, D. (1996). Beyond accuracy: What data quality means to data consumers. *Journal of Management Information Systems, 12*(4), 5–34.

Chapter 4
Opening the Closed World: A Survey of Information Quality Research in the Wild

Carlo Batini, Matteo Palmonari, and Gianluigi Viscusi

Abstract In this paper we identify and discuss key topics characterizing recent information quality research and their impact on future research perspectives in a context where information is increasingly diverse. The investigation considers basic issues related to information quality definitions, dimensions, and factors referring to information systems, information representation, influence of the observer and of the task. We conclude the paper by discussing how philosophical studies can contribute to a better understanding of some key foundational problems that emerged in our analysis.

4.1 Introduction

In the last decades, information systems of both private and public organizations have been migrating from a hierarchical/monolithic to a network-based structure, where the potential sources that single organizations or networks of cooperating organizations can use for the purpose of their activity is dramatically increased in size and scope. At the same time data representations have evolved from structured data, to semi-structured and unstructured text, to maps, images, videos and sounds. Now more than ever, information is available in different formats, media and resources and it is accessed and exploited through multiple channels. Each information is completely intertwined with the others, each contributing to the information assets of an organization. Among others, *data and information quality* (information quality in the following, IQ, for short), is becoming critical for human beings and organizations, referring to being able to define, model, measure and

C. Batini (✉) • M. Palmonari • G. Viscusi
Dipartimento di Informatica, Sistemistica e Comunicazione,
Università di Milano Bicocca, viale Sarca 336 – U14, 20037 Milan, Italy
e-mail: batini@disco.unimib.it; palmonari@disco.unimib.it; viscusi@disco.unimib.it

L. Floridi and P. Illari (eds.), *The Philosophy of Information Quality*,
Synthese Library 358, DOI 10.1007/978-3-319-07121-3_4,

improve the quality of data and information that are exchanged and used in everyday life, in business processes of firms, and administrative processes of public administrations.

However, it is our point that IQ issues are worth to be considered "in the wild", paraphrasing the title and the aims of the book by Hutchins (1995), where the terms "wild" referred to human cognition in its natural habitat, naturally occurring, and culturally constituted. As well as for cognition as investigated by Hutchins, we can consider the challenges and changes in the information quality paradigm when studied not only in the captivity of traditional database systems and IT units, but also in the everyday world of the information ecosystem produced by social networks and semantic information extraction processes. Accordingly, despite the relevance of the quality of information assets, the growing literature on information quality constructs and dimensions (Madnick et al. 2009; Wand and Wang 1996), we believe that a further clarification and formalization of their main concepts are required (Batini et al. 2012).

Thus, our aim in this paper is to make a comparative review of the recent literature on data and information quality, with the goal of providing several insights on recent developments along several dimensions. In Sect. 4.2 we critically discuss a recent standard that has been issued by the International Organization for Standardization (ISO). In Sect. 4.3 we introduce two coordinates that are used in the paper to survey the literature: *basic issues*, which concern founding features of IQ, and *influencing factors*, which represent aspects of information systems that have an influence on the interpretation and evaluation of information quality. Sections 4.4, 4.5 and 4.6 address the three basic issues, namely (a) IQ definitions (Sect. 4.4), (b) IQ dimensions (Sect 4.5), with specific reference to the accuracy and completeness dimensions, and c. IQ classifications (Sect. 4.6). Section 4.7 focuses on the relationships between IQ dimensions and the evolution of information systems, while Sects. 4.8 and 4.9 address the levels of semantic constraints and the evolution in the representation of data and knowledge from databases to web knowledge bases. Section 4.10 concludes the paper with a discussion focused on the relationships between IQ and philosophical issues.

4.2 Information Quality in the ISO Standardization Process

When attempting to formalize the concept of data and information quality, the first issue concerns the concepts of *data, information* and *quality*. Traditionally, international standard bodies are authoritative and knowledgeable institutions when definitional and classification issues are considered.

Luckily for our purposes, ISO has enacted in 2008 the standard ISO/IEC 25012:2008 (see Data Quality Model 2008), that defines data quality as the " degree to which the characteristics of data satisfy stated and implied needs when used under specified conditions", and provides "a general data quality model for data retained in a structured format within a computer system". The document presents:

Table 4.1 Data quality characteristics in the ISO standard

DQ characteristic	Definition (all definitions except for completeness and accessibility begin with: the degree to which data has attributes that…")
Correctness	Correctly represent the true value of the intended attribute of a concept or event in a specific context of use
Completeness	Subject data associated with an entity has values for all expected attributes and related entity instances in a specific context of use
Consistency	Are free from contradiction and are coherent with other data in a specific context of use
Credibility	Are regarded as true and believable by users in specific context of use
Currentness	Are of the right age in a specific context of use
Accessibility	Data can be accessed in a specific context of use, particularly by people who need supporting technology or special configuration because of some disability
Compliance	Adhere to standards, conventions or regulations in force and similar rules relating to data quality in a specific context of use
Confidentiality	Ensure that it is only accessible and interpretable by authorized users in a specific context of use
Efficiency	Can be processed and provide the expected levels of performance by using the appropriate amounts and types of resources in a specific context of use
Precision	Are exact or that provide discrimination in a specific context of use
Traceability	Provide an audit trail of access to the data and of any changes made to the data in a specific context of use
Understandability	Enable it to be read and interpreted by users, and are expressed in appropriate languages, symbols and units in a specific context of use
Availability	Enable it to be read and interpreted by users, and are expressed in appropriate languages, symbols and units in a specific context of use
Portability	Enable it to be installed, replaced or moved from one system to another preserving the existing quality in a specific context of use
Recoverability	Enable it to maintain and preserve a specified level of operations and quality, even in the event of failure, in a specific context of use

- a set of terms and definitions for concepts involved,
- two points of view that can be adopted when considering data quality *characteristics* (or *dimensions* (Batini and Scannapieco 2006), in the following),
  - the *inherent* point of view, that corresponds to intrinsic properties of data, and
  - the *system dependent* point of view, that depends on the system adopted to represent and mange data,
- a set of data quality characteristics and corresponding definitions, see Table 4.1.

When we look at the definitions of data and information proposed in the document, we discover that:

1. *data* is defined as "reinterpretable representation of information in a formalized manner suitable for communication, interpretation, or processing",

2. *information* is defined as "information-processing knowledge concerning objects, such as facts, events, things, processes, or ideas, including concepts that within a certain context have a particular meaning".

This choice is specular to the usual one in textbooks and scientific papers, where information is defined in terms of data (see e.g. Floridi 2011), and knowledge in terms of information in some definitions (e.g. in Merriam Webster). The ISO effort shows severe limitations, such as:

1. the flat classification adopted among characteristics (see Table 4.1 for the list of characteristics proposed and corresponding definitions), that contradicts e.g. the classification provided in the document "ISO/IEC 9126 Software engineering — Product quality, an international standard for the evaluation of software quality", where quality characteristics are expressed in terms of sub-characteristics;
2. several characteristics (e.g. completeness) depend on the model adopted for data representation, even though this dependence is not explicitly discussed;
3. data organized in models that neatly distinguish between instances and schemas are considered, e.g. the relational model, while schemaless data, such as e.g. textual documents, are ignored;
4. there is no attempt to distinguish between different types of data and information, from structured data to texts and images.

As a consequence of the above discussion, we can consider the ISO standard as a first standardization effort of the concept of data quality, which needs further investigation and elaboration.

In the rest of the paper, when we refer to *data quality*, we make reference to quality of structured data, while when we refer to *information quality*, we consider wider types of data represented according to different heterogeneous models, such as semi-structured data, texts, drawings, maps, images, videos, sounds, etc. This pragmatic distinction reflects a common use of these terms in the technical literature.

4.3 Information Quality Research Coordinates: Basic Issues and Influencing Factors

We define two coordinates to better formalize and analyze several aspects of IQ. One coordinate is represented by IQ *basic issues* and another coordinate is represented by IQ *influencing factors*, which have been both defined in the introduction. In the following we list a set of items for each of these two coordinates; we do not claim that these items provide an exhaustive coverage of the two concepts; rather they have to be seen as a first attempt to characterize and classify the issues discussed in the literature on IQ, following a classificatorial approach similar to the one adopted in a previous analysis of data quality methodologies (Batini et al. 2009). The basic issues considered in this paper are:

BI1. Definitions of IQ – How many different definitions exist of information quality?

BI2. IQ Dimensions – How many dimensions are considered in the literature to capture the multifaceted character of the concept of IQ?

BI3. IQ dimension classifications – In how many ways dimensions can be classified?

A list of significant factors influencing IQ is:

IF1. Type of information representation – As investigated in Batini et al. (2008), types of information representation can change significantly: if we want to emphasize the *visual perceptual character* of information, we can consider images, maps, graphical representations of conceptual schemas; if we want to emphasize the *linguistic character* of information, we can consider structured, unstructured and semi-structured types of text (specific types of semi-structured text that have been considered in the literature are e.g. laws and medical records). Another common distinction is the one among *structured data*, i.e. data having a rigid and pre-defined schema like relational databases, *unstructured data*, i.e., data having no schema like images and texts in natural language, and *semi-structured data*, i.e., data with a schema that is unknown, flexible or implicit like data in XML. In addition to the above mentioned types of data, we also consider data represented with languages such as RDF and JASON (Antoniou and van Harmelen 2008), called *weakly structured* data in this paper, which have a basic structure (e.g., RDF data have a graph structure) but have non-rigid, possibly changing and third-party schemas attached to the data. Considering the diversity of data to be considered, does the type of information representation influence IQ?

IF2. Life cycle of information – Information has usually a life cycle, made of acquisition (or imaging), validation, processing, exchange, rendering and diffusion. Does the life cycle of the different types of information representations influence IQ?

IF3. Type of information system – Information system architectures have evolved from hierarchical systems, where the information is highly controlled, to distributed, cooperative, peer to peer, web based information, where information flows are anarchic and undisciplined. How this evolution has influenced IQ?

IF4. Level of semantic constraints: binding vs. freedom in coupling data and schemas and open vs. closed world assumption – Data can undergo different levels of semantic constraints. In databases, data and schemas are tightly coupled, while other data, e.g. RDF data, can be loosely coupled with schema level constraints by means of metadata. Moreover, the closed world assumption (CWA) usually holds in data bases, meaning that any statement that is not known to be true is false. In knowledge bases, the open world assumption (OWA) states that any statement that is not known, cannot be predicated neither true nor false. Do the binding/freedom in coupling schemas and data and CWA/OWA influence IQ?

IF5. Syntax vs. semantics – How the syntax vs. the semantics of information play a role in IQ?

IF6. Objective vs. subjective assessment of IQ – With the term subjective we mean "evaluated by human beings", while the term objective means "evaluated by a measurement performed on real world phenomena". How the *objective vs. subjective quality evaluation* is related with IQ?

IF7. Influence of the observer – How IQ is influenced by the observer/receiver, human being vs. machine?

IF8. Influence of the task – IQ is intrinsic to information or it is influenced by the application/task/context in which information is used?

IF9. Topological/geometrical/metric space in visually perceived information – How the different spaces influence IQ?

IF10. Level of abstraction of information represented – The same real world phenomenon can be represented at different levels of abstraction (see Batini et al. 1993) where levels of abstractions are defined for conceptual database schemas). To give a simple example, a conceptual schema in the Entity Relationship model made of the two entities `Student` and `Course` and the relationship `Exam`, can be abstracted in terms of a schema made of the unique entity `Exam`, having as identifier the couple of identifiers of `Student` and `Course` in the refined schema. Is IQ influenced by (e.g. changes of) the level of abstraction?

IQ is a relatively new discipline in information sciences. As a consequence, a discussion on above basic issues and influencing factors can be made at the state of the art in terms of examples and counterexamples leading to observations, statements, conjectures that cannot be formally stated and validated. Conscious of these limitations and immaturity, in the rest of the paper we discuss (some) basic issues, influencing factors and relevant relationships between them.

4.4 Definitions of IQ

We first deal with one of the most controversial questions around IQ: is there an intrinsic information quality? Look at Fig. 4.1, and before reading the next paragraph, reply to this question: which is the most accurate/faithful image of Mars? Perhaps you said: the first one on the left…

The first image has been downloaded from a blog, while the second from the NASA site. Your judgments were probably based on your own model of Mars. Now that you have some ancillary data you could change your opinion…So, may we come to the conclusion that an intrinsic information quality does not exist? This conclusion seems too strong if we look at the two images of Fig. 4.2; they seem to represent the same flower, it is hard to say that the image on the left is of good quality.

The two previous examples show that in order to predicate the quality of a piece of information, sometimes (Fig. 4.1) we need a reference version of the information, other times we evaluate the quality according to perceptual and/or technological

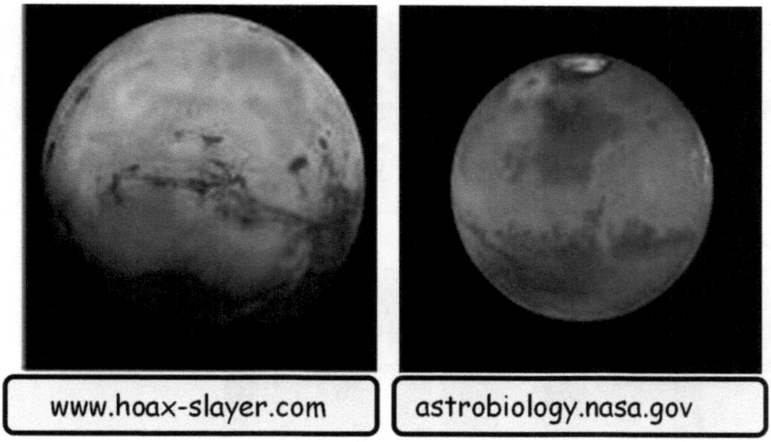

Fig. 4.1 Two pictures of Mars

Fig. 4.2 Two images of flowers

characteristics of information, that depend on the type of information representation (IF1), such as, in this case, the image resolution, that can be measured subjectively or else in terms of a metrics based on dots per inch.

As another example, Fig. 4.3 shows five different version of a photo, that make use of a decreasing number of dots per inch; looking at the 7 Kb version, we consider acceptable the rendering of the image with respect to the original, while in the 2 K case the resolution is not perceived as acceptable. So, we can conceive a

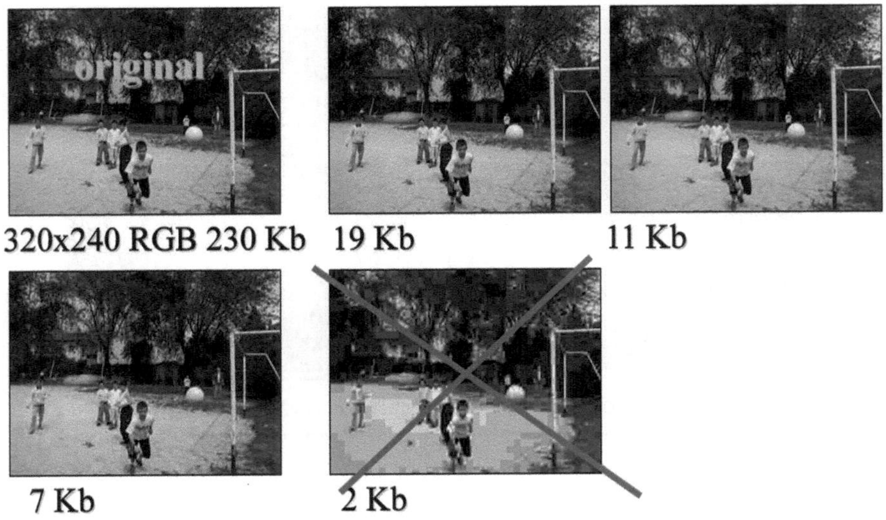

Fig. 4.3 Several representation of the same photo with decreasing amount of dots

concept of *minimal amount of data needed to represent a piece of information* over a threshold of reasonable quality. However we also observe that the context of use plays a role in defining this threshold; as an example, an image used as a web thumbnail is expected to be displayed at lower size (dpis and pixels) than the same image as a picture in a newspaper. We want now to investigate more in depth (see Table 4.2) the relationship between definitions of IQ in the literature and corresponding influencing factors shown in column 1 of the table.

Looking at columns, three different information representations are considered, (a) structured data, (b) images and (c) a specific type of semi-structured text, laws. We can define the quality of the image as the lack of distortions or artifacts that reduce the accessibility of its information contents. Some of the most frequent artifacts considered are: blurriness, graininess, blockiness, lack of contrast and lack of saturation. The definition referring to quality as a list of properties (BI2) is inspired by former contributions from the conceptual modeling research area (Lindland et al. 1994). Whereas the overall framework in Table 4.2 assumes the definition of data and information quality as based on the role of an information system as a representation (Wand and Wang 1996), and the consequent distinction between the internal and external views of an information system (Wand and Weber 1995). The internal view is use-independent, supporting dimensions of quality as intrinsic to the data; while the external view considered the user view of the real world system (the observer perspective), where possible data deficiencies happen (Wand and Wang 1996). Moreover, it is worth noting that most of the research effort in the literature on data quality has provided by far greatest attention to the design and production processes involved in generating the data as the main sources of quality deficiencies (Wand and Wang 1996). Notice also that the definition more closely influenced by

Table 4.2 Definitions of IQ and related issues and factors mentioned in definition

IF1 type of InfoR → Related issues/factors	Structured data	Images	Structured text: laws
IF2/IF6 Absence of defects		A perfect image should be free from all visible defects arising from digitalization and processing processes	
Adherence to the original			
BI2 – Quality as a list of properties	1. High quality data is accurate, timely, meaningful, and complete 2. The degree of excellence of data. Factors contributing to data quality include: the data is stored according to their data types, the data is consistent, the data is not redundant, the data follows business rules, the data corresponds to established domains, the data is timely, the data is well understood		
IF6/IF7 Impression of the observer		Impression of its merit or excellence as perceived by an observer neither associated with the act of photography, nor closely involved with the subject matter depicted (III Association 2007)	

(continued)

Table 4.2 (continued)

IF1 type of InfoR → Related issues/factors	Structured data	Images	Structured text: laws
IF8 Fitness for use/adequacy to the task	Data are of high quality "if they are fit for their intended uses in operations, decision making and planning	1. The perceptually weighted combination of significant attributes (contrast, graininess, etc.) of an image when considered in its marketplace or application 2. Degree of adequacy to its function/goal within a specific application field	Laws whose structure and performance approach those of "the ideal law": – It is simply stated and has a clear meaning – It is successful in achieving its objective – It interacts synergistically with other laws – It produces no harmful side effects – It imposes the least possible burdens on the people
Conformance…	…to requirements	Of match of the acquired/reproduced image with IF2 the original → fidelity IF7 viewer's internal references → naturalness	

the observer (third row) claims for a "third party" subjective evaluation, not influenced by the domain.

Coming to the fourth row, we see that *fitness for use*, that corresponds to IF9, Influence of the task, is the only common driving issue, while the impression of the observer (IF6) is typical of images, that are characterized by a high prevalence of subjective measures on objective ones (IF7). According to IF9, IQ can be expressed quantifying how it influences the performance of the task that uses it. Focusing on images (Batini et al. 2008):

- In the framework of medical imaging, an image is of good quality if the resulting diagnosis is correct.
- In a biometric system, an image of a face is of good quality if the person can be reliably recognized.
- In an optical character recognition system a scanned document has a good quality is all the words can be correctly interpreted.

Finally we comment the conformance definition, which in case of images may be associated:

(a) to the original, focusing in such a way on possible distortions during the processing life cycle (IF2), as a consequence subsuming the possibility to access to the original (Ciocca et al. 2009; Gasparini et al. 2012), or else
(b) to viewer's internal references (IF8), i.e. the perceived model in the user's mind of the image (Ciocca et al. 2009; Gasparini et al. 2012).

This last characteristic is typical of information representations such as images, that may influence emotions of human beings (Ciocca et al. 2009; Gasparini et al. 2012).

4.5 IQ Dimensions

Many possible dimensions and metrics can be conceived for IQ. Focusing on structured data in data bases, 13 methodologies for the assessment and improvement of data quality are listed in Batini et al. (2009), which mention a total of about 220 different dimensions with repetitions and about 70 without repetitions. In Batini and Scannapieco (2006) several examples of synonyms and homonyms existing in the literature among dimensions are shown.

Focusing on most frequently mentioned dimensions, namely accuracy, completeness, consistency, timeliness, currency, in Table 4.3 we see that multiple metrics are defined for each dimension, some of them objective and others subjective (IF6).

Coming to specific dimensions, we now investigate more in depth accuracy and completeness.

Table 4.3 Dimensions and related metrics

Dimensions	Name	Metrics definition
Accuracy	Acc1	Syntactic accuracy: it is measured as the distance between the value stored in the database and the correct one Syntactic accuracy = number of correct values/number of total value
	Acc2	Number of delivered accurate tuples
	Acc3	User survey – questionnaire
Completeness	Compl1	Completeness = number of not null value/total number of values
	Compl2	Completeness = number of tuples delivered/expected number
	Compl3	Completeness of web data = $(T_{max} - T_{current})^*$ $(completeness_{max} - completeness_{current})/2$
	Compl4	User survey – questionnaire
Consistency	Cons1	Consistency = number of consistent values/number of total values
	Cons2	Number of tuples violating constraints, number of coding differences
	Cons3	Number of pages with style guide deviation
	Cons4	User survey – questionnaire
Timeliness	Time1	Timeliness = (max (0;1-currency/volatility))
	Time2	Percentage of process executions able to be performed within the required time frame
	Time3	User survey – questionnaire
Currency	Curr1	Currency = time in which data are stored in the system – time in which data are updated in the real world
	Curr2	Time of last update
	Curr3	Currency = request time – last update
	Curr4	Currency = age + (delivery time – Input time)
	Curr5	User survey – questionnaire

4.5.1 Accuracy Dimension

Several methodologies investigated in Batini et al. (2009), see accuracy from two different points of view, syntactic and semantic (IF5). Figure 4.4 shows Italian first names, and compares them with the item "*Mrio*" that does not correspond to any of them. Semantic accuracy of a value *v* can be intuitively defined as closeness of the value *v* to the true value *v**; for a formal definition in the context of relational databases, the first order logic interpretation of the relational model can be adopted. Since semantic accuracy can be complex to measure and improve, a second type of accuracy, syntactic accuracy, measures the minimal distance between the value *v* and all possible values in the domain *D* of *v*. In our case, if we consider as distance the edit distance, the minimum number of character insertions, deletions, and replacements to convert "*Mrio*" to a string in the domain, the syntactic accuracy of "*Mario*", is 1. Notice that the string corresponding to "*Mrio*" is "*Mario*", but it could be possible that two errors have occurred so that the true value of "*Mrio*" is "*Maria*", another valid Italian name. To recognize this, we need more knowledge on the object represented by "*Mrio*", e.g. that is a male, or a female.

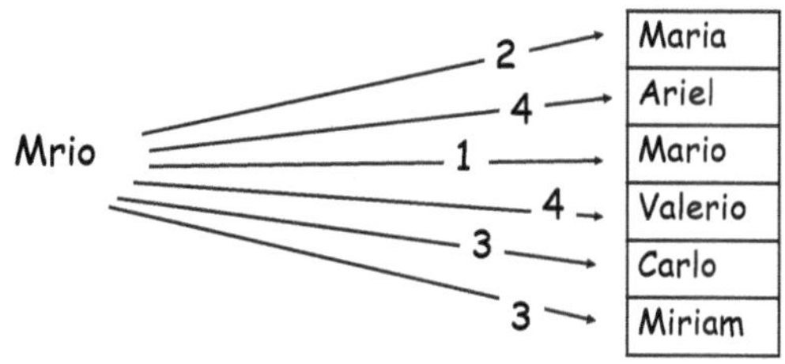

Fig. 4.4 Example of accuracy evaluation

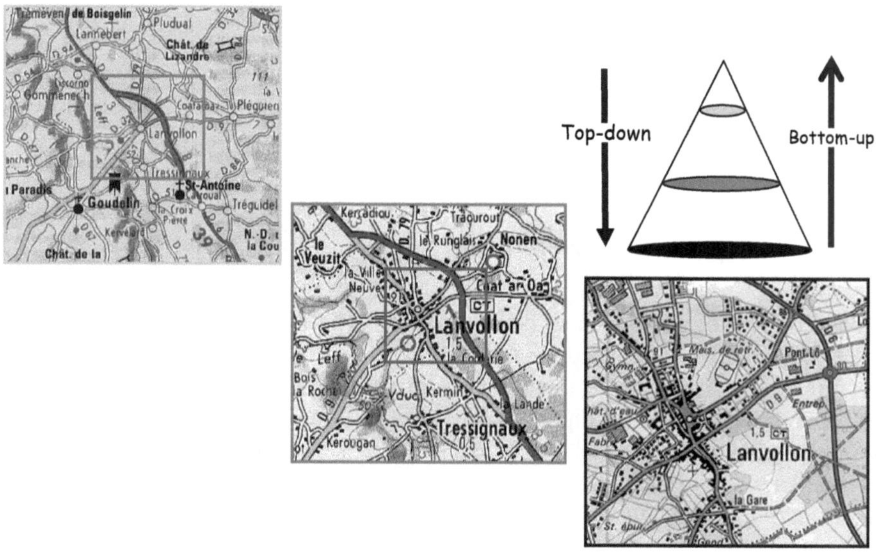

Fig. 4.5 The same geographic area represented at three abstraction levels

Another intriguing relationship to be investigated concerns accuracy and level of abstraction (IF10). Here we focus on maps. In our experience of visiting a city or making a travel by car, we need maps at different levels of detail. Cartographic generalization involves symbolizing data, and applying a set of techniques that convey the salient characteristics of that data. These techniques seek to give prominence to the essential qualities of the feature portrayed, e.g. that buildings retain their anthropogenic qualities – such as their angular form. In Fig. 4.5 we show the same geographic area around the town of Lanvollon in France represented at three abstraction levels.

As said in Encyclopedia of GIS (2010), "Different combinations, amounts of application, and different orderings of these techniques can produce different yet aesthetically acceptable solutions. The focus is not on making changes to information contained in the database, but to solely focus upon avoiding ambiguity in the interpretation of the image. The process is one of compromise reflecting the long held view among cartographers that making maps involves telling small lies in order to tell the truth!".

ID	Name	Surname	BirthDate	Email
1	John	Smith	03/17/1974	smith@abc.it
2	Edward	Monroe	02/03/1967	NULL
3	Anthony	White	01/01/1936	NULL
4	Marianne	Collins	11/20/1955	NULL

not existing
existing but unknown
not known if existing

Fig. 4.6 Completeness in relational tables

These considerations show that even a dimension such as accuracy, that is considered only from the inherent point of view in the ISO standard, is strongly influenced by the context in which information is perceived/consumed.

4.5.2 Completeness Dimension

The definition of completeness depends on the type of information representation (IF1), and is also influenced by the CWA/OWA (IF4). Let us consider the table reported in Fig. 4.6, with attributes `Name`, `Surname`, `BirthDate`, and `Email`. If the person represented by tuple 2 has no e-mail, tuple 2 is complete. If the person represented by tuple 3 has an e-mail, but its value is not known then tuple 3 presents incompleteness. Finally, if it is not known whether the person represented by tuple 4 has an e-mail or not, incompleteness may or may not occur, according to the two cases. In a model such as the relational model, in which only one type of null value is defined, these three types of incompleteness are collapsed into one.

Further, relation completeness, i.e., the number of tuples w.r.t. to the total number of individuals to be represented in the table, depends on the validity of the CWA or else of the OWA. Usually it is assumed that the closed world assumption holds in data bases, in this case a relation is always complete. Instead semantic data are usually considered under the OWA; if we adopt this assumption for our table, then we cannot compute completeness, unless we introduce the concept of reference relation, i.e. a relation that is considered complete, and used as a reference for measuring the completeness of other relations representing the universe, for details see Batini and Scannapieco (2006).

4.6 IQ Dimension Classifications

Several classifications of dimensions are considered in the literature, we shortly mention them, while their comparison is outside the scope of the paper. In Lee et al. (2002), a two ways classification is proposed based on

(a) conforms to specification vs. meets or exceeds consumer expectations (here we find an influence from IF6),
(b) product quality vs. service quality.

Wang and Strong (1996) proposes an empirical classification of data qualities, based on intrinsic, contextual, representations, accessibility qualities. The approach of Liu et al. (2002), is based on the concept of evolutional data quality, where the data life cycle is seen as composed of four phases:

- *Collection*, data are captured using sensors, devices, etc.
- *Organization*, data are organized in a model/representation.
- *Presentation*, data are presented by means of a view/style model.
- *Application*, data are used according to an algorithm, method, heuristic, model, etc.

Qualities that in other approaches are generically attached to data, here are associated to specific phases, e.g. accuracy to collection, consistency to organization. A theory in Liu et al. (2002) is a general designation for any technique, method, approach, or model that is employed during the data life cycle. E.g. when data in the Organization phase is stored, a model is chosen, such as a relational or object-oriented model to guide the data organization. Due to the attachment of data to theories, when defining quality, we need to consider how data meet the specifications or serve the purposes of a theory. Such a concept of quality is called *theory-specific*; e.g., in the relational model, theory specific qualities are normal forms and referential integrity.

In order to investigate the influence of the type of information representation (IF1) on the analysis of several quality dimensions, we use adopt in the following the classification of dimensions proposed in Batini et al. (2008), where dimensions are empirically included in the same cluster according to perceived similarity. Clusters concern:

1. *Accuracy/correctness/precision* refer to the adherence to a given reference reality.
2. *Completeness/pertinence* refer to the capability to express all (and only) the relevant aspects of the reality of interest.
3. *Currency/volatility/timeliness* refer to temporal properties.
4. *Minimality/redundancy/compactness* refer to the capability of expressing all the aspects of the reality of interest only once and with the minimal use of resources.
5. *Readability/comprehensibility/usability* refer to ease of understanding and fruition by users.
6. *Consistency/coherence* refer to the capability of the information to comply with all properties of the membership set (class, category,...) as well as to those of the sets of elements the reality of interest is in some relationship.
7. *Credibility/reputation*, information derives from an authoritative source.

In Table 4.4 we relate dimensions cited in the literature with the above dimension classification (BI3) and with a set of types of information representation (IF1).

Table 4.4 Comparative analysis of quality dimensions for diverse information representations

Quality dimension cluster	Structured data	Geographic maps	Images	Unstructured texts	Laws and legal frameworks
Correctness/accuracy/precision	**IF4** Schema accuracy w.r.t requirements w.r.t. the model	Instance **IF9** Spatial accuracy Relative/absolute Relative inter layer	**IF8** Accuracy Syntactic Semantic "Reduced" semantic	IF8 Accuracy IF5 Syntactic IF5 Semantic IF4 Structural similarity	Accuracy Precision Objectivity Integrity
	IF4 Instance accuracy	Locally increased r.a.	Genuineness		Correctness
	IF5 Syntactic **IF5** Semantic **IF8** Domain dependent (ex. Last Names, etc.)	External/internal Neighbourhood a. Vertical/horizontal/ height Attribute accuracy	Fidelity Naturalness Resolution		Reference accuracy
		IF8 Domain dependent accuracy (ex. traffic at critical intersections, urban vs rural areas, etc.) Accuracy of raster representation	Spatial resolution **IF2** Scan type		
Completeness/pertinence	Schema Completeness Pertinence **IF5** Instance Value C. Tuple C. Column C. Relation C. Database C.	Completeness (btw different datasets) Pertinence	Completeness	Completeness	Objectivity Completeness

Temporal	Currency IF8 Timeliness, volatility	Recency/temporal accuracy/temporal resolution			
Minimality/redundancy/compactness/cost	Schema Minimality Redundancy	Redundancy	Minimality		For a law: Conciseness For a legal framework: **Minimality, redundancy**
Consistency/coherence/interoperability	Instance Intrarelational Consistency Interrelational Consistency Interoperability	**IF9** Consistency Object consistency Geometric consistency Topological consistency Interoperability	Interoperability	**IF5** Cohesion Referential Temporal Locational Causal Structural **IF5** Coherence Lexical Nonlexical	**Coherence** Consistency among laws Consistency among legal frameworks
Readability/comprehensibility/usability/usefulness/interpretability	Schema **IF7** – Diagrammatic readability Compactness Normalization	Instance Readability/legibility Clarity Aesthetics	**IF5** – Readability, lightness, brightness, uniformityness, sharpness, hue chroma reproductionness usefulness	**IF5** – Readability Comprehensibility **IF5** Cultural readability	**IF6** Clarity Simplicity

Several dimensions in the table are associated with corresponding influencing criteria. Notice:

(a) the great variability of the accuracy cluster with the type of information representation,
(b) the clear distinction between schema and instance related dimensions in the "Structured data" column,
(c) the differentiation in the "Laws and Legal framework" column between qualities of single laws and qualities for the legal framework.

After these general considerations, we discuss more in depth the influence of type of information representation (IF1) on specific dimension clusters listed in the table.

1. *Accuracy* is often considered as an intrinsic IQ dimension (IF9), and its quality level is measured either by comparison with the "true" value (IF5, semantics) or else by comparison with a reference table (IF5, syntax).
2. *Accuracy* for structured data is defined both at the schema level and at the instance level, while for unstructured texts is defined at the instance level, with reference to a weaker property called *structural similarity* (IF4), referring in the word "structural" to the latent internal organization of the text.
3. *Accuracy* for structured data has different metrics for different definition domains. We may focus here on (a) surnames of persons, that are made of one word item (e.g. Smith), or else (b) names of businesses, that may involve several word items (e.g. AT&T Research Labs). When data values are typically composed of one single word, distance metrics are adopted that compare the two words seen as strings of characters, without any further internal structure considered. When data values consist of groups of items, then distance metrics consider the total number of items in data values, and the number of common items (Jaccard's distance), or variants of metrics that are based on the internal structure of values. Even in case of single words, metrics are sensitive to the average length of words in the universe of discourse; so that they change when, e.g., consider surnames in United States and in Asia, where surnames in certain populations are very long.
4. *Spatial accuracy* for maps refers to a bidimensional or tridimensional metric space (IF9).
5. *Consistency* for geographic maps is defined both in the topological space and in the geometric space (IF9).
6. *Cohesion* and *coherence* are proposed for unstructured texts. Both cohesion and coherence represent how words and concepts conveyed in a text are connected on particular levels of language, discourse and world knowledge. Cohesion is considered an objective property (IF6) of the explicit language and text, and is achieved by means of explicit linguistic devices that allow expressing connections (relations) between words, sentences etc. These cohesive devices cue the

reader on how to form a coherent representation. Coherence results from an interaction between text cohesion and the reader. The coherence relations are constructed in the mind of the reader (IF7) and depend on the skills and knowledge that the reader brings to the situation. Coherence is considered a characteristic of the reader's mental representation, and as such is considered subjective (IF6). A particular level of cohesion may lead to a coherent mental representation from one reader but an incoherent representation for another (IF7).

7. *Diagrammatic readability* is usually expressed in terms of the achievement of several aesthetic criteria such as:

   (a) Minimize crossings
   (b) Use only horizontal and vertical lines
   (c) Minimize bends in lines
   (d) Minimize the area of the diagram
   (e) Place most important concept in the middle
   (f) Place parent objects in generalization above child objects.

   Notice that criteria a, b, c and d can be considered syntactic criteria, while e and f are semantic criteria (IF5). Applying such criteria to the two semantically equivalent Entity Relationship diagrams in Fig. 4.7, we may come to the conclusion that the diagram on the right is more readable than the diagram on the left. Unfortunately (or fortunately) this is not a universal conclusion, since about 30 years ago one of the authors was invited to visit Beda University at Peking, and Chinese professors preferred the diagram on the left, claiming that they liked asymmetry and sense of movement (IF7).

8. *Readability of unstructured texts* and *cultural accessibility* refer to the readability/comprehensibility cluster. Readability is usually measured by using a mathematical formula that considers *syntactic features* of a given text, such as complex words and complex sentences, where e.g. complex words are evaluated on the basis of shallow syntax, such as number of syllables. *Cultural readability* refers to difficult (to understand) words, so they are related to the understanding of the word meaning, and as such can be considered more semantic oriented (IF6).

9. Concerning the relationship between IQ dimensions in the different representations vs. objective/subjective measures (IF6), we have produced some figures in the past that confirm the validity of the following intuitive statement in the literature: the less the information is structured, from a restricted domain to a totally unstructured domain, the more subjective measures prevail on objective measures.

In Fig. 4.8 we show two types of information representations, relational tables and diagrams, and three measures of IQ quality, respectively for *accuracy* of data for relational tables, and *readability* for diagrams addressed in previous point 8. It is clear (also recalling the previous example on Chinese professors) that objective measures can be conceived for diagrams, but only to a certain extent, after that we have to deal with human being perceptions.

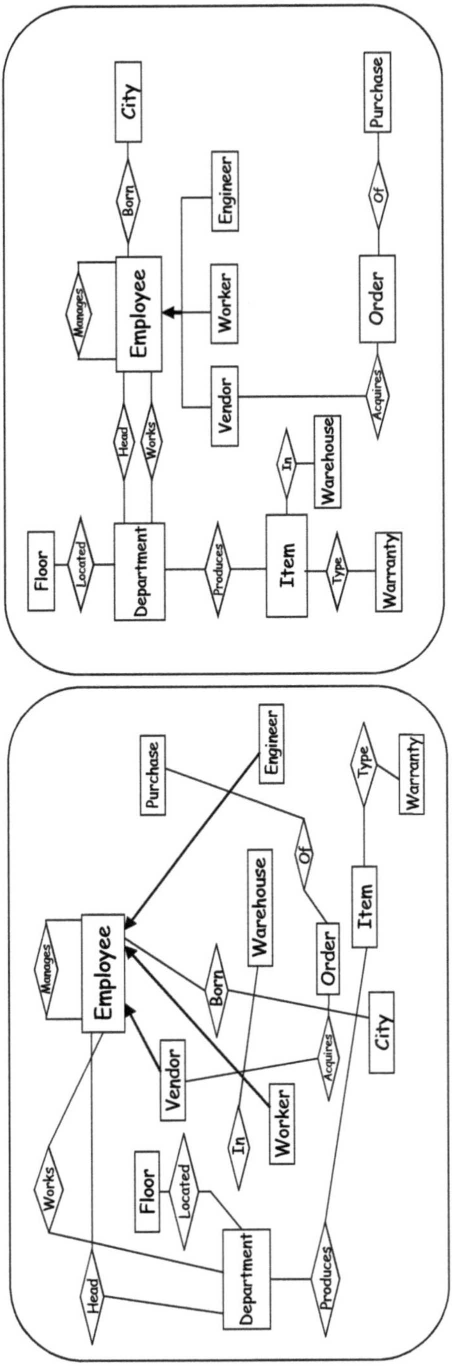

Fig. 4.7 Two semantically equivalent entity relationship diagrams

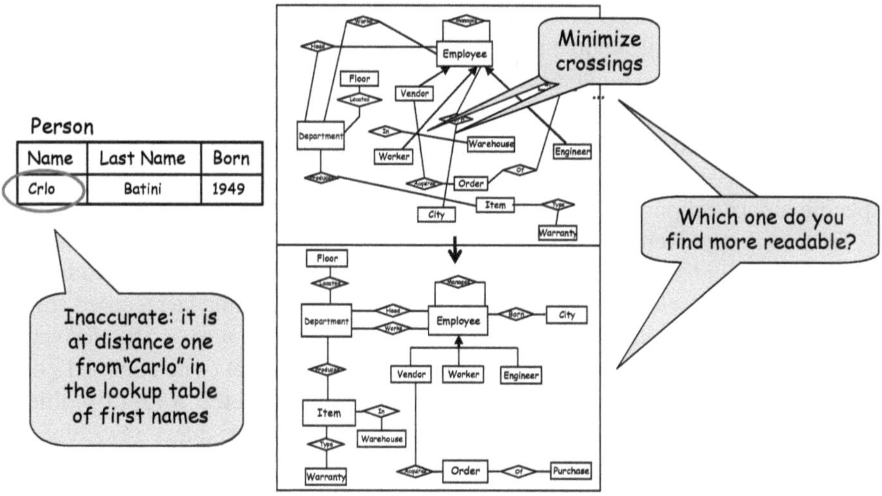

Fig. 4.8 comparison of IQ measures for relational tables and diagrams

4.7 IQ Dimensions and Types of Information Systems (IF3)

We now investigate the relationships between IQ dimensions and the evolution of types of information systems, enabled by the evolution of ICT technologies. The shift from centralized and tightly coupled distributed systems to loosely coupled, distributed and peer to peer systems, and from "controlled" sources to the unrestrainable web results both in bad and in good news from the point of view of IQ. From one side, the overall quality of the information that flows between networked information systems may rapidly degrade over time if both processes and their inputs are not themselves subject to quality control. On the other hand, the same networked information system offers new opportunities for IQ management, including the possibility of selecting sources with better IQ, and of comparing sources for the purpose of error localization and correction, thus facilitating the control and improvement of data quality in the system.

Peer to Peer data management (P2P) Systems, typical of many application areas such as the ones found in the domain of biological databases, differently from centralized and strongly coupled distributed systems do not provide a global schema of the different sources. P2P systems are characterized by their openness, i.e. a peer can dynamically join or leave the system, and by the presence of mappings usually relating pairs of schemas. In P2P systems (and even more in the web) new quality dimensions and issues have to be considered such as *trustworthiness* and *provenance*.

The evaluation of the trustworthiness (or confidence) of the data provided by a single peer is crucial because each source can in principle influence the final, integrated result. A common distinction is between the reputation of a source, which refers to the source as a whole, and the trust of provided data, e.g., the trust of the mapping that the source establishes with the other sources in a P2P system. While several trust and reputation systems have been proposed in the literature (see Josang et al. (2007) for a survey), there is still the need to characterize the trust of a peer with respect to provided data and use such information in the query processing step. Effective methods for evaluating trust and reputation are needed, with the specific aim of supporting decisions to be taken on result selection.

Information provenance describes how data is generated and evolves with time going on, which has many applications, including evaluation of quality, audit trail, replication recipes, citations, etc. Generally, the provenance could be recorded among multiple sources, or just within a single source. In other words, the derivation history of information could take place either at schema level (when defined), or at instance level. Even if significant research has been conducted, a lot of problems are still open. For the schema level, the most important are query rewriting and schema mappings including data provenance, and for the instance level, we mention relational data provenance, XML data provenance, streaming data provenance (Buneman and Tan 2007). Moreover another important aspect to be investigated is dealing with uncertain information provenance for tracking the derivation of information and uncertainty.

4.8 IQ Dimensions and Levels of Semantic Constraints (IF4)

Influencing factor IF4 deserves special attention in the context of this book. We address in this section the discussion on levels of semantic constraints and the adoption of OWA vs. CWA, while next section details the changes in perspective when moving form databases to ontologies and knowledge bases.

As we anticipated in Sect. 4.3, different levels of semantic constraints can be imposed to data. In databases, data and schemas are tightly coupled; schemas pre-exist to data and control methods implemented by database management systems can enforce data to comply to the schema, which, even if poorly, defines their semantics. As an example, normal forms in relational databases are defined at the schema level, and are expressed in terms of properties of functional dependencies defined in relational schemas. A relational database whose relation schemas are in normal form, has relation instances free of redundancies and inconsistencies in updates, since every "fact" is represented only once in the database.

The coupling of data and schemas in semi-structured or weakly structured data is way looser. Even when languages for semi-structured or weakly structured data are accompanied with languages for describing data schemas, e.g., XML-Schema for XML, RDFS and OWL2 for RDF (Antoniou and van Harmelen 2008), schemas are not required to pre-exist to data and the enforcement of the compliance of data to a schema at publishing time is weaker (it is left to the data publisher). Data in these cases are associated with schemas by means of annotation mechanisms. Finally, the use of metadata, e.g., based on folksonomies, or other annotation schemes, can be seen as a way to associate data with schema-level information that provides data with semantics. However, the maximum freedom achieved by these representation approaches leads to a yet weaker coupling of data and schemas.

As an example, let us focus on semantic data represented in RDF, which is also accompanied with expressive languages for the representation of schemas. A schema for RDF data can be defined by a RDFS vocabulary; however, there is no mechanism to enforce data to be compliant to the schema; even using reasoning, RDFS is not expressive enough to detect inconsistencies, because of its deductive semantics (the schema is used to make inference, not to constraint their meaning) and the lack of expressivity (concept disjointness and cardinality restrictions cannot be modeled in RDFS) (Antoniou and van Harmelen 2008); although counterintuitive inferences can be considered a measure of poor compliance between data and schemas (Yu and Heflin 2011), no inconsistencies can be detected, making a quality dimension such as *soundness* difficult to assess.

In addition, the adoption of CWA or OWA has an influence on this discussion; OWA has an impact on the difficulty of defining and evaluating the compliance between data and schemas: a relation between two instances can hold even if the schema does not model such relation between the concepts the instances belong to; conversely, we cannot conclude that a relation between two concepts of different schemas does not hold because it is not represented in the data instances.

4.9 The Impact of the Information Representation Model Flexibility on IQ

4.9.1 From Databases to Knowledge Bases on the Web

Considering the remarks in the previous section, we can conclude that the more types of information are considered, and the more diverse and decentralized information management models and architectures are, the more we are in need of rethinking the perspective through which we look at information quality (in computer science). An interesting perspective on the role that diversity of information objects can play in IQ emerges if we investigate how the IQ perspective changes when moving from data bases to web knowledge bases (KBs), i.e., knowledge bases published, shared and accessible on the web. Web KBs are, in fact, diverse and often decentralized information sources.

In general, we can see a web KB as composed of a terminological component and an assertional component (Staab and Studer 2004). The terminological component of a web KB, usually called *ontology*, conveys general knowledge about a domain in terms of logical constraints that define the meaning of the concepts (and relations) used in the language (e.g. "*every Cat is an Animal*"); ontologies for web KBs (web ontologies for short) are represented with web-oriented formal languages like OWL, RDFS, and SKOS (Antoniou and van Harmelen 2008).

The assertional component of a web KB expresses facts in terms of properties of individuals, i.e., instances of ontology concepts, and relations holding between them (e.g. "*Fritz is a Black Cat*"; "*Fritz is friend of Joe*"). We remark that the distinction between the two components in a KB can be more or less sharp depending on the language used to represent the KB and the ontology, but it can be adopted without loss of generality for our purposes.[1] Also, terminological and assertional components can be independent (see the Sect. 4.9.2) and several ontologies that are not designed for specific assertional components exist, e.g., consider an upper-level ontology such as DOLCE.[2]

In the following we focus on IQ as investigated in the field of ontologies because they represent a fundamental aspect of web KBs.[3]

4.9.2 Some Issues Arising from the Investigation of IQ for Ontologies: Semiotic, Diversity, Reuse

We concentrate on three main characteristics of ontologies, each of which shed light on significant aspects of IQ when considered in an open information spaces.

4.9.2.1 Ontologies Are Semiotic Objects

One of the first works that addressed the problem of evaluating (the quality of) ontologies exploited a framework based on a semiotic model (Burton-Jones et al. 2005). A similar approach appears in a model that describes the relationship between ontologies as formal (externalized) specifications, (mental) conceptualization and the "real world" (Gangemi et al. 2006). Within this cognitive-flavored semiotic approach, several quality dimensions, and metrics have been defined on top of these frameworks. Gangemi et al. (2006) distinguishes between quality dimensions and evaluation principles.

Three types of dimensions under which it is possible to evaluate an ontology are discussed. The *structural dimension* focuses on syntax and formal semantics, i.e. on ontologies represented as graphs (context free metrics). The *functional dimension*

[1] The use of lexical resources such as WordNet or other taxonomies represented in SKOS in KBs is widespread. Although these resources are used for annotation purposes in the assertional components of KBs, they are very often referred to as *ontologies* in the community (Manaf et al. 2012) and share likewise terminological components of KBs define semantic relations between concepts in a domain.

[2] http://www.loa.istc.cnr.it/DOLCE.html

[3] Most of these approaches explicitly consider ontologies as KB terminologies represented in web-compliant formal languages. Some of the approaches use a even broader definition of ontology which includes instances and relations among instances and is equivalent to our definition of web KB.

is related to the intended use of a given ontology and of its components, i.e. their function in a context. The focus is on the conceptualization specified by an ontology. The *usability-profiling dimension* focuses on the ontology profile (annotations), which typically addresses the communication context of an ontology (i.e. its pragmatics). Then several principles (or evaluation-driven dimensions) are introduced, namely: *cognitive ergonomics, transparency, computational integrity and efficiency, meta-level integrity, flexibility, compliance to expertise, compliance to procedures for extension, integration, adaptation, generic accessibility*, and *organizational fitness*.

Following the cognitive flavor of this point of view, a quite recent approach studied a measure of cognitive quality based on the adequacy of represented concept hierarchies w.r.t. the mental distribution of concepts into hierarchies according to a cognitive study (Evermann and Fang 2010). These cognitive approaches clarify an important issue that has been central in the research about IQ in the ontology domain: ontologies are knowledge objects that are used by someone and for some specific goals; the evaluation of the quality of ontology should consider ontology in its semiotic context.

4.9.2.2 Ontologies as Diverse Knowledge Objects

As it can be captured from the broad definition of ontology given at the beginning of this paragraph, ontologies are very different one from another. Some ontologies are flat, while some others consist in deep concept hierarchies; some ontologies are deeply axiomatized, while others, e.g. Geonames,[4] look more like database schemas (Cruz et al. 2012, 2013). Moreover, often ontologies cannot be modified but are reused and eventually extended. Some metrics defined for evaluating an ontology can be adopted to provide a value judgment about an ontology. Other metrics proposed so far are more intended as analytic dimensions to profile an ontology, and to understand its structure and its properties. As an example, one of the first unifying framework proposed to assess ontology quality distinguishes between syntactic, semantic, pragmatic and social qualities (see Table 4.5) (Burton-Jones et al. 2005).

Although lawfulness and interpretability clearly lead to a value judgment (positive vs. negative), metrics such as richness and history can be hard to be associated with a value judgment. In other frameworks such as the one proposed by (Gangemi et al. 2006; Tartir et al. 2005), which put a lot of focus on the computability of the defined metrics, most of the metrics are more aimed at profiling an ontology, rather than at assessing its quality from a value perspective. The idea is that these quality metrics can be used to summarize the main property of an ontology and their evaluation can be used by third party applications. As an example, a machine learning method that takes advantage of fine-grained ontology profiling techniques (extended from Tartir et al. (2005)) to automatically configure an ontology matching system

[4] http://www.geonames.org/

Table 4.5 Types of qualities and dimensions in Burton-Jones et al. (2005)

Dimension	Metrics	Definition
Syntactic quality	Lawfulness	Correctness of syntax
	Richness	Breadth of syntax used
Semantic quality	Interpretability	Meaningfulness of terms
	Consistency	Consistency of meaning of terms
	Clarity	Average number of word senses
Pragmatic quality	Comprehensiveness	Number of classes and properties
	Accuracy	Accuracy of information
	Relevance	Relevance of information for a task
Social quality	Authority	Extent to which other ontologies rely on it
	History	Number of times the ontology has been used

has been recently proposed (Cruz et al. 2012). These approaches, which consider ontologies also as computational resources (see point above), differ from early works on ontology quality that were based on philosophical (metaphysical) principle to establish the quality of an ontology as a conceptual model, but whose analytical principles are more difficult to be made computable.

4.9.2.3 Ontologies as (Reusable) Computational Resources

A key aspect of ontologies is that they are expected to be reused by other ontologies, applications, or, more generically, third party processes. It is often the case that one has to select an ontology to reuse it in a given domain. Ontologies can be used to support search or navigation. Different aspects of an ontology can be more or less amenable depending on the task an ontology is aimed to support. Approaches that evaluate ontologies on a task basis (Yu et al. 2007; Lei et al. 2007; Strasunskas et al. 2008) seem to have received more attention, recently, than previous approach based on metaphysical and philosophical considerations (Guarino and Welty 2002), which better fit the use of ontologies as conceptual models, rather than as computational objects.

4.10 Conclusive Remarks

In this paper we have discussed the main issues considered in data quality and information quality research, identifying several factors influencing them. According to a quite common use of the terms in the technical literature published by the data management community, we referred to data quality when structured data where addressed, and to information quality when information represented according to other data models is considered. However, the consideration of information digitally represented by different types of data and organized according to different data models has definitely a deep impact on the most relevant issues considered in information quality, including the definition itself. The more heterogeneous the considered information is, the more a comprehensive theoretical framework defining in a general way the mutual relationship between several crucial concepts in the definition and assessment of information quality (e.g., data, information, information carrier, observer, task, and so on) is needed. Recent works in the field of ontology evaluation framed the (information) quality problem within a broader semiotic and cognitive framework (see Gangemi et al. (2006) and Evermann and Fang (2010)). A similar concern can be found in several works on information quality coming from the Information Systems community (see Wand and Weber (1995, 1990) and Wand and Wang (1996)). These approaches can provide important contributions to a theoretical clarification of the common use of information quality core concepts and issues, in a context where the amount and the degree of complexity, diversity, and interconnection of the information managed in ICT is constantly increasing.

One problem that we believe particularly interesting is tightly related to the influencing factor IF4 addressed in this paper, which considers the impact on information quality of the degree of coupling between data and schemas (where available), and the difference in the semantics associated with structured and other types of data (e.g., schemaless data such as texts, images, sounds). An interesting research question concerns the extent to which information quality is affected by the degree of coupling between data and schemas, or, more in general, the role played by semantics defined by data models and schemas in the definition of information quality. This issue tightly relates to the relationship between data, information and *truth* in information systems. In this case, information quality faces the dualism of scheme and content, of organizing systems and something waiting to be organized, as criticized by Davidson as the third dogma of empiricism (Davidson 1974). If schema-driven data can be easily interpreted as carriers of factual information and interpreted according to a semantic theory of truth (Kirkham 1992) (e.g., through mapping to First-Order Logic), the connection between other types of information representations (e.g., maps, images, sounds) and factual information has been less investigated and results more obscure. Texts can be taken as borderline examples from this point of view: most of textual documents are clearly carriers of factual information to a human reader, but their digital representation is by no means related to any factual interpretation (hence, investigations in the field of natural language processing, knowledge extraction, and so on).

Moving to the conceptual challenges to be faced in the future, as also shown by the above reference to the work of Davidson, it is our point that contributions from philosophy can bring some theoretical clarification to IQ basic issues and influencing factors. Otherwise, we argue that these challenges are going to be tangled by dichotomies such as the ones implied in the discussion carried out in previous sections. As an example, consider factual information, which is represented both in structured and semi-structured information. Some of the quality dimensions proposed in the literature pose the question of adherence of a certain representation to real world (see for example IF6, and BI2 as for clusters of dimensions such as accuracy or consistency). As for these issues, considering (IF4), the critical question here is whether information qualities pertain to facts of sense or rather to laws of logic or, else, whether IQ is a matter of synthetic rather than analytic knowledge (e.g., are there truly intrinsic quality dimensions?). This and other issues related to IQ and discussed in the paper recall in fact philosophical disputes about the two dogmas of empiricism, against which Quine provided arguments, in favor of a holistic perspective. On the one hand, Quine rejected the distinction between truths independent from facts, and truths grounded in facts; on the other hand, he contrasted reductionism as the theory according to which the meanings of statements come from some logical construction of terms, exclusively referring to immediate experience (Quine 1951).

An example is the current debate among scholars and practitioners about the use of quality dimensions coming from practice in specific domains (Embury et al. 2009), instead of well-established (often academic) ones. Furthermore considering practitioners' debate on Linkedin Groups (e.g., the IAIDQ – Information/Data Quality Professional Open Community) where some members argue the need for better definition of "data quality" as different from "data qualities",[5] and of "dimensions",[6] likewise. As for a lesson learned by the work of Quine, this issue may require challenging the ontology anchoring data quality since (Wand and Wang 1996). In particular, we believe that the following assumptions are worth being challenged when conducting research on information quality "in the wild":

- *the quality of the data generated by an information system depends on the design of the system*: this assumption is grounded in a closed perspective on the information system design as bound by an organization requirements; whereas today we assist to an ecosystems of information systems, providing information to both businesses and lay users, often in an open and bidirectional way, actually having different design requirements for intended use by different organizations and target users.
- *The internal/external views of an information system*: strictly related to the previous assumption, this dichotomy leads to a focus on the internal view considered as use-independent, and the identification of a set data quality dimensions comparable across applications and viewed as intrinsic to data. This perspective is based on the idea that systems requirements capture the true intentions of the users (Wand and Wang 1996) As said above, today it is difficult to identify the true intentions of the users, due to the variety, heterogeneity, and the openness of the information systems, thus questioning the internal view assumption: "issues related to the external view such as why the data are needed and how they a used are not part of the model" (Wand and Wang 1996, p. 11).
- *The definition of data deficiency as inconformity* between a view of a real-world system mediated by a representing information system, and a reference view of

[5] See IAIDQ discussion "Do data quality dimensions have a place in assessing data quality?", 2nd July 2013.

[6] See IAIDQ discussion "Do data quality dimensions have a place in assessing data quality?", 9th July 2013.

a real-world system obtained by direct observation. Again, today information systems are not designed "from scratch" and are composed both by legacy systems and a (often) dynamic configuration of external systems for information search, retrieval, and production (social networks, internet of things, etc.). Thus, *inconformity* between views of real-world is actually difficult to ascertain, being today probably a rule rather than an anomaly of information systems "in the wild" (as for this issue, the arguments by Quine on the indeterminacy of translation and the meaning of the expressions of one's own language (Weir 2008; Quine 1960) may provide insights to information quality research).

Furthermore, the above issues may also be related to the problem of knowledge of things *by acquaintance* (e.g. in the case of images) and *by description* (e.g. in the case of structured data), as stated for example by Bertrand Russell: "We shall say that we have acquaintance with anything of which we are directly aware, without the intermediary of any process of inference or any knowledge of truths" (Russell et al. 1910). Thus, differently from knowledge by acquaintance, knowledge by description connects the truths (carried by data, in our case) with things with which we have acquaintance through our direct experience with the world (*sense-data*, in the Russell perspective).

As an example of the role of factual information carried by data in information quality, consider the above discussion pointing out data and information quality pose the question of adherence of a certain representation to real world (see for example, clusters of dimensions such as *Accuracy/correctness/precision* or *Completeness/pertinence*). This question points to one of the most controversial issues discussed in philosophy so far. Significantly, Russell discusses this issue using the term *data*, and in particularly distinguishing between *hard data* and *soft data:* "The hardest of hard data are of two sorts: the particular facts of sense, and the general truths of logic" (Russell 1914/2009, p. 56).

Indeed, from the above discussion we could ask ourselves to which extent information quality (and specific quality dimensions) may pertain to the domain of both hard and soft data. Thus, the critical question is if information quality pertains to facts of sense or rather to laws of logic, which play a fundamental role both at the data model level (e.g., relational algebra for relational databases) and at the schema level (e.g., all persons are identified by their Social Security Number). Again, what can we say about data that are not straightforwardly associated with any truth-based semantics (e.g. images)? Finally, we mention that the role of the processes and tasks that are supported by an information system has to be considered when investigating the above research questions (the number of papers focusing on task-oriented evaluation of information quality is in fact increasing, see, e.g. Yu et al. (2007), Lei et al. (2007), Strasunskas et al. (2008)).

We believe that the above insights should be considered working constructs, with the aim of investigating whether philosophical research can help to clarify significant relationships between basic issues and influencing factors of IQ, too often narrowly considered under a technical perspective in computer science and information systems areas. In particular, the previously cited well known contributions from philosophy may help bending IQ basic issues and influencing factors towards a holistic or else pragmatist perspective (Rorty 1982); this latter being suitable to challenge what we have described in the introduction as the current wild landscape in which information is published, processed and consumed.

Acknowledgments We acknowledge Raimondo Schettini and his research group for providing insights and some of the figures in the paper, with specific reference to image quality.

References

Antoniou, G., & van Harmelen, F. (2008). *A semantic web primer*. Cambridge: MIT Press.

Batini, C., & Scannapieco, M. (2006). *Data quality: Concepts, methodologies and techniques*. Berlin/Heidelberg: Springer.

Batini, C., Di Battista, G., & Santucci, G. (1993). Structuring primitives for a dictionary of entity relationship data schemas. *IEEE Transactions on Software Engineering, 19*, 344–365.

Batini, C., Cabitza, F., Pasi, G., & Schettini, R. (2008). *Quality of data, textual information and images: A comparative survey*. Tutorial at the 27th International Conference on Conceptual Modeling (ER 2008), Barcelona, available on request to batini@disco.unimib.it.

Batini, C., Cappiello, C., Francalanci, C., & Maurino, A. (2009). Methodologies for data quality assessment and improvement. *ACM Computing Surveys, 41*, 16:1–16:52.

Batini, C., Palmonari, M., & Viscusi, G. (2012, July 2–6). The many faces of information and their impact on information quality. In P. Illari & L. Floridi (Eds.), *Information quality symposium at AISB/IACAP World Congress*, Birmingham (pp. 5–25). The Society for the Study of Artificial Intelligence and Simulation of Behaviour.

Buneman, P., & Tan, W. (2007). *Provenance in databases*. SIGMOD Conference (pp. 1171–1173), ACM Press.

Burton-Jones, A., Storey, V. C., Sugumaran, V., & Ahluwalia, P. (2005). A semiotic metrics suite for assessing the quality of ontologies. *Data & Knowledge Engineering, 55*, 84–102.

Ciocca, G., Marini, F., & Schettini, R. (2009). Image quality assessment in multimedia applications. *Multimedia Content Access Algorithms and Systems III, SPIE*, Vol. 7255, 72550A.

Cruz, I. F., Fabiani, A., Caimi, F., Stroe, C., & Palmonari, M. (2012). Automatic configuration selection using ontology matching task profiling. In *ESWC 2012* (pp. 179–194).

Cruz, I. F., Palmonari, M., Caimi, F., & Stroe, C. (2013). Building linked ontologies with high precision using subclass mapping discovery. *Artificial Intelligence Review, 40*(2), 127–145.

Davidson, D. (1974). On the very idea of a conceptual scheme. In J. Rajchman & C. West (Eds.), *Proceedings and addresses of the American Philosophical Association*, Vol. 47 (1973–1974), pp. 5–20. JSTOR.

Embury, S. M., Missier, P., Sampaio, S., Greenwood, R. M., & Preece, A. D. (2009). Incorporating domain-specific information quality constraints into database queries. *Journal of Data and Information Quality, 1*, 11:1–11:31.

Encyclopedia of GIS. (2010). *Encyclopedia of geographical information systems*. Springer.

Evermann, J., & Fang, J. (2010). Evaluating ontologies: Towards a cognitive measure of quality. *Information Systems, 35*, 391–403.

Floridi, L. (2011). Semantic conceptions of information. The Stanford Encyclopedia of Philosophy.

Gangemi, A., Catenacci, C., Ciaramita, M., & Lehmann, J. (2006). Modelling ontology evaluation and validation. In Y. Sure & J. Do-mingue (Eds.), *ESWC* (pp. 140–154), Vol. 4011 of Lecture Notes in Computer Science, Springer.

Gasparini, F., Marini, F., Schettini, R., & Guarnera, M. (2012). A no-reference metric for demosaicing artifacts that fits psycho-visual experiments. *EURASIP Journal on Advances in Signal Processing, 2012*, 4868–4873.

Guarino, N., & Welty, C. A. (2002). Evaluating ontological decisions with OntoClean. *Communications of the ACM, 45*, 61–65.

Hutchins, E. (1995). *Cognition in the wild*. Cambridge: MIT Press.

ISO/IEC FDIS 25012. (2008). Software engineering – Software product quality requirements and evaluation – Data quality model.

Josang, A., Ismail, R., & Boyd, C. (2007). A survey of trust and reputation systems for online service provision. *Decision Support Systems, 43*, 618–644.

Kirkham, R. (1992). *Theories of truth: A critical introduction* (pp. xi, 401). Cambridge, MA: The MIT Press.

Lee, Y. W., Strong, D. M., Kahn, B. K., & Wang, R. Y. (2002). AIMQ: A methodology for information quality assessment. *Information and Management, 40*, 133–146.

Lei, Y., Uren, V. S., & Motta, E. (2007). A framework for evaluating semantic metadata. In D. H. Sleeman & K. Barker (Eds.), *K-CAP* (pp. 135–142). ACM.

Lindland, O. I., Sindre, G., & Solvberg, A. (1994). Understanding quality in conceptual modeling. *IEEE Software, 11*, 42–49.

Liu, L., & Chi, L. (2002). Evolutional data quality: A theory-specific view. In *The 6th International Conference on Information quality*, Boston.

Madnick, S. E., Wang, R. Y., Lee, Y. W., & Zhu, H. (2009). Overview and framework for data and information quality research. *Journal of Data and Information Quality, 1*, 1–22.

Manaf, N. A. A., Bechhofer, S., & Stevens, R. (2012). The current state of SKOS vocabularies on the web. In *ESWC 2012* (pp. 270–284). Berlin/Heidelberg: Springer-Verlag.

Merriam Webster. Knowledge. http://www.merriam-webster.com/dictionary/knowledge

Quine, W. V. O. (1951). Two dogmas of empiricism. *Philosophical Review, 60*, 20–43.

Quine, W. V. O. (1960). *Word and object*. Cambridge: MIT Press.

Rorty, R. (1982). *Consequences of pragmatism: Essays, 1972–1980*. Minneapolis: University of Minnesota Press.

Russell, B. (1910). Knowledge by acquaintance and knowledge by description. In *Proceedings of the Aristotelian Society* (New Series), Vol. XI (1910–11), pp. 108–128.

Russell, B. (1914/2009). *Our knowledge of the external world*. London/New York: Routledge.

Staab, S., & Studer, R. (Eds.). (2004). *Handbook on ontologies*. Berlin: Springer.

Strasunskas, D., & Tomassen, S. L. (2008). Empirical insights on a value of ontology quality in ontology-driven web search. In R. Meersman & Z. Tari (Eds.), *OTM Conferences (2)*, Vol. 5332 of Lecture Notes in Computer Science (pp. 1319–1337). Springer.

Tartir, S., Arpinar, I. B., Moore, M., Sheth, A. P., & Aleman-Meza, B. (2005). *OntoQA: Metric-based ontology quality analysis*. IEEE Workshop on Knowledge Acquisition from Distributed, Autonomous, Semantically Heterogeneous Data and Knowledge Sources.

Wand, Y., & Wang, R. Y. (1996). Anchoring data quality dimensions in ontological foundations. *Communications of the ACM, 39*, 86–95.

Wand, Y., & Weber, R. (1990). An ontological model of an information system. *IEEE Transactions on Software Engineering, 16*, 1282–1292.

Wand, Y., & Weber, R. (1995). On the deep structure of information systems. *Information Systems Journal, 5*, 203–223.

Wang, R. Y., & Strong, D. M. (1996). Beyond accuracy: What data quality means to data consumers. *Journal of Management Information Systems, 12*, 5–33.

Weir, A. (2008). Indeterminacy of translation. In E. Lepore & B. C. Smith (Eds.), *The Oxford handbook of philosophy of language* (pp. 233–249). Oxford: Oxford University Press.

Yu, Y., & Heflin, J. (2011). Extending functional dependency to detect abnormal data in RDF graphs. In L. Aroyo, C. Welty, H. Alani, J. Taylor, & A. Bernstein (Eds.), *The 10th International Conference on The semantic web – Volume Part I (ISWC'11)* (pp. 794–809). Berlin/Heidelberg: Springer Verlag.

Yu, J., Thom, J. A., & Tam, A. (2007). Ontology evaluation using Wikipedia categories for browsing. In *The Sixteenth ACM Conference on Information and knowledge management, CIKM '07* (pp. 223–232). New York: ACM.

Chapter 5
What Is Visualization Really For?

Min Chen, Luciano Floridi, and Rita Borgo

Abstract Whenever a visualization researcher is asked about the purpose of visualization, the phrase "gaining insight" by and large pops out instinctively. However, it is not absolutely factual that all uses of visualization are for gaining a deep understanding, unless the term insight is broadened to encompass all types of thought. Even when insight is the focus of a visualization task, it is rather difficult to know what insight is gained, how much, or how accurate. In this paper, we propose that "saving time" in accomplishing a user's task is the most fundamental objective. By giving emphasis to "saving time", we can establish a concrete metric, alleviate unnecessary contention caused by different interpretations of insight, and stimulate new research efforts in some aspects of visualization, such as empirical studies, design optimization and theories of visualization.

5.1 Introduction

Visualization was already an overloaded term, long before it has become a fashionable word in this era of data deluge. It may be used in the context of meditation as a means for creative imagination, or in sports as a means for creating a heightened sense of confidence. If one considers the term literally, as Robert Spence said,

M. Chen (✉)
University of Oxford, Oxford, UK
e-mail: min.chen@oerc.ok.ac.uk

L. Floridi
Oxford Internet Institute, University of Oxford,
1 St Giles, Oxford OX1 3JS, UK
e-mail: luciano.floridi@oii.ox.ac.uk

R. Borgo
Swansea University, Swansea, UK
e-mail: r.borgo@swansea.ac.uk

L. Floridi and P. Illari (eds.), *The Philosophy of Information Quality*,
Synthese Library 358, DOI 10.1007/978-3-319-07121-3_5,

"visualization is solely a human cognitive activity and has nothing to do with computers" (Spence 2007).

In this article, we focused on visualization in computing, which may be referred to technically as *Computer-supported Data Visualization*. In this context, the process of visualization features data, computer and human users. The first two essential components differentiate this technological topic from those above-mentioned contexts. In the remainder of this article, we will simply refer to "computer-supported data visualization" as "visualization".

Visualization is intrinsically related to *information quality*. Firstly, a visualization image is a form of data and conveys information. Hence the quality of the visualization image is at least one of the significant metrics of the quality of information being conveyed. Secondly, the process of visualization always involves transforming one data representation to another, for instance, from a table of numbers to a bar chart, and from a stack of x-ray images to a geometric surface. Hence, it is most likely that the visualization process also alters the quality of information after the transformation, for "better" hopefully. However, any interpretation or measurement of "better" is fundamentally underpinned by the definition of visualization.

Scott Owen (1999) compiled a collection of definitions and rationale for visualization, most of which are still widely adopted or adapted today. These definitions were intended to define the two questions, namely what is visualization and what is it for?

> The goal of visualization in computing is to gain *insight* by using our visual machinery. (McCormick et al. 1987)

> Visualization is a method of computing. It transforms the symbolic into the geometric, … Visualization offers a method for seeing the unseen. It enriches the process of scientific discovery and fosters profound and unexpected *insights*. (McCormick et al. 1987)

> Visualization is essentially a mapping process from computer representations to perceptual representations, choosing encoding techniques to maximize human understanding and communication. (Owen 1999)

> Visualization is concerned with exploring data and information in such a way as to gain understanding and *insight* into the data. The goal … is to promote a deeper level of understanding of the data under investigation and to foster new *insight* into the underlying processes, relying on the humans' powerful ability to visualize, (Earnshaw and Wiseman 1992)

> The primary objective in data visualization is to gain *insight* into an information space by mapping data onto graphical primitives. (Senay and Ignatius 1994)

Most of the above definitions were made in the context of *Scientific Visualization* (a subfield of visualization), where data traditionally features some spatial or geometrical information. Nevertheless, the definitions for *Information Visualization* (another subfield of visualization), which often deal with non-spatial data, bear a high level of resemblance in terms of placing an emphasis on *gaining insight*. As an addition to Scott Owen's collection, here are other commonly cited definitions, many of which were written specifically for information visualization, while some were intended to encapsulate both spatial and non-spatial data visualization.

> Visualization facilitates "the use of computer-supported, interactive, visual representations of abstract data to amplify cognition." (Card et al. 1999)
>
> The purpose of information visualization is to amplify cognitive performance, not just to create interesting pictures. Information visualizations should do for the mind what automobiles do for the feet. (Card 2008)
>
> Graphics reveal data. Indeed graphics can be more precise and revealing than conventional statistical computations. (Tufte 2001)
>
> Information visualization helps think. (Few 2009)
>
> Information visualization utilizes computer graphics and interaction to assist humans in solving problems. (Purchase et al. 2008)
>
> The goal of information visualization is to translate abstract information into a visual form that provides new *insight* about that information. Visualization has been shown to be successful at providing *insight* about data for a wide range of tasks. (Hearst 2009)
>
> The goal of information visualization is the unveiling of the underlying structure of large or abstract data sets using visual representations that utilize the powerful processing capabilities of the human visual perceptual system. (Berkeley 2010)
>
> "Visual representations of data enable us to communicate a large amount of information to our viewers." In comparison with texts, they "can encode a wealth of information and are therefore, well suited to communicate much larger amounts of data to a human." (Marty 2009)
>
> "A following is a summary of visualization benefits:" "answer a question", "pose new questions", "explore and discover", "support decisions", "communicate information", "increase efficiency", and "inspire". (Marty 2009)
>
> The purpose of visualization is to get *insight*, by means of interactive graphics, into various aspects related to some processes we are interested in … (Telea 2008)

In the above definitions, there are many references to *gaining insight*, or likewise phrases such as *amplifying cognition*, *seeing the unseen*, *unveiling structure*, *answering questions*, *solving problems*, and so forth. It is unquestionable that these are the benefits that visualization can bring about in many occasions. There has been an abundance of evidence to confirm that such goals are achievable. However, *insight* is a non-trivial concept. It implies "accurate and deep intuitive understanding" according to many dictionaries. While this may be what everyone who creates or uses visualization is inspired to achieve, it is an elusive notion and rather difficult to measure, evaluate, or validate objectively.

Perhaps it is also because of its vagueness, it is relatively easier for people to interpret the term *insight* differently. The charged debate about chart-junks a few years ago was perhaps partly caused by the diverse interpretation of what *insight* to be gained from visualization.

The debate started with a paper by Bateman et al. (2010), which reported an empirical study on the effects of using visual embellishments in visualization. They compared conventional plain charts with highly embellished charts drawn by Holmes (1984). The findings of the study suggest that embellishment may aid memorization. Following this work, Hullman et al. (2011) proposed a possible explanation that "introducing cognitive difficulties to visualization" "can improve a user's understanding of important information." Obviously this was a major departure from the traditional wisdom of avoiding chart-junks in visualization. For example, in Tufte (2001), some of Holmes's visual designs were shown as counter examples of this wisdom.

These two pieces of work attracted much discussion in the blogosphere. Stephen Few, the author of several popular books on visualization (e.g., Few 2009), wrote two articles. On Bateman et al. (2010), he concluded:

> At best we can treat the findings as suggestive of what might be true, but not conclusive. (Few 2011a)

Few was much more critical on Hullman et al. (2011):

> If they're wrong, however, which indeed they are, their claim could do great harm. (Few 2011b)

In many ways, the two sides of the debate were considering different types of insight to be gained in different modes of visualization. The difficulty in defining insight resulted in different assessment of the quality of visualization. We will revisit this debate later in Sects. 5.2.5 and 5.3.3.

5.2 A Story of Line Graph

Before we attempt to answer the question what visualization is really for, let us examine some examples of visualization. We start with one of the simplest form of visualization, *line graph*, which is also referred to as *line chart* and *line plot*.

Figure 5.1 shows a line graph created by an unknown astronomer in the tenth (or possibly eleventh) century, depicting the "inclinations of the planetary obits as a function of the time" (Funkhouser 1936). More line graphs were found in the seventeenth century records, noticeably the plot of "life expectancy vs. age" by Christiaan Huygens in 1669, and the plot of "barometric pressure vs. altitude" by Edmund Halley in 1686 (Friendly 2008). The eighteenth and nineteenth centuries saw the establishment of statistical graphics as a collection of charting methods, attributed to William Playfair, Francis Galton, Karl Pearson and others (Cleveland 1985). The invention of coordinate papers in the eighteenth century also helped make line graph a ubiquitous technique in science and engineering.

Today, digitally stored data, which captures or exhibits a functional relationship $y=f(x)$, is everywhere. For example, there are thousands or millions of real time data feeds of financial information. Weather stations and seismic monitors around the world generate an overwhelming amount of data in the form of $y=f(t)$. In fact, every pixel in a video results in a time series (if it is in grey scale), or three time series (if it is in color). There are usually about a million of pixels in each video frame. In some cases, we still use line graphs for visualization, and in other cases, we do not. What has been the *most fundamental factor* that makes visualization users choose one visual representation from another? Is it a more quantifiable factor, such as the

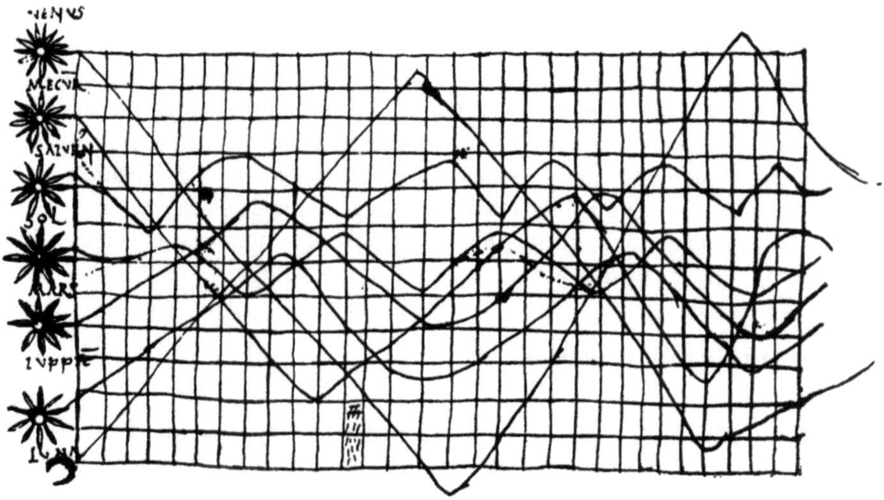

Fig. 5.1 This is the earliest line graph found in the literature. It divides the 2D plane onto some 30 temporal zones across the x-axis and uses horizontal lines to indicate zodiac zones across the y-axis. Seven time series were displayed in this chart (Source: Funkhouser 1936)

number of data series, the *number of data points per data series*, or another data-centric attribute? Is it a less quantifiable factor such as the *amount or type of insight*, the *amount of cognitive load required*, the *level of aesthetic attraction*, the *type of judgment to be made*, or any other human-centric attribute?

Let us consider why a seismologist uses a seismograph, which is a type of line graph that depicts the measured vibrations over time. (For the convenience of referring, we use the female pronoun for the seismologist.) The main task supported by a seismograph is for a seismologist to *make observation*. Her first priority is simply to see, or to know, the data stream in front of her, so she can confidentially say "I have seen the data". She may wish to observe some signature patterns of a potential earthquake, relationships among several data series measured at different locations, anomalies that may indicate malfunction of a device, and so on. The seismologist also uses seismographs as a mechanism of *external memorization*, since they can "remember" the data for her. In real time monitoring, she does not have to stare at the seismometer constantly and can have a break from time to time. In offline analysis, she does not need to remember all historical patterns, and can recall her memory by inspecting the relevant seismographs. Viewing seismographs *simulates various thoughts*, such as hypotheses. After observing a certain signature pattern in a seismograph, she may hypothesize that the vibrations would become stronger in the next few hours. While the seismograph advances with newly arrived data, she *evaluates her hypothesis* intuitively. When discussing with her colleagues, she draws their attention to the visual patterns on the seismograph, and explains her hypothesis and conclusion. In other words, she uses the seismograph to *aid her communication* with others.

Perhaps the seismologist does not have to use seismographs. The vibration measures could simply be displayed as a *stream of numbers*; after all viewing these numbers would be more accurate than viewing the wiggly line on a seismograph. Alternatively, to make more cost-effective use of the visual media, the stream of number could be *animated* in real time as a dot moving up and down, accompanied by a precise numerical reading updated dynamically. One would expect that it is intuitive to visualize temporal data using time (i.e., animation). Let us have a close look at the advantages of a seismograph over a stream of numbers or an animation.

5.2.1 Making Observation

It is not difficult for most people to conclude that viewing a stream of numbers is much slower than viewing a line graph such as a seismograph. Numerous studies in psychology have confirmed this (e.g., Styles 2006). The latter often facilitates pre-attentive processing, allowing information to be obtained from a visual medium or environment unconsciously. One might suggest that line graphs make better use of space than a stream data. This is certainly true, but space optimization cannot be the fundamental factor, as line graphs are not the optimal in terms of space usage in static visualization (Chen and Jänicke 2010). Furthermore, the *animation* of dots and numbers would offer much better space utilization.

The difficulty of using *animation* to support tasks of making observations is due to its excessive demand for various cognitive capabilities, including attention and memory. While watching such an animation, it is difficult for a viewer to pay attention to a specific temporal pattern, and almost impossible to have a photographic memory to record a specific set of numbers. Of course, one could view the same animation repeatedly, and would eventually work out interesting patterns in the movement of the dot and the variations of the numbers. It would no doubt take much more time than viewing a line graph.

When considering a video capturing a real world scene, however, it would not be a good idea to display the signal at each pixel as a line graph. There would be about a million of wiggly lines to observe. Figure 5.2 shows three common approaches for dealing with such temporal data. Although watching videos requires little learning, it has a high demand for time, attention and memory. In particular, the cost of human time is usually unaffordable in many video surveillance applications.

Several video visualization techniques were developed to reduce this time cost by transforming a video to a static visualization image. For example, Chen et al. (2006) proposed a method to summarize a video using a horseshoe shape. The temporal axis follows the curved from left to right, and each video frame is placed orthogonally along this curved axis. Those pixels showing no significant change from the background are not displayed. Hence the green shape in the middle of the horseshoe shows that some non-background objects have occupied those pixels in the video segment. The strong intensity variation at a pixel in a short period suggests that there may be something in motion. Those detected motions are high-

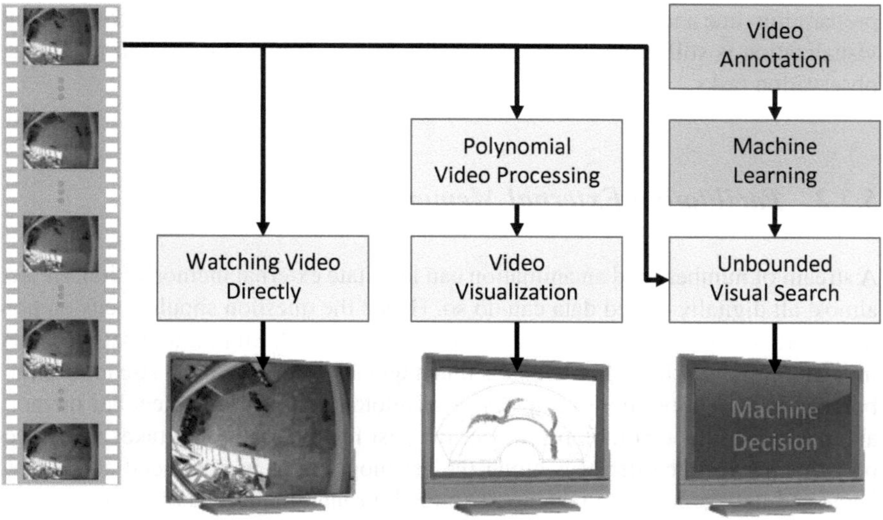

Fig. 5.2 Three common approaches for dealing with video data. *Left*: A video may be watched directly by a human viewer. This is usually a costly affair. *Right*: An idealized computer vision system may replace humans in making observation in the future. *Middle*: A visualization summarises the events in a video that features a person entering the foyer of a building, leaving a bag on the floor, then walking out of the scene, a few minutes later re-entering the scene, picking up the bag and continuing walking. This visualization allows a CCTV operator to make observation quickly, without the need for staring at the video monitor constantly. It thus saves time (Source: Chen et al. 2006)

lighted with dark dots on the green shape. Using simple logic reasoning, one can observe that an object in motion appeared in the video on the left. It then broke up into two objects, one with motion and one without. The one with motion then disappeared from the camera view. Sometime later, an object in motion appeared in the scene, merged with the object not in motion, and continued their movement. Chen et al. (2006) also conducted an empirical study confirming that human observers can learn to recognize signature patterns in such visualization after a short training (30 min). Using this method, the burden upon human observers is partly transferred to the computer. As most basic video processing and visualization techniques require only polynomial computation time, the summary visualization of a video can be generated in seconds.

One may consider an ideal alternative for the computer to make observation on behave of humans. Tsotsos (1989) discovered that many vision tasks involved unbound visual search, and proved that such tasks belong to a class of intractable computational problems (i.e., NP-complete problems). This partly explains why there are still many unsolved problems in computer vision. Furthermore, in order for machine learning to produce a reliable learned system, there must be a good set of training data. The more complex the system is, the larger and the more comprehensive the training data has to be. Not many observation tasks can afford the costly preparation time and human resources for such training data. Hence, at the moment, visualization is still providing humans with cost-effective assistance in many data observation tasks.

5.2.2 Facilitating External Memorization

A stream of numbers and an animation can facilitate external memorization. In fact, almost all digitally-stored data can do so. Hence the question should focus on how fast a form of visual display can facilitate memory recall. Similar to what discussed in Sect. 5.2.1, viewing a line graph is much quicker than viewing a stream of numbers, or an animation. For example, if a seismologist tries to recollect her memory about some events that took place over the past few hours, it only takes a few seconds for her to trace her eyes along the seismograph to be reminded about what happened before. It would take hours to read through thousands of numbers, or to watch the animation repeatedly.

Interestingly, inspired by seismographs, Botchen et al. (2008) developed a visual representation for summarizing videos, which facilitates better external memorization than the horseshoe design in Fig. 5.2. They referred to this new design as *VideoPerpetuoGram* (VPG), implying “endless video visualization”.

5.2.3 Stimulating Hypotheses and Other Thoughts

While some visualization tasks may involve routine, and perhaps sometimes mundane, observations, others can be highly analytical, involving various aspects of thought process, such as data comprehension, facts deliberation, experience reflection, hypothesis generation, hypothesis evaluation, opinion formulation, and decision making. Many aspects of human thinking are stimulated by visual signals. If we compare a seismograph with a stream of numbers or an animation of dots and numbers, we are interested in which type of visual signal can stimulate more thought, or stimulate a specific aspect of thought faster. To our knowledge, there is yet any reported study on such questions. However, with the availability of technologies, such as functional magnetic resonance imaging (fMRI), we hope that there will be more conclusive answers in the near future.

Nevertheless, there have been many anecdote evidences suggesting that when visualization is appropriately designed to convey overviews and is supported by interaction for details-on-demand exploration, it can simulate hypotheses more effectively (e.g., Teoh et al. 2003; Kehrer et al. 2008).

Figure 5.3 shows a visualization designed for depicting the advancing and retreating patterns of some 200 calving glaciers in Greenland over a 10 year period. In other words, there are some 200 time series in the data set. The conventional line graphs were initially used to plot the aggregated time series or a small set of time series for individual glaciers (usually 8–12 time series per chart). Because of the

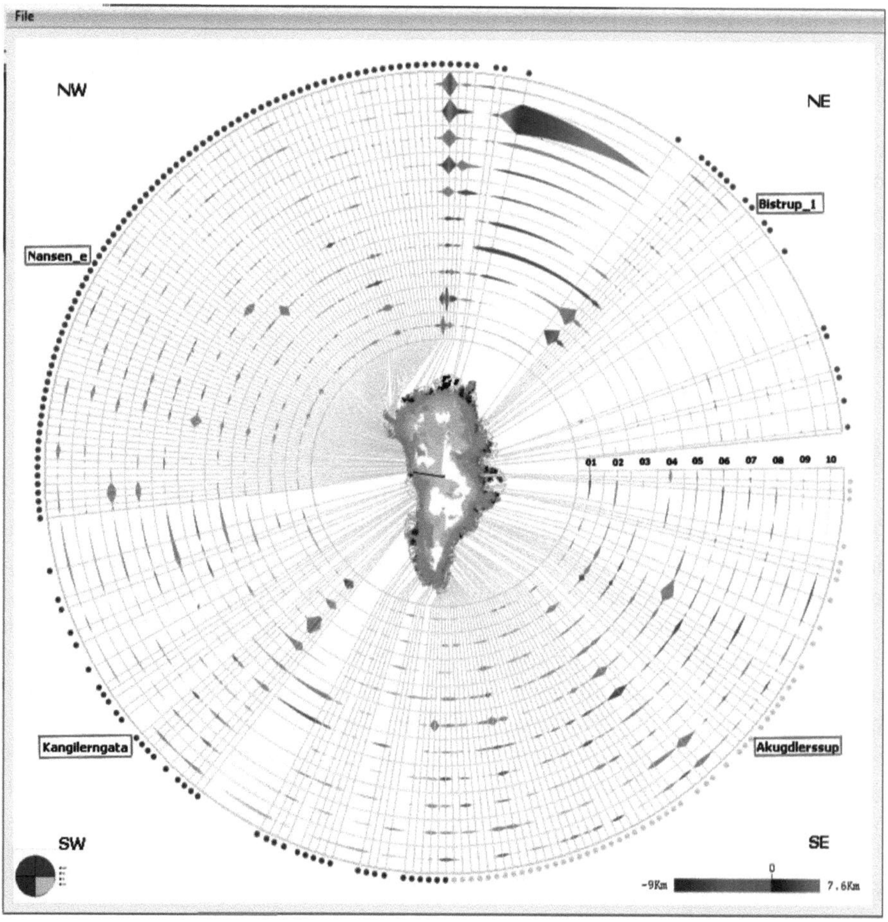

Fig. 5.3 The radial visualization shows 199 time series, each representing the relative frontal positions of a calving glacier in Greenland over 10 years (rings). The design facilitates rapid overview of the whole data set, which has been difficult with conventional line graphs previously used by the domain experts (Source: Drocourt et al. (2011), reprinting with permission granted by Eurographics Association and John Wiley & Sons, Inc)

lack of an overview of the spatial and temporal information together, it was not easy to formulate hypotheses from isolated observations. After observing the glaciologists at work, Drocourt et al. (2011) realized that this group of glaciologists knew the geography of Greenland extremely well, so displaying the actual map of Greenland was not essential. They thus transported the position of each glacier to a point on a circular ring while maintaining some geographical information such as primary orientation, neighborhood, and relative proximity. By using only 1D angular coordinates for the locations of glaciers, they were able to use the radial coordinates for different time steps. This provides an effective overview of both spatial and temporal information. When the glaciologists saw the visualization for the very first time, they immediately spotted some unusual patterns in the visualization, and many hypotheses were proposed. One of the hypotheses led to the discovery of a serious error in the data curation process.

Whereas some hypotheses or thoughts stimulated by visualization may qualify as *insight*, or *deep understanding*, currently this is not an everyday phenomenon happening to all visualization users. In fact, the endeavor of *Visual Analytics*, a relatively new subfield of visualization (Thomas and Cook 2005), is to enable humans to gain more *insight* from data through integrated uses of different technologies, including mathematical analysis, visualization, data mining, machine learning and human-computer interaction. This however raises a new question about which component really contributes to an *insight* gained by a human.

5.2.4 Evaluating Hypotheses

There is no doubt that hypothesis testing is a critical process in scientific investigation. Whenever applicable and feasible, one should always utilize scientific methods, such as statistical hypothesis testing and Bayesian hypothesis testing.

However, such scientific methods often require a non-trivial amount of time and effort for collecting, processing and analyzing data. In practice, visualization is often used as an intuitive form of hypothesis evaluation. For example, in scientific computation, to evaluate a simulation model (which is essentially a hypothesis), scientists visualize the results of a simulation run and visually compare the results with some ground truth data. Such intuitive evaluation is based on the principles of default reasoning and counterfactual reasoning. It is not in any way unscientific. It saves time. In the visualization literature, there have been many case studies that confirm the use of visualization as a tool for hypothesis evaluation (e.g., Hsu et al. 2010).

5.2.5 Disseminating Knowledge

Visualization is used extensively for disseminating knowledge (Gomez et al. 2012). In fact, this is often mistaken as the main or only function of visualization. In such situations, visualization is a tool for assisting a scientist or scholar in delivering a collection of messages to an audience. These messages may consists of data being visualized, background information to be appreciated, concepts to be comprehended, and opinions to be accepted. Clearly, the person presenting the visualization would like to direct the audience to receive the intended messages as fully, and as fast, as possible. Such intended messages may encode a piece of knowledge known to the presenter, the insight gained by the presenter in analyzing the data, or a subjective opinion of the presenter.

There is a subtle difference between this visualization task and those in the above four sections (5.2.1, 5.2.2, 5.2.3, and 5.2.4). Assessing how well a task is performed is generally difficult in many practical situations. For example, consider a task of making seismological observation. If a visual pattern of a potential risk was not noticed in a seismograph during routine monitoring, unless the risk is actualized, it always seems debatable as to such pattern should be noticed or not. The same paradox can be suggested for external memorization, hypothesis simulation and hypothesis evaluation. In knowledge dissemination, however, as the person presenting the visualization usually has a set of defined criteria for measuring task performance, he/she can assess the audience to determine whether the intended messages were received. Meanwhile, the time is more a constraint rather than a quality metric, since, for instance, a presentation, a meeting or a lecture is usually time-limited. In such a situation, visualization is often embellished in order to "energize" the messages intended by the presenter.

5.3 Saving Time in Different Modes of Visualization

Some of the visualization tasks mentioned in the above discussions have an objective for gaining insight by the user, but some do not. Nevertheless all visualization tasks feature an objective related to saving time. Table 5.1 summarizes the main objectives of these five types of visualization tasks. The above story of line graph highlights the importance of *saving time* in performing a user's tasks.

Of course, this is not a new discovery. Amid many "insight-based" definitions, some appreciated the purpose of saving time:

> Today's researchers must consume ever higher volumes of numbers … If researchers try to read the data, … they will take in the information at snail's pace. If the information is rendered graphically, however, they can assimilate it at a much *faster* rate. (Friedhoff and Kiley 1990)

> One of the greatest benefits of data visualization is the sheer quantity of information that can be *rapidly* interpreted if it is presented well. (Ware 2004)

Table 5.1 The main objectives of using visualization in relation to different tasks

Visualization tasks	Gaining insight	Saving time	Others
Making observation	Not an essential requirement	View data quickly	With sufficient accuracy
Facilitating external memorization	Not an essential requirement	Recall information quickly	Recall more information
Stimulating hypotheses	Gain good quality or "correct" hypotheses	Stimulate hypotheses quickly	
Evaluating hypotheses	Evaluate hypotheses accurately	Evaluate hypotheses quickly	
Disseminating knowledge	Pass on one's insight to others correctly	Pass on one's insight to others quickly	Attract attention, stimulate curiosity, help memorization

> Visual representations and interaction technologies provide the mechanism for allowing the user to see and understand large volumes of information *at once*. (Thomas and Cook 2005)

> Information visualization promises to help us *speed* our understanding and action in a world of increasing information volumes. (Card 2008)

5.3.1 What Is Visualization Really For?

Based on the reasoning and evidence in Sect. 5.2 as well as the insightful remarks by Friedhoff and Kiley (1990), Ware (2004), Thomas and Cook (2005) and Card (2008), we are ready to answer the question of "what is visualization really for?"

Definition Visualization (or more precisely, computer-supported data visualization) is a study of transformation from data to visual representations in order to facilitate effective and efficient cognitive processes in performing tasks involving data. The fundamental measure for effectiveness is sufficient correctness and that for efficiency is the time required for accomplishing a task.

Note that we choose the verb "accomplish" to emphasize that the task has to be performed to a certain degree of satisfaction before the measure of efficiency becomes meaningful. When the correctness has reached a satisfactory level, or becomes paradoxically difficult to assess (as discussed in Sect. 5.2.5), the time required to perform a visualization task becomes the most fundamental factor. Such time is a function of three groups of variables:

(a) *data centric attributes*, such as the size of a dataset, the number of multivariate dimensions, the entropy of the data space, etc.
(b) *human-centric attributes*, such as the amount or type of insight to be gained, the type of judgment to be made, the amount of cognitive load required, the level of aesthetic attraction, etc.
(c) *information delivery attributes*, such as the type of medium, the properties of the display device, the type of visual representations, the type of exploration, etc.

In most real world applications, there is usually little flexibility with (a) and (b). Hence choosing the appropriate *information delivery attributes* can be critical to accomplish a task efficiently.

5.3.2 Modes of Visualization

Visualization serves as a medium and a tool for human-human interaction. Let us refer to those who create visualization as **visualization producer** and those who view visualization in order to gain an insight as **visualization consumer**.

Table 5.2 Common modes of human participation in visualization processes

Mode	Participants	Example scenarios
(1)	A	An analyst works alone
(2)	$A_1, A_2, \ldots, A_k$	A team of analysts conduct visual data mining collaboratively
(3)	A, V	A personal visualization assistant and a boss
(4)	P, V	A personal tutor and a student
(5)	$P, V_1, V_2, \ldots, V_n$	A presenter (or a teacher) and an audience (or students)
(6)	$A_1, A_2, \ldots, A_k, V_1, V_2, \ldots, V_n$	A team of analysts carry out visual data mining in real time, while a panel of onlookers eagerly observe the process
(7)	$A, P, V_1, V_2, \ldots, V_n$	An analyst works for a domain expert who needs to disseminate his/her research to others

In some cases, the producer differs from the consumer. For example, a business analyst, who has a good understanding of a financial data set, creates a collection of charts for a company board meeting; or a teacher, who has a good understanding of a concept, creates an illustration to disseminate his or her knowledge to students. In many cases, the producer is also the consumer. For example, in a visual data mining process, an analyst, who has difficulties to comprehend a complex data set by simply reading the textual or numerical data, interactively explores various visual representations of the data, in order to gain an overview about the data, or make a discovery.

Let us consider three types of visualization users: **analyst** A, who is a producer as well as a consumer, **presenter** P, who is a producer but not a consumer, and **viewer** V, who is a consumer but not a producer. Different combinations of analysts, presenters and viewers in visualization processes will usually lead to different styles of human-human interaction. Table 5.2 lists several typical operational modes of visualization processes.

Most analytical tasks (e.g., *Making Observation, Stimulating Hypotheses and Other Thoughts*, and *Evaluating Hypotheses*) are likely to be conducted in modes (1), (2), and (3). Only the tasks of *Knowledge Dissemination* are normally conducted in modes (4) and (5). Mode (6) is relatively rare, but one can easily imagine that some visualization tasks during disaster management may be performed in this mode. On the other hand, mode (7) is rather common, but often has conflicting requirements between the knowledge dissemination task and those analytical tasks.

5.3.3 Reasoning About Visual Embellishment

Let us revisit the debate about visual embellishment discussed in Sect. 5.1. Borgo et al. (2012) reported a study on visual embellishment, which used more conservative stimuli than (Bateman et al. 2010). It shows that visual embellishments may help information retention in terms of both accuracy of and time required for memory recall. However, this is at the expenses of an increase in the time required for visual search, which is an activity typically taking place in tasks of making observation, hypothesis generation and evaluation. Their study also indicates that visual embellishment may help viewers to grasp concepts that a presenter would like to disseminate, especially when such concepts are embedded among complex information.

If we divide the notion of *insight* into two categories, the insight to be gained from data by visualization consumers, and that by a visualization producer, we can easily see that visual embellishments only help the former. Visual embellishments can assist in improving memory recall of the consumers, directing their attention, and enabling them to grasp quickly the key messages to be conveyed by the visualization. The producer can usually influence what *insight* to be gained by the consumers. On the other hand, when a visualization producer performs his/her analytical tasks, the *insight* to be gained, if any, would be unknown or uncertain beforehand. The effort for creating any visual embellishment would only cost extra unnecessary time.

Borgo et al. (2012) also pointed out in their explanation that the finding about the negative impact on visual search tasks provides scientific evidence to indicate some disadvantages of using visual embellishments. In other words, visual embellishment is unlikely to save time for an analyst in performing tasks such as making observation, hypothesis generation and evaluation. They also pointed out that the positive impact on memory and concept grasping should not be generalized to situations where visualizations are created by data analysts for their own use. In other words, the positive impact is relevant mainly to the above-mentioned modes (4) and (5). In addition, they made a connection between their findings and the information theoretic framework of visualization (Chen and Jänicke 2010).

Visual communication between a presenter and one viewer or a group of viewers in modes (4) and (5) is "noisier" and less reliable than that between a data analyst and himself/herself in mode (1). The reliability of visualization modes (2) and (3) is somewhere in-between. In many ways, the minimalistic design principle echoes Shannon's source coding theorem (Shannon 1948) that places emphasis on time saving in communication. In general, there is much less need for a data analyst to help himself/herself to grasp the key concepts or to remember a visualization by using embellishments, though data analysts can benefit from other forms of redundancy in visualization (e.g., lines that join dots in time series, and domain-specific metaphors).

On the other hand, the use of embellishments in visualization echoes Shannon's noisy-channel coding theorem (Shannon 1948) that places emphasis on making use of redundancy to facilitate automatic error detection and correction at the receiver's end. Visual embellishments can thus be viewed as redundancy, which strengthens "signals" in visualization, hence helping memorization and concept grasping. Improved memorization has a direct impact on the time required to recall acquired information as well as the accuracy of recall. Ultimately, it may save the viewers' time in gaining the messages that the presenter intended to send, though some of such messages may only be the opinions of a presenter and may not truly reflect what the data shows.

Table 5.3 Characteristics of analytical tasks and dissemination tasks in visualization

	Analytical tasks	Dissemination tasks
Modes of visualization	Producer (may) = consumer	Producer ≠ consumer
Saving time	Producer and consumer's time	Producer and consumer's time
Gaining insight	For producer to gain	For consumer to gain
Assessing correctness	Relatively difficult	More feasible
Using embellishment	Usually not helpful	Can be helpful
Information theory	Source encoding	Channel encoding

Consider those visualization tasks discussed in Sect. 5.2. There are analytical tasks, and dissemination tasks. Table 5.3 summarizes the main characteristics of these two groups of visualization tasks. Since the majority of work in visualization concerns about analytical tasks, the notion of "saving time" must not sit on the backbench in the definition of visualization. As it implicitly implies the completion of task, it encapsulates the notion of "gaining insight" to a large degree, but not vice versa. Hence, "saving time" is more fundamental.

5.4 How "Saving Time" Makes a Difference?

One might wonder whether bringing the "saving time" emphasis to the frontbench in the definition of visualization has a different implication from those existing definitions given in Sect. 5.1. There are indeed some fundamental differences.

5.4.1 Measurement

Firstly, time is much easier to measure and quantify than insight, knowledge or cognitive load, especially in the case of analytical tasks. In many ways, time may also be easier to measure than information, that is, the quantitative measures used in information theory. For example, any measurement in terms of Shannon entropy relies on the knowledge about the probability distribution in the whole data space. Such distribution is often unavailable and may vary from time to time, from one place to another. Of course, ideally we should be able to measure the amount of time in conjunction with the amount of new insight gained, or the amount of cognitive load imposed upon the user. While the measurement about insight or cognitive load may be undertaken in a laboratory condition, it is usually far too intrusive for a practical environment. Such a measurement would be uncertain as the measurement introduces a significant amount of artefacts and distortion to a normal cognitive process of gaining insight.

5.4.2 Empirical Studies

Most empirical studies involved measurement of accuracy and response time. It is comforting to know such measurements are not only meaningful, but also fundamental. While we encourage and experiment with other studying methods, it is important not to underestimate the measurement of time.

It is necessary to recognize the limitation of empirical studies in assessing "insight gained", especially when domain-specific knowledge is required. "Insight gained" depends on data as well as existing knowledge of participants. When such knowledge varies dramatically from one person to another, the study results have to be treated with care.

Hence empirical studies should focus on fundamental questions in visualization, and have to minimize variables, especially those hard-to-observe and hard-to-control variables such as a priori knowledge and insight to be gained. There is a limited amount of research resources in the field of visualization. We should devote more time to empirical studies for more fundamental scientific investigation, while controlling our appetite for evaluating everything, especially those hard-to-measure properties such as the amount of insight gained by domain experts.

5.4.3 Design Optimization

Measuring the time taken to perform a task can often be done seamlessly by a system, subject to the necessary ethical consideration and user consensus. This provides a metric for guiding the optimization of the design of a visual representation, a visualization system, or a visual analytics workflow. In comparison with the existing metrics (e.g., abstraction [Cui et al. 2006], salience [Jänicke and Chen 2010], reconstructability [Jänicke et al. 2011]), the time required to perform a task is undoubtedly the most important. It is easier to measure, more objective and more generic to all types of data, visual designs, systems, tasks and users.

In addition, the optimization of visualization also applies to analytical algorithms, interaction methods, automation mechanisms, and animation designs, since they all have significant impact on the time required to accomplish a task.

5.4.4 Theory of Visualization

The visualization community has not yet found a theory of visualization that most would agree to be fundamental. The good news is that many researchers are inspired to find such a theory, and some frameworks have been proposed (e.g., Chen and Jänicke 2010). Any theory of visualization should try to account for the impact of time required for performing or accomplishing visualization tasks.

5.4.5 *A Practical Wisdom*

Most visualization researchers have had some experience of engaging with scientists or scholars in different disciplines, or potential users from industrial or governmental organizations. Many of us had encountered difficulties in persuading potential collaborators about the merits of using visualization, or the need for developing advanced visual designs and visualization systems. After demonstrating some visualization techniques, typically conversations between a visualization researcher and a potential user might flow like that:

Potential user (engagingly): These pictures are very pretty. We are interested in having such techniques. I wonder how I can justify the costs for developing the system that you proposed.

Visualization researcher (enthusiastically): As you can see from the demo, visualization enables you to gain new insights from the data. This very much outweighs the development costs.

Potential user (doubtfully): Really, what kind of insights are we talking about?

Visualization researcher (anxiously): Patterns. (Pause, trying to recollect some definitions of visualization.) Interesting patterns, such as various anomalies, complex associations, warning signs, and potential risks.

Potential user (hopefully but cautiously): Can those pictures tell me all these automatically?

Visualization researcher (truthfully but uneasily): Not quite automatically. The mapping from data to visual representations will enable you see these patterns more easily and help you to make decisions.

Potential user (disappointedly): I can understand my data with no problem. I could not imagine how these pictures can help me make better decisions.

After a while, some of us learned a wisdom, i.e., never suggesting to potential collaborators that visualization could offer them insight. It is much better to state that visualization could save their time. As this paper has shown, visualization can indeed save time.

5.5 Conclusion

In this paper, we have presented a reasoned argument that it is more appropriate to frame the main purpose of visualization as saving time in accomplishing a task. This framing encompasses a variety of visualization tasks, including many basic operations that are not quite at the level of gaining insight, such as making routine observations, monitoring dynamic data streams, and relying on visualization as a means of external memorization.

We reasoned about our thesis with a close examination of different visualization tasks in the context of line graph and its contemporary adaption in video visualization and geo-spatial data visualization. We identified a number of typical visualization tasks, where insight has different bearings. We analyzed the roles of analysts, presenters and viewers in visualization with the aid of categorization of visualization producer and consumer. We further examined the notions of "gaining insight" and "saving time" in different modes of visualization. This led us to conclude that the purpose of "saving time" and that of "gaining insight" ought to swap their prominence in the definition of visualization. Building upon the rationale, we discussed the impact of drawing attention to "saving time" to different aspects of visualization.

"Gaining insight" has been an elusive purpose of visualization for several decades. It is perhaps the time to invigorate visualization as a scientific discipline by shining the spotlight on a more concrete purpose, that is, *to save the time required for accomplish a visualization task.* On the central theme of this book, *information quality*, we can now state the followings:

The most important metric for measuring the quality of a visualization process or a visual representation is its ability to save the time required for a user(s) to accomplish a data handling task.

References

Bateman, S., Mandryk, R. L., Gutwin, C., Genest, A., McDine, D., & Brooks, C. (2010). Useful junk?: The effects of visual embellishment on comprehension and memorability of charts. In *Proceedings of ACM* CHI, pp. 2573–2582.

Borgo, R., Abdul-Rahman, A., Mohamed, F., Grant, P. W., Reppa, I., Floridi, L., & Chen, M. (2012). An empirical study on using visual embellishments in visualization. *IEEE Transactions on Visualization and Computer Graphics, 18*(12), 2759–2768.

Botchen, R. P., Bachthaler, S., Schick, F., Chen, M., Mori, G., Weiskopf, D. & Ertl, T. (2008). Action-based multi-field video visualization. *IEEE Transactions on Visualization and Computer Graphics*, *14*(4), 885–899.

Card, S. (2008). Information visualization. In A. Sears & J. A. Jacko (Eds.), *The human-computer interaction handbook.* Mahwah: Lawrence Erlbaum Assoc Inc.

Card, S., Mackinlay, J., & Shneiderman, B. (1999). *Readings in information visualization: Using vision to think.* San Francisco: Morgan Kaufmann Publishers.

Chen, M., & Jänicke, H. (2010). An information-theoretic framework for visualization. *IEEE Transactions on Visualization and Computer Graphics, 16*(6), 1206–1215.

Chen, M., Botchen, R. P., Hashim, R. R., Weiskopf, D., Ertl, T., & Thornton, I. M. (2006). Visual signatures in video visualization. *IEEE Transactions on Visualization and Computer Graphics, 12*(5), 1093–1100.

Cleveland, W. S. (1985). *The elements of graphic data.* Pacific Grove: Brooks Cole.

Cui, Q., Yang, J., Ward, M., & Rundensteiner, E. (2006). Measuring data abstraction quality in multiresolution visualizations. *IEEE Transactions on Visualization and Computer Graphics, 12*(5), 709–716.

Drocourt, Y., Borgo, R., Scharrer, K., Murray, T., Bevan, S. I., & Chen, M. (2011). Temporal visualization of boundary-based geo-information using radial projection. *Computer Graphics Forum, 30*(3), 981–990.

Earnshaw, R. A., & Wiseman, N. (1992). An introduction to scientific visualization. In K. W. Brodlie et al. (Eds.), *Scientific visualization, techniques and applications.* Berlin/New York: Springer.

Few, S. (2009). *Now you see it.* Oakland: Analytics Press.

Few, S. (2011a). *The chartjunk debate: A close examination of recent findings.* http://www.perceptualedge.com/articles/visual_business_intelligence/the_chartjunk_debate.pdf

Few, S. (2011b). *Benefitting InfoVis with visual difficulties? Provocation without a cause*. http://www.perceptualedge.com/articles/visual_business_intelligence/visual_difficulties.pdf
Friedhoff, R. M., & Kiley, T. (1990). The eye of the beholder. *Computer Graphics World, 13*(8), 46.
Friendly, M. (2008). A brief history of data visualization. In *Handbook of Data Visualization* (pp. 15–56). Springer Handbooks of Computational Statistics. Heidelberg: Springer.
Funkhouser, H. G. (1936). A note on a tenth century graph. *Osiris, 1*, 260–262.
Gomez, S. R., Jianu, R., Ziemkiewicz, C., Guo, H., & Laidlaw, D. H. (2012). Different strokes for different folks: Visual presentation design between disciplines. *IEEE Transactions on Visualization and Computer Graphics, 18*(12), 2411–2420.
Hearst, M. A. (2009). *Search user interfaces*. Cambridge/New York: Cambridge University Press.
Holmes, N. (1984). *Designer's guide to creating charts and diagrams*. New York: Watson-Guptill Publications.
Hsu, C.-H., Ahrens, J. P., & Heitmann, K. (2010). Verification of the time evolution of cosmological simulations via hypothesis-driven comparative and quantitative visualization. In *Proceedings of IEEE Pacific Visualization Symposium*, pp. 81–88.
Hullman, J., Adar, E., & Shah, P. (2011). Benefitting InfoVis with visual difficulties. *IEEE Transactions on Visualization and Computer Graphics, 17*(12), 2213–2222.
Jänicke, H., & Chen, M. (2010). A salience-based quality metric for visualization. *Computer Graphics Forum, 29*(3), 1183–1192.
Jänicke, H., Weidner, T., Chung, D., Laramee, R. S., Townsend, P., & Chen, M. (2011). Visual reconstructability as a quality metric for flow visualization. *Computer Graphics Forum, 30*(3), 781–790.
Kehrer, J., Ladstädter, F., Muigg, P., Doleisch, H., Steiner, A., & Hauser, H. (2008). Hypothesis generation in climate research with interactive visual data exploration. *IEEE Transactions on Visualization and Computer Graphics, 14*(6), 1579–1586.
Marty, R. (2009). *Applied security visualization*. Upper Saddle River: Addison-Wesley.
McCormick, B. H., DeFanti, T. A., & Brown, M. D. (Eds.). (1987). Visualization in scientific computing. In *ACM SIGGRAPH Computer Graphics*, New York, Vol. 21, p. 6.
Owen, G. S. (1999). *HyperVis – Teaching scientific visualization using hypermedia*. ACM SIGGRAPH Education Committee. http://www.siggraph.org/education/materials/HyperVis/hypervis.htm
Purchase, H. C., Andrienko, N., Jankun-Kelly, T. J., & Ward, M. (2008). Theoretical foundations of information visualization. In A. Kerren et al. (Eds.), *Information visualization: Human-centered issues and perspectives* (Lecture notes in computer science, Vol. 4950, pp. 46–64). Berlin/New York: Springer.
Senay, H., & Ignatius, E. (1994). A knowledge-based system for visualization design. *IEEE Computer Graphics and Applications, 14*(6), 36–47.
Shannon, C. E. (1948). A mathematical theory of communication. *Bell System Technical Journal, 27*, 379–423.
Spence, R. (2007). *Information visualization: Design for interaction*. Harlow: Pearson.
Styles, E. (2006). *The psychology of attention*. Cambridge: Psychology Press.
Telea, A. C. (2008). *Data visualization, principles and practice*. Wellesley: A K Peters.
Teoh, S. T., Ma, K.-L., & Wu, S. F. (2003). A visual exploration process for the analysis of Internet routing data. In *Proceedings of IEEE Visualization*, 523–530.
Thomas, J. J., & Cook, K. A. (Eds.). (2005). *Illuminating the Path: The Research and Development Agenda for Visual Analytics*. National Visualization and Analytics Center. http://vis.pnnl.gov/pdf/RD_Agenda_VisualAnalytics.pdf
Tsotsos, J. K. (1989). The complexity of perceptual search tasks. In *Proceedings of 11th International Conference on Artificial Intelligence*, Detroit, pp. 1571–1577.
Tufte, E. R. (2001). *The visual display of quantitative information*. Cheshire: Graphics Press.
U. C. Berkeley, School of Information. (2010). *Information visualization showcase*, Events Archive (also in course information for i247). http://www.ischool.berkeley.edu/newsandevents/events/20100510infoviz
Ware, C. (2004). *Information visualization: Perception for design*. San Francisco: Morgan Kaufmann.

Chapter 6
Object Matching: New Challenges for Record Linkage

Monica Scannapieco

Abstract Record Linkage is the problem of identifying pairs of records coming from different sources and representing the same real world object. Available techniques for record linkage provide a satisfying answer when data are "traditional" records, that is well-structured information with clearly identified metadata describing values. When this latter condition does not hold, record linkage is most properly called Object Matching. In this paper, we will focus on objects that have "some degree of structure", which is the case of most part of the data available on the Web. We will describe challenges of Object Matching when objects have this latter meaning, and we will provide several examples of techniques that permit to face some of these challenges.

6.1 Introduction

Record Linkage (RL) is the problem of identifying pairs of records coming from different sources and representing the same real world object. Since the earliest contributions, dated back to Dunn (1946), Newcombe et al. (1959) and Fellegi and Sunter (1969), there has been a proliferation of different approaches to record linkage, coming from different scientific areas, including (just to cite the most significant ones) databases, statistics and machine learning. However, available techniques for record linkage provide a satisfying answer when data are "traditional" records, that is well-structured information with clearly identified metadata describing values. When this latter condition does not hold, record linkage is most properly called

M. Scannapieco (✉)
Information and Communication Technology Directorate,
ISTAT, Italian National Institute of Statistics, Via Balbo 16, 00185 Roma, Italia
e-mail: scannapi@istat.it

L. Floridi and P. Illari (eds.), *The Philosophy of Information Quality*,
Synthese Library 358, DOI 10.1007/978-3-319-07121-3_6,

Object Matching (OM). Given this definition the scope of OM is very huge, going from images (*image matching*) to completely unstructured data like documents (*document matching*). When the object to match has very specific features (like it happens for images or documents), it is often the case that dedicated techniques are devised for them.

In this chapter, we will focus on objects that have "some degree of structure", which is the case of most part of the data available on the Web. Objects in this latter meaning pose the following challenges with respect to traditional record linkage approaches:

- Size of data. This has been a challenge for traditional record linkage too. Indeed, record linkage between two data sets, each of size equals to n records, requires n^2 comparisons. When n is of the order of 10^4 or greater, the space of all the comparison pairs must be reduced in order to permit its practical exploration. Turning to Web data, the size of data sets can be order of magnitude greater, hence dedicated procedures must be devised. Section 6.3.1 expands this point.
- Time variance of data. In traditional record linkage settings, it is typically well-know the time to which the snapshot of the real world captured by a database refers. In the case of Web sources: (i) this time instant may be defined but unknown (e.g. Web sites that can have competitiveness problems in releasing the update frequency of their data, like E-commerce sites); (ii) this time instant may be not defined at all (e.g. social platforms where data are updated in a continuous and uncontrolled way). However, when it is available, it is a very relevant information that OM techniques should take into account. More details on this are provided in Sect. 6.3.2.
- Data quality. When data sources are not directly controlled, like it often happens when using Web data that belong to various providers, data are likely to have poor quality. OM plays a major role in this context by providing a necessary "basic" step to permit Web sources integration. It is however harder to be carried out, because poor quality reflects into a complication of the decision phase of the OM process. Sect. 6.3.3 details this issue.
- Schema information is typically very poor, or, in other words, the degree of structure is low. As an example of the impact of this condition, it may be the case that available attribute values cannot provide enough information to come up with a decision on the status of Match/Unmatch of objects. Hence, dedicated techniques that take explicitly into account various degree of structure of data need to be investigated. Sections 6.3.4 and 6.3.5 are dedicated to this point.

The paper will illustrate the challenges of OM for Web data with a specific focus on Linked Data and Deep Web data (see Sect. 6.2 for an introduction to basic notions). Moreover, some techniques for OM will be illustrated in order to exemplify the specificity of the OM process to this context and to highlight current and future research challenges in the field.

6.2 Background

6.2.1 Record Linkage

The term record linkage is mentioned for the first time in Dunn (1946) and it is the process that decides if two (or more) records refer to the same real world entity or not. Record linkage is also known as duplicate detection, record matching, instance identification, entity resolution and with several other synonyms (see the survey (Elmagarmid et al. 2007)). In order to show why RL is difficult, let us refer to Fig. 6.1.

In the figure, there are several issues that need to be solved in order to perform a RL task:

- A surrogate identifier (`ID`) is present in both tables but it is differently codified. Social Security Number (`SSN`) could be used as an identifier but has missing values.
- The two tables have different schemas. Hence, schema-level matching should be done in order to identify that the attribute `Lastname` and `Surname` match, as well as to individuate the attributes common to the two tables on which RL can be performed.
- Assuming that RL is performed on `Name`, `Surname/Lastname` and `SSN`, a value-level comparison, supported by approximate string distance functions, must be performed in order to match `Jhn` with `John`.

RL has been subject of research from different research communities, including the statistics community, the database community and the machine learning community.

In the statistics community, Fellegi and Sunter were the first to formalize the notion of record linkage by means of a theoretical model (Fellegi and Sunter 1969). Such a model is considered as the reference one for formalizing the problem of record linkage.

NAME	LASTNAME	BIRTHYEAR	SSN	ID
Marie	Gold	2000	322Y	***A1***
Jhn	Smith	1974	***455X***	***A2***

NAME	SURNAME	TELNUM	SSN	ID
Marie	Gold	999555	322Y	***B1***
John	Smith	222444		***B2***

Fig. 6.1 Example of RL issues

Given two sets of records A and B, it considers the cross product $A \times B = \{(a, b)|a \in A \text{ and } b \in B\}$. Two disjoint sets M and U can be defined starting from $A \times B$, namely, $M = \{(a, b)|a \equiv b, a \in A \text{ and } b \in B\}$ and $U = \{(a, b)|a ! \equiv b, a \in A \text{ and } b \in B\}$, where the symbol $\equiv$ means that the records a and b represent the same real world entity (and $! \equiv$ if they do not). M is named the *matched* set and U is named the *unmatched* set. The record linkage procedure attempts to classify each record pair as belonging to either M or U. A third set P can be also introduced representing *possible matches*. Beyond the formalization of RL, Fellegi and Sunter also proposed a solution to the problem by means of a probabilistic approach based on: (i) modeling the comparison vector **y** among the candidate pairs as a random variable, (ii) modeling the distribution `m(y)` of matches and `u(y)` of unmatches in dependence of the comparison vector, and (iii) estimating parameters related to the distribution of `m` and `u` in order to minimize the probability that a candidate pair belong to the set of *possible matches* (i.e., where neither a matching nor a non-matching decision is taken).

In the database community, most of the research effort on RL regarded the efficient computation of similarity joins (e.g., Gravano et al. 2001; Chaudhuri et al. 2005; Sarawagi and Kirpal 2004). In these works, the focus is on carrying out the comparison among candidate pairs within a database management system in order to optimize efficiency performances.

Machine learning approaches also provided a significant contribution to RL research. Indeed, in some applications it is useful to have an a priori sample of data for which it is known whether they match or not; such a sample is called labeled data, while unlabeled data are data for which the matching status is unknown. Labeled data can be used effectively to learn probabilities, distance functions, or knowledge used in the different techniques. Supervised learning techniques assume the presence of a labeled training set for which knowledge is available on matching/unmatching pairs (e.g., Bilenko et al. 2003), while unsupervised learning techniques does not rely on training data (e.g., Zardetto et al. 2010).

6.2.2 Object Matching and Web Data: Linked Data and Deep Web

Object Matching generalizes the notion of record to the one of *object*: it aims at identifying pairs of data-objects that represent the same real world object (see Fig. 6.2).

Given this definition the scope of OM is very huge, going from images (*image matching*) to completely unstructured data like documents (*document matching*).

As stated in the Introduction, the focus of this paper is more restricted, and specifically, we will consider two categories of Web data, namely Linked Data and Deep Web data. Each of these two categories will be briefly defined and described in the following.

Linked Data (LD) (2006) is among the major initiatives of the so-called Web of Things (Fortuna and Grobelnik 2011). Indeed, not only does the Web of Things

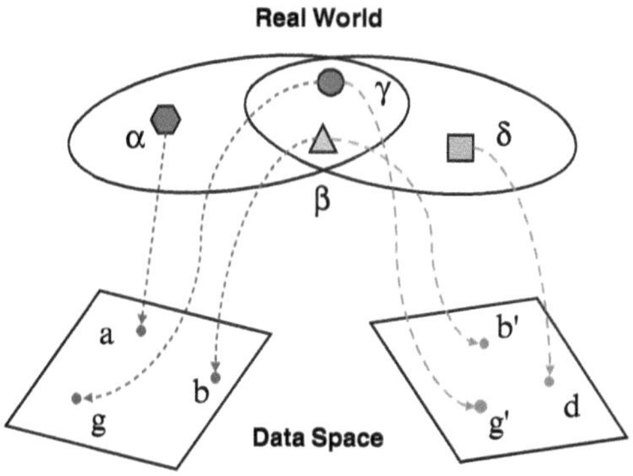

OM Problem

a ↔ b'	no
a ↔ g'	no
a ↔ d	no
b ↔ b'	yes
…………	…
…………	…
g ↔ g'	yes
g ↔ d	no

Fig. 6.2 OM problem

need access to data, but relationships among data should be made available, too, in order to actually interconnect sources otherwise separated. The collection of interrelated datasets on the Web is referred to as Linked Data. A relevant case of a large linked dataset is DBPedia (2007), which makes the content of Wikipedia infoboxes available in Resource Description Framework (RDF) (2002). RDF uses URIs[1] to name the relationship between things as well as the two ends of the link (this is usually referred to as a "triple"). This simple model allows representation and sharing of structured and semi-structured data. The linking structure forms a directed, labeled graph, where the edges represent the named link between two resources, represented by the graph nodes.

One of the most critical activities in order to prepare data to open publication is the discovery/definition of connections among different URIs dealing with the same entity. Hence, OM is a particularly significant issue for LD.

Deep Web indicates that part of the Web that is not directly indexed by standard search engines. A huge amount of information on the Web is sunk on dynamically generated sites, and traditional search engines cannot access their content as those pages do not exist until they are created dynamically as the result of a specific search. Most Web sites are interfaces to databases, including E-commerce sites, flight companies sites, online bibliographies, etc. The Deep Web includes all these sites and thus it is estimated that its size is several orders of magnitude larger than the surface Web (Bergman 2001).

6.3 Object Matching of Web Data

From an OM perspective LD and Deep Web data share some relevant features and are instead different with respect to some other features.

[1]A uniform resource identifier (URI) is a string of characters used to identify a name of a web resource.

Specifically, the features *shared* by these two subcategories of Web data are:

- Big, that is data sets size can be several order of magnitude greater than the size of traditional non-Web data.
- Time dependence, related to the time variability of data.
- Quality, in terms of both intrinsic data quality dimensions (such as accuracy, completeness, etc.) and external data quality dimension (such as source reputation, credibility, trustworthiness, etc.).

Instead, the feature that is really *different* for LD and Deep Web data is:

- Structure, related to the inherent structure of objects that can be, for instance, a record structure or a tree structure or a graph one or something else. Structure is well-defined for LD and loosely-defined for Deep Web data

In the following, for each of the above listed feature, we will illustrate the impact on the OM process, and some examples of research works that address the OM problem with respect to the specific feature under analysis.

6.3.1 Object Matching and Big

The term Big is used for data whose size has order of Peta (10^15) or greater. Web data, in the meaning of this paper, can actually reach such a Big size, though it is not said that they are natively characterized by it.

However, when Web data are involved in a OM process, the Big size is often achieved due to the inherent complexity of the phase named "search space creation", i.e. the phase that creates the space of all the possible object pairs to be compared in order to decide their Matching or Non-Matching status.

More specifically, OM between two data sets, each of size equals to n objects, requires n^2 object comparisons. When n is of the order of 10^4 or greater, the space of all the comparison pairs must be reduced in order to permit its practical exploration. To the scope, it is possible to adopt search space reduction techniques (see Elmagarmid et al. 2007) for a review of these techniques).

Kolb et al. (2012a) is an example of a work that adapts a traditional search space reduction technique to take into account the Big data feature of Web data. The search space reduction technique considered in Kolb et al. (2012a) is Sorted neighborhood (Hernandez and Stolfo 1998), which is one of the most popular approaches for the reduction task: it sorts all entities using an appropriate blocking key and only compares entities within a predefined distance window. In the work, it is shown how to use MapReduce for the parallel execution of Sorted neighborhood, demonstrating a highly efficient OM implementation. In such an implementation, the "Map" function determines the blocking key for ordering records. The map output is then distributed across to multiple "Reduce" tasks, each implementing the sliding window strategy. The authors of Kolb et al. (2012a) also implemented a tool called Dedoop (Deduplication with Hadoop) for MapReduce-based entity resolution of large datasets (Kolb et al. 2012b).

6.3.2 Object Matching and Time Dependence

The relationship between time and Web data has two main aspects. The first one is data volatility, i.e. a temporal variability of the information the data are meant to represent: there are data that are highly volatile (e.g. stock options), other which exhibit some degree of volatility (e.g. product prices), and some which are not volatile at all (e.g. birth dates). The second aspect is more generally related to the time features of the data generating mechanism. For instance, some Web data spring up and get updated in an almost unpredictable fashion, so that their time dimension is not available in a direct way, but does need to be re-constructed, if wishing to use those data in any meaningful analysis.

From an OM perspective, the data volatility aspect has the direct implication that manual tasks are not anymore possible during the OM process, that is the process should be fully automated. Decision models for OM are often supervised or semi-supervised, or, in other words, selected record pairs (typically the more difficult to classify) are sent to be clerically reviewed and training set of prelabeled record pairs can be prepared. Implementations of the Fellegi and Sunter model (1969) are often classified as unsupervised methods for learning the status of Matching or Non-Matching of object pairs. However, such implementations are not actually fully automated, as it would be necessary in a OM process on Web data. As an example, several implementations of Fellegi and Sunter rely on the Expectation Maximization (EM) algorithm (Dempster et al. 1977) for the estimation of the parameters of the model. However, in these techniques, manual intervention is required due to (i) the need of setting thresholds for identifying Matching and Unmatching pairs; (ii) possible unsuccessful parameter estimation via the EM algorithm (that may happen for instance if the size of the search space is too huge or too much limited). An example of fully automated technique that can fit the timely requirement of Web data is provided in (Zardetto et al. 2010), where a statistical approach based on mixture models is adopted. Zardetto et al. (2010) structures an OM process into two consecutive tasks: first, mixture parameters are estimated by fitting the model to observed distance measures between pairs; then, a probabilistic clustering of the pairs into Matches and Unmatches is obtained by exploiting the fitted model.

Let's consider the second aspect related to OM and time dependence, i.e. the possible availability of a timestamp for Web data. The OM matching process does need to be aware of this specific kind of information, and indeed there are some preliminary works that actually take explicitly into account the temporal information. As an example, in Pei et al. (2012), an approach that leverages temporal information with linkage is presented. The approach takes into account cases in which as time elapses, values of a particular entity may evolve; for example, a researcher may change affiliation or email. On the other hand, different objects are more likely to share the same value(s) with a long time gap. Thus the concept of decay is defined, with which the penalty for value disagreement is reduced and, at the same time, the reward for value agreement over a long period is reduced as well. Moreover, temporal clustering algorithms are proposed that explicitly consider time order of records in order to improve linkage results.

6.3.3 Object Matching and Quality

The characterization and evaluation of the quality of LD and of Deep Web data is a current research area. Data quality dimensions that characterize traditional non-Web data, including accuracy, timeliness, consistency, completeness, cannot be applied to Web data in a direct way. As an example, even before the Web 2.0 era, preliminary works on a dedicated characterization and evaluation of the quality of Web data were published (e.g. Pernici and Scannapieco 2003).

With respect to the relationship between OM and Web data quality, it can't be denied that a poor quality implies extensive OM activities. Unfortunately, preliminary works on assessing the quality of Web data actually reveal that the overall quality is poor. In Li et al. (2012), an assessment of the quality of Deep Web data from Stock (55 sources) and Flight (38 sources) domains is presented. The results of the assessment report a bad quality in terms of inconsistency (for 70 % data items more than one value is provided) and of correctness (only 70 % correct values are provided by the majority of the sources). Every data integration task that aims to use such sources in a unified way does need to match data and improve their quality. Hence a first step could be the OM one, followed by a fusion step (see the survey Bleiholder and Naumann 2008), in which activities are actually carried out in order to improve the quality and facilitate the data integration task.

6.3.4 Object Matching and Structure: LD

As anticipated, this feature is different for LD and Deep Web data. In this section, we will deal with LD, while in Sect. 6.3.5, we will deal with Deep Web data.

LD have a well-defined structure, as they are specified on the basis of the RDF family of languages. More specifically, RDF Schema (2004) and Ontology Web Language (OWL) (2004) enable the description of an ontology, i.e. a formal representation of domain concepts that permits modeling, querying and reasoning tasks.

Hence, OM can exploit such a high degree of structure: the implementation of the OM task can rely on extensional information (concepts, attributes, well-defined relationships, etc.) in order to match objects. However, despite the fact that OM can exploit the modeling efforts of the LD community, OM is really the central task for interlinking data sets belonging to different sources in an automated way. With respect to such a role, the ability of "discovery" links among objects becomes a significant feature for OM procedures enabling LD creation.

The OM task really needs a specific definition in the LD domain. Indeed, identity is commonly expressed using the standard `owl:sameAs` property. In OWL, two resources connected by the `owl:sameAs` property are considered identical in the sense that the subject and object of this statement share all their properties. Such interpretation of identity, however, appears too strong in many cases. The authors of Halpin et al. (2010) distinguished weaker varieties of identity beyond the canonical one. In Datalift (2011), different notions of identity are proposed, namely:

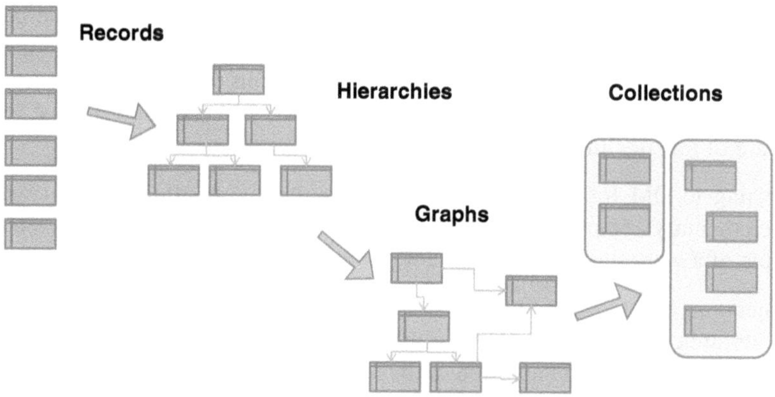

Fig. 6.3 Degree of structure of objects

1. Ontological identity, that occurs when the properties of two objects are the same.
2. Logical identity, considering that two descriptions are identical when they can be substituted one to the other in a logical expression without changing the meaning of the expression.
3. Formal identity, considering the fact that identity is superimposed, e.g. by imposing the same URI.

6.3.5 Object Matching and Structure: Deep Web Data

The structure of data retrieved from the Deep Web can go from highly structured *records* to *collections*, in which the degree of structure can be lower (see Fig. 6.3). Hierarchies and graphs can be considered as objects having an intermediate level of structure.

As remarked in the Introduction, the linkage of records is a research field explored since more than four decades with several results that can be effectively exploited. Instead, when objects to match have a different data structure, though there are already some relevant results, there is however still need for dedicated research efforts.

Considering OM for hierarchical data, most of the work is related to matching XML objects (see e.g. Milano et al. (2006), in which tree distances are used). Indeed, OM for XML data has two major challenges:

- Identification of the objects to compare. While, in RL it is well defined that the objects to compare are records, the delimitation of the boundaries of the elements to compare within an XML document is an issue.
- Flexibility of the XML data model. As a semistructured data model, XML permits the definition of an element in multiple ways, and allows the definition of

optional attributes. For instance, an element person can be defined as <!ELEMENT person(name;surname)|(surname;DoB)>, meaning that it has two both valid definitions, the first in terms of name and surname, and the latter in terms of surname and data of birth. OM techniques must thus be aware of this flexibility.

Turning to OM for graph-based data structures, the relationships between the most elementary components of the data structure are not anymore hierarchical. An example of a work dealing with OM for graphs is Chen et al. (2005). In this work:

- Objects in a database are the nodes of the graph and relationships are edges. Nodes are compared by features similarity, leading to similarity edges, that are a further kind of edge, in addition to the relationships.
- Partitions induced by similarity edges form Virtual Connected Subgraphs and only nodes belonging to such subgraphs are considered for comparison.
- A connection strength is computed by assigning weights to relationships.
- Decision on the matching status is taken on the basis of connection strengths.

In collections, the objects to be matched can be grouped according to particular relationships. As an example, in On et al. (2007) a dedicated set of techniques is proposed for the cases where entities can be represented as groups of relational records (sharing a group ID), rather than individual relational records; an example is given by an author in a digital bibliography that consists of a group of publications, where each publication is represented by a relational record. In this work, two groups can be linked if:

- High enough similarity between "matching" pairs of individual records that constitute the two groups.
- A large fraction of records in the two groups form matching record pairs; not all records in each of the two groups need to match.

The proposed techniques is a generalization of the Jaccard similarity metric $J(A,B)=|A \cap B|/|A \cup B|$ based on bipartite graph matching.

6.4 Conclusions

OM is a relevant issue for performing integration of Web data. The hardness of this latter task is undeniable, as proven by big failures like, e.g., Google Squared launched by on Google Labs on June 2009 and shut down on September 2011. However, initiatives like LD, DBPedia and more recently Wikidata (2012) are significant drivers that can give a new sprint to the process of integrating Web data.

In this paper, we have outlined that OM of Web data is a research field that provides several interesting and important challenges to tackle and have fun with. Specifically the following set of methods and techniques for OM do need further investigation: (i) dealing with the Big data size; (ii) fully automating the OM process; (iii) extracting and managing the time dimension of data; (iv) performing in a satisfactory way even in bad quality contexts, (v) permitting discovery of links among different objects, and (vi) dealing with various degrees of structure of objects.

References

Bergman, M. K. (2001). The deep web: Surfacing hidden value. *The Journal of Electronic Publishing, 7*(1), 1–17.

Bilenko, M., Mooney, R., Cohen, W., Ravikumar, P., & Fienberg, S. E. (2003). Adaptive name matching in information integration. *IEEE Intelligent Systems, 18*(5), 16–23.

Bleiholder, J., & Naumann, F. (2008). Data fusion. *ACM Computing Surveys, 41*(1), 1–41.

Chaudhuri, S., Ganti, V., & Motwani, R. (2005). Robust identification of fuzzy duplicates. In *Proceedings of the International Conference on Data Engineering (ICDE 2005)*, Tokyo.

Chen, Z., Kalashnikov, D. V., & Mehrotra, S. (2005). Exploiting relationships for object consolidation. In *Proceedings of the International Workshop on Information Quality in Information Systems (IQIS)*, Baltimore.

Datalift. (2011). Deliverable 4.1 methods for automated dataset interlinking.

DBPedia. (2007). DBPedia. Retrieved February 15, 2013, from http://dbpedia.org/About

Dempster, A. P., Laird, N., & Rubin, D. B. (1977). Maximum-likelihood from incomplete data via the em algorithm. *Journal of the Royal Statistical Society, 39*(1), 1–38.

Dunn, H. L. (1946). Record linkage. *American Journal of Public Health, 36*, 1412–1416.

Elmagarmid, A. K., Ipeirotis, P. G., & Verykios, V. S. (2007). Duplicate record detection: A survey. *IEEE Transactions on Knowledge Data Engineering, 19*(1), 57–72.

Fellegi, I. P., & Sunter, A. B. (1969). A theory for record linkage. *Journal of the American Statistical Association, 64*, 1183–1210.

Fortuna, C., & Grobelnik, M. (2011). Tutorial: The web of things. In *Proceedings of the World Wide Web Conference*, Hyderabad.

Gravano, L., Ipeirotis, P. G., Jagadish, H. V., Koudas, N., Muthukrishnan, S., & Srivastava, D. (2001). Approximate string joins in a database (almost) for free. In *Proceedings of Very Large Data Base (VLDB* 2001), Rome.

Halpin, H., Hayes, P. J., McCusker, J. P., McGuinness, D. L., & Thompson, H. S. (2010). When owl:sameas isn't the same: An analysis of identity in linked data. In *Proceedings of the 9th International Semantic Web Conference (ISWC)*, Shanghai.

Hernandez, M. A., & Stolfo, S. J. (1998). Real-world data is dirty: Data cleansing and the merge/purge problem. *Data Mining and Knowledge Discovery, 2*(1), 9–37.

Kolb, L., Thor, A., & Rahm, E. (2012a). Multi-pass sorted neighborhood blocking with MapReduce. *Computer Science – Research and Development, 27*(1), 45–63.

Kolb, L., Thor, A., & Rahm, E. (2012b). Dedoop: Efficient deduplication with Hadoop. In *Proceedings of the 38th International Conference on Very Large Databases (VLDB)/Proceedings of the VLDB Endowment* 5(12), 1878–1881.

Li, P., Dong, X. L., Maurino, A., & Srivastava, D. (2012). Linking temporal records. *Frontiers of Computer Science, 6*(3), 293–312.

Li, X., Luna Dong, X., Lyons, K. B., & Srivastava, D. (2012). Truth finding on the deep web: Is the problem solved? *PVLDB, 6*(2), 97–108.

Linked Data. (2006). Retrieved February 15, 2013, from http://linkeddata.org/

Milano, D., Scannapieco, M., & Catarci, T. (2006). Structure-aware XML object identification. *IEEE Data Engineering Bulletin, 29*(2), 67–74.

Newcombe, H. B., Kennedy, J. M., Axford, S., & James, A. (1959). Automatic linkage of vital records. *Science, 130*(3381), 954–959.

On, B.W., Koudas, N., Lee, D., & Srivastava, D. (2007). Group linkage. In *Proceedings of the International Conference on Data Engineering*, Istanbul.

OWL. (2004). Ontology web language. Overview. Retrieved February 15, 2013, from http://www.w3.org/TR/2004/REC-owl-features-20040210/

Pernici, B., & Scannapieco, M. (2003). Data quality in web information systems. *Journal of Data Semantics, 1*, 48–68.

RDF. (2002). Resource description framework. Retrieved February 15, 2013, from http://www.w3.org/RDF/

RDF Schema. (2004). RDF vocabulary description language 1.0: RDF schema. Retrieved February 15, 2013, from http://www.w3.org/TR/rdf-schema/

Sarawagi, S., & Kirpal, A. (2004). Efficient set joins on similarity predicates. In *Proceedings of ACM SIGMOD International Conference on Management of Data (SIGMOD'04)*, Paris.

Wikidata. (2012). Wikidata. Retrieved February 15, 2013, from http://www.wikidata.org/wiki/Wikidata:Main_Page

Zardetto, D., Scannapieco, M., & Catarci, T. (2010). Effective automated object matching. In *Proceedings of the International Conference on Data Engineering (ICDE 2010)*, Longbeach.

Chapter 7
Algorithmic Check of Standards for Information Quality Dimensions

Giuseppe Primiero

Abstract An important aspect of defining IQ standards is that sound information conforming to a specification should be error-free. We propose to assess information quality dimensions and check their standards by way of an algorithmic procedure. We design an effective procedural way to determine if and where IQ standards fail and to establish algorithmic resolution and evaluation methods that provide a metric appropriate to our quality checking system. This model is largely inspired by systems for quality standard assessment of software production, but it assumes a very high abstraction level. Our claim is that any information processing system, also not necessarily software based ones, can be designed after (some variations of) our model. A detailed formal translation of the definitions involved in our model is available in a machine-checked code.

7.1 Introduction

An important aspect of defining IQ standards is that sound information conforming to a specification should be error-free (Kahn et al. 2002). This often further unqualified requirement has to be properly addressed. While much work is devoted to the classification and proper definition of information dimensions, less is done to establish how to check their standards. We do not enter the long and articulated debate on *how* to classify IQ dimensions, for which we refer to any of the standard studies (e.g. Wand and Wang 1996; Batini and Scannapieco 2006). Rather, our focus here is on *failures* of IQ dimensions: we define an algorithmic check procedure to identify where a given dimension fails and what kind of errors cause the failure. To do so, we first propose a general categorization of errors based on the full taxonomy introduced in Primiero (2013), where each family is determined by the condition breach

G. Primiero (✉)
Department of Computer Science, Middlesex University, London, UK
e-mail: G.Primiero@mdx.ac.uk

L. Floridi and P. Illari (eds.), *The Philosophy of Information Quality*,
Synthese Library 358, DOI 10.1007/978-3-319-07121-3_7,

induced and a mode of occurrence. Secondly, we define an information processing system as model for our checking algorithm. This model is largely inspired by systems for quality standard assessment of software production (ISO/IEC 25010: 2011; Abran et al. 2008; Suryn et al. 2003), but it assumes a very high abstraction level. Our claim is that any information processing system, also not necessarily software based ones, can be designed after (some variations of) our model. Then we proceed in mapping error kinds to each stage in this model, resulting in a detailed analysis of the possible breaches of correctness. On this basis, we design an algorithm for error resolution and one for assessment of a quality metric without resolution. Finally, all errors definitions and the algorithm formulation are translated in the machine checkable language of the CoQ proof assistant, to verify formal and syntactic correctness.

7.2 Methodology

Our approach to the problem of evaluating standards for IQ dimensions is characterized as an *algorithmic negative approach*: we design an algorithm to establish precisely which information quality standards fail, to which extent does the failure affects the system and if the failure is considered reversible. Our first step consists therefore in defining the nature of possible errors. Without providing a detailed categorization of error families and their definition, we present here a simplified schema of error cases[1] (Table 7.1):

The uppermost row categorizes three main requirements that can be breached when an error occurs:

- *validity requirements*: the set of conditions established by the logical and semantic structure of the process defined for the given purpose; e.g. contradictory or undefined requirements;
- *correctness requirements:* the syntactic conditions for the same process; e.g. incomplete or incorrect definitions;
- *physical requirements*: the purely contextual conditions in which the information processing is executed; e.g. malfunctioning routines.

Table 7.1 Categorization of errors

	Validity	Correctness	Physical
Conceptual	Mistake	Failure	X
Material	X	Failure/slip	Malfunctions
Executive	X	X	Slip

[1]The full taxonomy of errors applied to this schema for algorithmic check is originally developed in Primiero (2013).

In the leftmost column we list three aspects or modes in which errors occur at a given stage of the process:

- *conceptual:* all aspects related to configuration and design of the information process;
- *material:* all aspects related to implementation of the process;
- *executive:* all aspects related to successful execution and use of the process.

By criss-crossing error conditions and error forms, we define four main cases:

1. *Mistakes:* errors caused by breaching validity requirements at the configuration and design levels;
2. *Failures:* errors caused by breaching correctness requirements at the configuration, design or implementation levels;
3. *Malfunctions:* errors caused by breaching physical requirements at execution level[2];
4. *Slips:* errors caused by either breaching correctness requirements at the implementation level; or by breaching physical requirements at the level of system use.

Given this general categorization, we proceed first by locating IQ dimensions in a model of information processing where all intermediate steps are qualified by translation to a model of software production. Our aim is to identify for each stage in this information processing model the corresponding IQ dimensions and establish where and why they may fail. Given the general nature of the model, stages will also be referred to as Levels of Abstraction (LoA) for information processing. We admit the possibility that the same dimension appears at different LoAs. Provided our categorization of conditions breaches, a different possible error is induced at each such stage.

Second, we propose a metric of standards for such dimensions. The standards will be located on a scale of their own, from low- to high-level ones, and matched to LoAs. Whatever specific scale is used to assess standards, this should allow for an accommodation of lower and higher standards. We suggest that such a scale is obtained as a by-product of evaluating errors: for an error at a lower LoA resulting from breaching a low-profile requirement, the overall assessment of the standard should be comparably softer than in the case of an error at a higher LoA resulting from breaching a high-profile requirement.

Finally, we operationalize the present conceptual analysis by formulating an inductive algorithm that is meant to proceed on the obtained scale to assess standards and evaluate the overall quality of the system. The algorithm is then translated to machine-checkable code to be verified correct.

---

[2]We deviate here from the original taxonomy presented in Primiero (2013), where malfunctions are characterized as a sub-family of failures.

7.3 Model

Our abstract model of information flow is defined by the following standard stages of processing (as shown in Fig. 7.1):

- at *functional specification level* (FSL) the Architect formulates general requirements for the system's purpose (e.g. the request to develop a system that collects, distributes and manage orders for a restaurant);
- at *design specification level* (DSL) the System Designer translates the requirements formulated at FSL into an informal description (a directed graph of all possible input/outputs relations, from order to delivery);
- at *algorithm design level* (ADL) the Algorithm Designer translates the informal instructions coming from DSL into the appropriate and correct formal instructions of an algorithm (the set of formal IF/THEN instructions translating the above relations);
- at *algorithm implementation level* (AIL), the Engineer further pins down the algorithm into the appropriate formulation in terms of a given language (translation to an appropriate programming language or other relevant implementation);

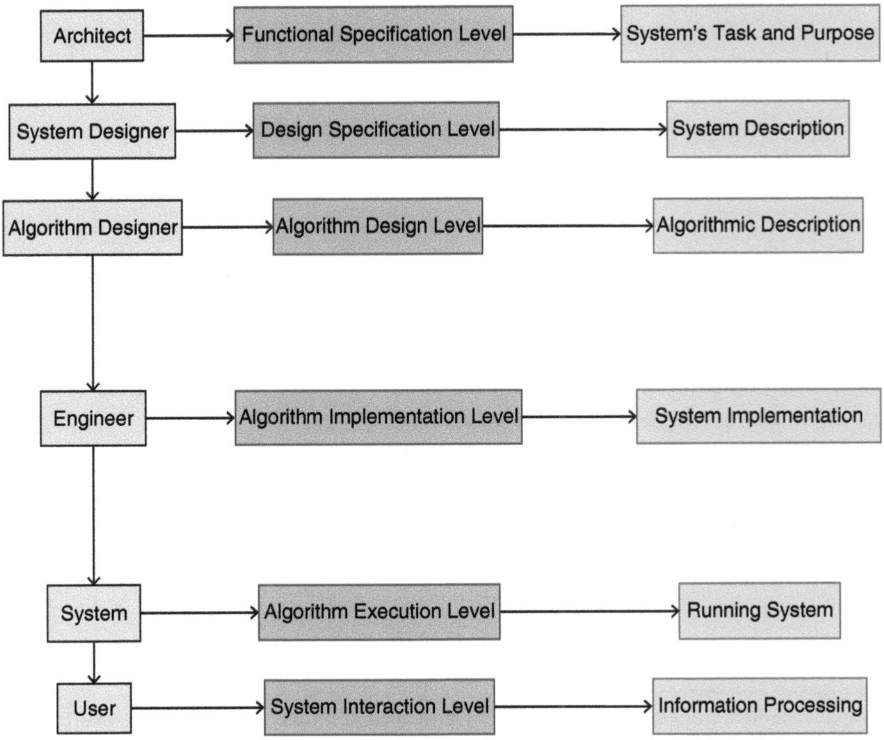

Fig. 7.1 A schema of the information processing structure

Table 7.2 Matching LoAs, agents and acts

LoA	Agent	Information act
FSL	Architect	Semantic purpose definition
DSL	System designer	Operational representation
ADL	Algorithm designer	Syntactical representation
AIL	Engineer	Translation to supported language
AEL	System	Data manipulation
SIL	User	Semantic information manipulation

- at *algorithm execution level* (AEL), the System processes information (orders are placed, routines get triggered and executed);
- at *system interaction level* (SIL), the User interacts and makes use of the output of the information processing from the System (requests are collected, orders fulfilled).

We associate to each LoA of this model an agent and the task fulfilled. The sequence of LoAs constitutes the structure through which information flows and gets transformed: the kind of information at stake is determined by the corresponding act the agent performs at the assigned LoA. Such procedural description of information flow is sketched in Table 7.2:

- at FSL, the Architect acts as purpose-giving entity and determines the semantics of the computation. The relevant question here is: *what do we want the system to do?* The information involved at this stage, though embedded in a certain verbal, graphical or otherwise formulated data substrate, fulfils a purely semantic task;
- at DSL, the System Designer interprets the semantic purpose, giving it a procedural representation. The relevant question here is: *how do we want the system to do what we want it to do?* The information involved at this stage is translated to a functional structure;
- at ADL, the Algorithm Designer provides an appropriate syntactic representation to the semantic purpose of the system. The relevant question here is: *which processes do we need the system to perform and on which data?* The information involved at this stage translates the procedural task to its syntactic representation;
- at AIL, the Engineer translates data into a proper support. The relevant question here is: *which language does the system need to be implemented in to deal with the given processes and data?* The information involved at this stage translates the syntactic representation back into instructional information;
- at AEL, the System acts as data manipulator. The relevant question here is: *which Input/Output relations do data and instructions instantiate?* The information involved at this stage is symbol manipulation;
- at System Interaction Level, the User interprets the Input/Output relations. The relevant question here is: *which purpose gets instantiated by these Input/Output relations? And how can I use them?* The information involved at this stage is semantic information, where data is matched back to meaning by recollecting the system's purpose. At this stage, evaluation of the overall quality system is possible.

The next step consists in matching to each LoA of this information processing schema one or more IQ dimensions and the related kind(s) of error(s).

Table 7.3 Matching LoAs, agents, actions and errors

Agent	LoA	Action	Breach	IQ dimensions	Error
Architect	FSL	Purpose Definition	Invalid	Consistency (reqs)	Mistake
			Incorrect	Accuracy (specs)	
			Incorrect	Completeness (specs)	
System Designer	DSL	Procedure Definition	Invalid	Consistency (design)	Mistake
			Incorrect	Completeness (routines)	Failure
			Incorrect	Accuracy (data)	Failure
			Incorrect	Accessibility (data)	Failure
Algorithm Designer	ADL	Algorithm selection	Invalid	Consistency (processes)	Mistake
			Invalid	Completeness (design)	Mistake
			Invalid	Relevance (design)	Mistake
			Invalid	Accuracy (design)	Mistake
Engineer	AIL	Algorithm implementation	Incorrect	Access (data)	Failure
			Incorrect	Security (routines)	Failure
			Incorrect	Flexibility/scalability (data)	Failure
			Incorrect	Precision (I/O)	Failure
			Incorrect	Efficiency (task)	Failure
			Incorrect	Reliability (task)	Failure
			Incorrect	Sufficiency (design)	Failure
System	AEL	Execution	Unusable	Usability	Malfunction
			Unusable	Usefulness	Malfunction
			Unusable	Accessibility (data)	Malfunction
User	SIL	Use	Unusable	Understandability	Malfunction
			Unusable	Efficiency	Malfunction
			Unusable	Precision (system)	Malfunction
			Unusable	Precision (user)	Slip
			Invalid	Relevance (purpose)	Mistake
			Incorrect	Completeness	Failure/slip

Table 7.3 extends Table 7.2 with the condition or requirement that is breached in the action at hand: validity, correctness or physical requirement. The fifth column matches the IQ dimensions relevant to that particular LoA: our selection is extracted from a standard list and might be further extended. The last column identifies the type of error corresponding to the breaching of a particular requirement for a certain IQ dimension at a given LoA.

Our analysis offers thus the following match of breached conditions and errors for IQ dimensions:

- Breaching of validity conditions induces a conceptual error, i.e. a mistake, which can occur related to:
  - Consistency of the requirements at FSL
  - Consistency of procedure definition at DSL
  - Consistency of selected processes at ADL

- Completeness of selected processes with respect to design at ADL
- Accuracy of the specification description with respect to purpose at FSL
- Accuracy of selected processes with respect to specification at SDL
- Accuracy of selected routines with respect to design at SDL
- Relevance of selected processes with respect to design at ADL
- Relevance of system use with respect to purpose at SIL

- Breaching of correctness conditions induces a material error, i.e. a failure, which can occur related to:

  - Completeness of selected routines at DSL
  - Accuracy of selected input data at DSL
  - Accessibility of selected input data at DSL
  - Accessibility of selected input data at AIL
  - Security of selected routines at AIL
  - Flexibility/Scalability of selected input data at AIL
  - Precision of Input/Output relation at AIL
  - Efficiency of task execution at AIL
  - Reliability of task execution at AIL
  - Sufficiency of task execution with respect to design at AIL
  - Completeness of use with respect to purpose at SIL

- Breaching of physical conditions at the material level induces errors of functioning, i.e. a malfunction, which can occur related to:

  - Accessibility of data for the system at AEL (due to Design Failure)
  - Usability of system at AEL (due to Design Failure)
  - Usefulness of system at AEL (due to Conceptual Error)
  - Understandability of the system by the user at SIL (due to Design Failure)
  - Efficiency of the system at SIL (due to Design Failure)
  - Precision of the system at SIL (due to Design Failure)

- Breaching of physical conditions at the executive level induces errors of use, i.e. a slip, which can occur related to:

  - Precision of the user at SIL
  - Completeness of execution procedures by the user at SIL

  Let us now briefly consider our interpretation of the most relevant dimensions.

7.3.1 *Consistency*

This dimension involves the semantics of requirements description, system design and algorithm design. These in turn define integrity constraints at each lower LoA.[3] Failure of this dimension is caused by inconsistent requirements at function description; or by inconsistent operations at design, or inconsistent routines at algorithmic translation.

[3] This characterization agrees with the one offered in Batini and Scannapieco (2006, pp. 30–32).

7.3.2 *Completeness*

This dimension involves the procedural representation of the purpose for the system. This dimension can fail either at design specification level when incomplete processes are given for requirements representation, or at algorithm design level when incomplete routines are used for design representation.

7.3.3 *Accuracy*

This dimension is affected both in terms of semantics and of syntax. At the semantic level one measures the distance of the given specification with the intended one; at the syntactic level, one measures the closeness of data used to those actually occurring in the domain of the intended design.[4] Its failure, accordingly, might occur at various levels: at functional specification level, when requirements import semantic errors; at design specification level, when selected representations for the specifications import semantic errors or when the related data input of those representation import syntactic errors; finally, at algorithm design level, when selected routines for the algorithmic translation of the design import syntactic errors.

7.3.4 *Relevance*

This dimension is affected at levels of semantics and use. It can fail at algorithm design level when processes are selected to represent procedures defined at design specification level; and when the system is put to use, relevance can fail at system interaction level.

7.3.5 *Accessibility*

This dimension is typically syntactical or related to use. While it is usually characterized as an ability of the user,[5] we consider it as a production act at level of design, algorithm implementation and system execution. Hence, in our model accessibility can fail with respect to data accessibility when the Designer has to configure the informal procedures; for algorithm implementation when the Engineer has to define access of both software and hardware data; and for algorithm execution for the material accessibility of data (e.g. in the case of mobile or distributed systems).

[4] See also Batini and Scannapieco (2006, pp. 20–22).

[5] See Batini and Scannapieco (2006, p. 34).

A number of other dimensions can all be grouped together according to their possible failure at the level of implementation: *Security*; *Flexibility/Scalability*; *Precision*; *Efficiency*; *Reliability*; *Sufficiency*. They all concern syntactic requirements that have to be fulfilled when procedures are translated in a chosen language and with respect to data input, routines, I/O algorithms, task execution. Of these, *Efficiency* and *Precision* also occur as possible cases of failure at the interaction level either from system or from user end. *Usability*, *Usefulness*, *Understandability* concern the system use, hence they fail at algorithm execution level.

This list does not include all the possible dimensions, and it might be extended in view of other error categories, matching other information qualities. On the other hand, our model restricts the list of applicable dimensions at given LoAs: for example, it seems that a dimension such as *Believability* could be located in this model only at SIL level (and we have actually avoided considering it for the time being). Also, some dimensions are here implicitly given by other higher-order dimensions: for example, *Clarity* of data does not appear in our list, but it could be argued that it is indirectly obtained by consistent, complete and accurate design and implementation. Direct and detailed definitions of all dimensions and the corresponding failures are formulated formally in the code presented in the Appendix.

7.4 Metric from Errors

A metric for IQ standards is typically used to make a quantitative statement over how much certain information is suited for a given purpose. In the following, we shall define such an abstract metric, based on how many failures occur in the information processing, and at which LoAs do they occur. In defining such a metric, we start from the viewpoint that the higher the LoA at which errors occur and the more important the requirement breached, the greater the loss of quality. Vice versa, for errors occurring at a lower LoA, or involving less important requirements breaches, the less IQ one loses. This order is used in two ways: first, to establish the error-check order; second, to establish which errors are more costly in terms of information quality. The scale of ordered IQ dimensions can be given as the graph obtained by accommodating the various dimensions we have taken into account as Cartesian coordinates over a X-axis representing conditions ordered by increasing relevance and a Y-axis representing LoAs ordered from the lowest to the highest (Fig. 7.2).

Starting from the highest point in the graph, we look at values that can be assessed in a *stand-alone metric*, namely values related to dimensions produced by the Architect alone; proceeding backwards on the vector, we move towards *context-dependent* values of information content, where the information user is more and more present and the standard of the dimension is generated by way of

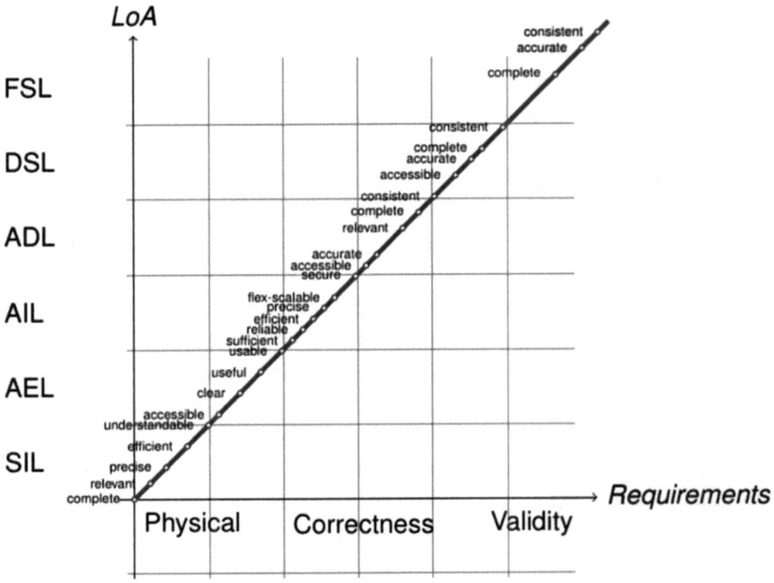

Fig. 7.2 Ordering IQ dimensions at LoAs and requirements

interacting agents. Using this vector as a scale, we intend to provide an understanding of IQ metric according to an agile checking method,[6] such that the evaluation and correction mechanisms are well-definable and scalable. The evaluation of the IQ metric for the whole system is a function of the evaluation at each point in the graph: an inductive check proceeds on each dimension at each of the corresponding LoA/Requirement pair and assesses if a failure occurs. In the following, one strategy for the resolution of errors and one for the system's evaluation are presented.

7.4.1 First Strategy: Resolve Errors

The first strategy requires to identify errors, resolve them, then move back to the above level in order to proceed with the improved evaluation before moving again down to the next dimension. This method assumes therefore that "(e)ach processing step transforms one or more inputs into a new data source, with a new set of IQ

[6] Here the terms "agile" refers to the technical understanding of software quality checking procedure at each different LoA, as opposed to the usual counterpart, the "waterfall" method. See Royce (1970), AA.VV (The Agile Manifesto).

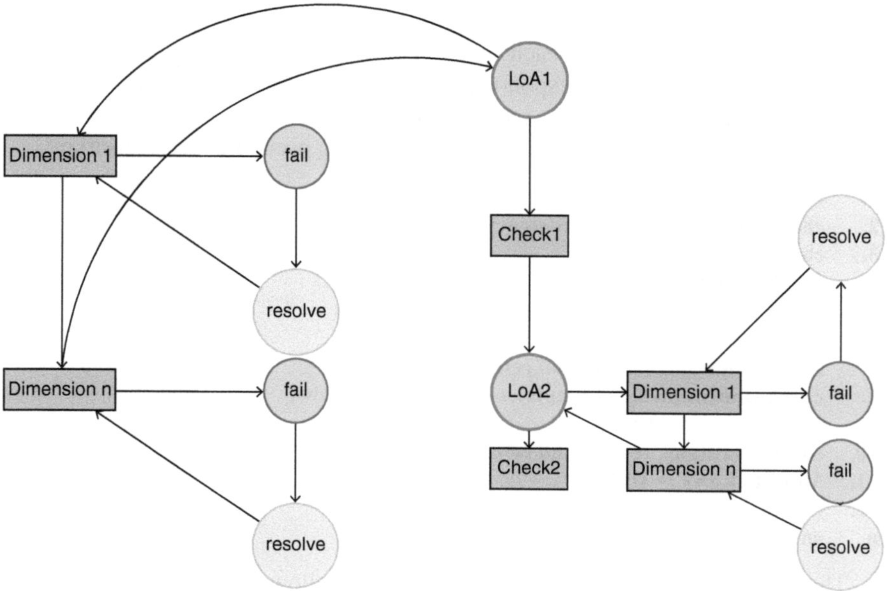

Fig. 7.3 Resolve method

metrics".[7] The process is described in the directed graph from Fig. 7.3. The graph gives a procedural description of a resolution method on a simple information processing with two LoAs and *n* dimensions: an arrow from a dimension to a "fail"-node is always followed by an arrow to a resolution step, followed by a re-initiation of the procedure. In this case, one starts from the highest LoA, and moves by increasing indices down on the scale of LoAs:

$$LoA1 := FSL;\ LoA2 := DSL;\ LoA3 := ADL; \ldots$$

The IQ dimensions are ordered in a downward scale as well, first according to the quality requirement family they satisfy:

$$IQReq1 := Validity;\ IQReq2 := Correctness;\ IQReq3 := Use;$$

then according to the mode of error that occurs when one of the above requirement is breached:

$$Mode1 := Conceptual;\ Mode2 := Material;\ Mode3 := Physical.$$

[7] Wand and Wang (1996, p. 3).

This gives us an ordered list of dimensions to check, which corresponds to the list given in the Table 7.3. For each such dimension, the procedure consists in looking for the occurrence of corresponding errors, namely by checking the satisfaction of use, correctness and validity requirements, resolve where errors occur and re-initiate. The directed graph instantiates the following pseudo-coded algorithm:

1. Start LoA1 := FSL;
2. check Dimension 1 := Consistency of Requirement;
3. check = yes, move to 5;
4. check = mistake, resolve and return to 2;
5. check Dimension 2 := Accuracy of Specifications;
6. check = yes, move to 8;
7. check = mistake, resolve and return to 5;
8. check Dimension 3 := Completeness of Specifications;
9. check = yes, move to 11;
10. check = mistake, resolve and return to 8;
11. Move to LoA2 := DSL;
12. ...;
n end.

The list of LoAs and the categorization of errors allows us to know precisely for which criteria should we check, and in which form. To exemplify:

1. Start with System Specification;
2. check mistake in dim1 = are the requirements presented consistent? ;
3. check = yes, move to 5;
4. check = contradictory req found, resolve and return to 2;
5. check mistake in dim2 = are the requirements presented accurate? ;
6. check = yes, move to 8;
7. check = unclear req found, resolve and return to 5;
8. check mistake in dim3 = are the requirements presented complete? ;
9. check = yes, move to 11;
10. check = incomplete req found, resolve and return to 8;
11. Move to LoA2 := DSL;
12. check mistake in dim1 = are the procedures designed consistent?
13. check = yes, move to 15;
14. check = contradictory proc found, resolve and return to 12;
15. check failure in dim2 = are the routine complete?
16. ...;
n end.

The complexity of the routine is determined by the corresponding complexity of each LoA. For *FSL*, this means that the length of the resolution depends from how many parameters are given as requirements; for *DSL*, from how many are given in design; and so on. We are requiring that for each error found by the algorithm, a resolution procedure is started and the procedure does not move on unless such a resolution has been performed. This algorithmic check corresponds therefore to an evaluation for an *optimal metric*, i.e. where the system is required to be optimally evaluated with respect to standards of dimensions.

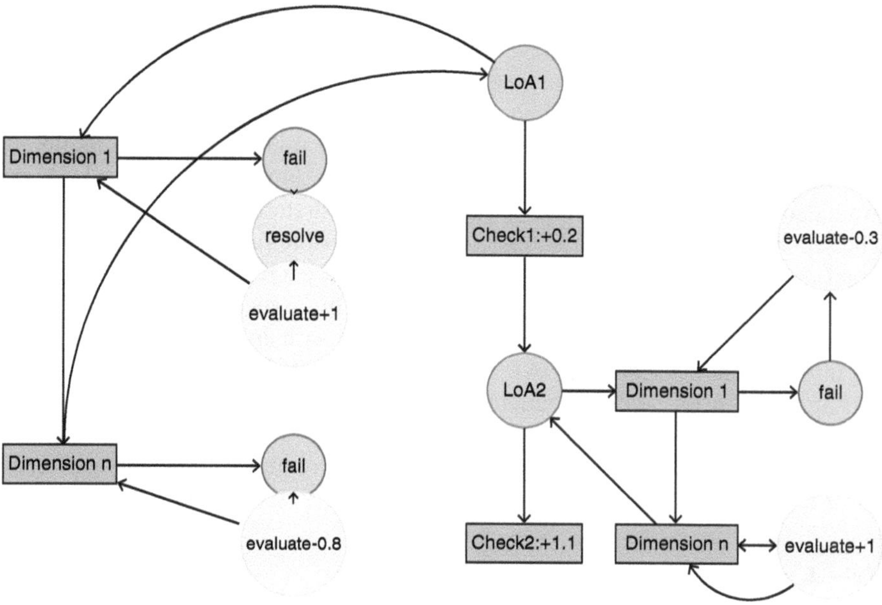

Fig. 7.4 Resolve and evaluate methods combined

7.4.2 *Second Strategy: Evaluate Errors*

Our algorithmic approach is not only easily scalable, but it is also very adaptable. In the second method, we do not proceed by resolution, but by evaluation: the specification on the algorithm is now that the assessment of a dimension's standard might not require a reset of the procedure, but only a lowering of the overall evaluation. This means allowing to move down on the LoAs without executing any resolution strategy. Such procedure might be permitted on some dimensions only (i.e., combining it with the previous strategy for most fundamental dimensions at certain LoAs) and it allows different standards metrics at different stages. Dimensions can be listed in any desired priority order and with any desired assignment of evaluations. In other words, the evaluation can map to any positive check a given positive integer or rational value, which does not need to be the same for each stage. Similarly, a negative value can be assigned to any non-positive check. For example: passing check on consistent requirements can be evaluated to 1.0; passing check on accurate requirements can be evaluated to 0.8; passing check on complete requirements can be evaluated to 0.6 (or for that matters to 0.9, as the logical order of checks does not need to correspond to ordered values); passing check on consistent procedures can be evaluated again to 1.0; passing check on complete routines should be evaluated to $n < 1.0$, as it moves down in the scale of requirements from validity to correctness; similar operations can be defined on negative checks, by assigning negative values. The values are of course better chosen depending on the application. The previously given directed graph is now modified as in Fig. 7.4 (with arbitrary evaluations assigned).

In this example, the first dimension at the first LoA is considered of highest relevance and it is required to have a resolution method in the case of failure, followed by an evaluation assignment (+1). The *n*th dimension of the first LoA is considered not of maximal relevance as it allows for an evaluation without resolution; nonetheless it is obviously considered of some high relevance, as its failure is evaluated with high loss of IQ (−0.8). After all dimensions of one LoA are evaluated, the check on LoA1 evaluates the current status of the system (+0.2). The second LoA presents an unresolved but admissible failure at the first dimension (i.e. the system does not stop, nor does it require a resolution, but lowers the overall evaluation with a −0.3 score), followed by a successful nth dimension (score +1). After which the system again proceeds with an overall evaluation taking into account the score at LoA1 (total: +1.1). It is obviously an issue of context and application to establish which range of values are considered admissible, which are positive and which are not. In the above pseudo-code we have imposed very stringent conditions on the satisfaction of requirements, namely by giving an all-or-nothing approach:

if check = yes; then move;,
if check = error found; then resolve and return;.

In this new model, we change the algorithm accordingly. For example, one might want to just reduce the overall metric of the system when recognizing the processing makes use of routines that are known to be less secure than others, but still efficient (while keeping the latter entirely independent from the former). Then one could improve the algorithm accordingly at AIL:

```
n          Move to LoA4 := AIL;
n+m        ...;
n+m+1      check failure in dim2 = are the procedures used secure?
n+m+2      check = yes, move to [n+m+4];
n+m+3      check = hackable proc found; evaluate to -0.8 and move to
           [n+m+4];
n+m+4      check failure in dim3 = are the routines effective?
n+m+5      check = yes, return to [n+m+2] force = yes;
n+m+6      check = no output proc, evaluate to -1.0 and return to LoA3
           := ADL;
n+m+7      ...
```

By this algorithm: when encountering an hackable procedure, the system gets a lowering of the overall quality score, but it moves further to check if they still provides the required output; if so, it forces a positive answer to the security check; otherwise, it lowers further the quality score and moves back to the above LoA. The latter represents a conservative, security move to proceed again with all possible mistakes and failure checks. Similarly, various other modifications can be easily implemented: skip the check on precision (for the I/O relation that are known to be precise up to a certain range of values); give low scores to process-clarity and high scores to system-usability (respectively: for processing at the lower levels, such as machine language processes that remains hidden to the user; and for interface processes); or viceversa (when considering programming processes). And so on, again depending on appropriate parameters chosen on context and application.

7.5 Conclusions

An algorithmic treatment of information quality dimensions offers a novel perspective to improve systems' design and its technical and epistemological traits. The negative account here introduced allows determining where and how IQ dimensions fail. The quantitative aspect is also crucial to establish thresholds of reliability. Issues of security and trust in systems design can profit from such an approach, by allowing to identify minimal criteria to be satisfied in order for the system to be considered of sufficient quality. This has of course great repercussions on the impact of technology on everyday life, from education to communications. Of particular interest is the applicability of the model here presented to issues of program safety in the context of distributed computing.

Appendix

A machine-checkable formal translation of the checking algorithm for IQ dimensions standards given above is provided. In this code we provide all definitions for errors and how they match to information quality standards in the language of the Calculus of Constructions. By means of such formal translation we satisfy a double task:

1. we give a precise formulation for each error case as it gets applied to an Information Quality Dimension;
2. we are able to check the syntactic correctness of such definitions and of the algorithmic check for the errors resolution strategy.

To help readability, we introduce the different parts of the code by a brief explanation.

We start by providing standard definitions for Boolean values, processes interpreted as characteristic functions of propositions, and some set-theoretic relations reduced to corresponding defining processes.

```
Coq <
Coq < Check Prop : Type.
Prop:Type
: Type
Coq < Variable A B C : Prop.
A is assumed
Warning: A is declared as a parameter because it is at a global level
B is assumed
Warning: B is declared as a parameter because it is at a global level
C is assumed
Warning: C is declared as a parameter because it is at a global level
Coq <
Coq < Inductive tm : Prop :=
        | tm true : tm
        | tm false : tm
        | tm if : tm > tm > tm > tm.
tm is defined
tm ind is defined
Coq <
Coq <
Coq < Inductive bvalue : tm > Prop :=
        | b true : bvalue tm true
        | b false : bvalue tm false.
bvalue is defined
bvalue ind is defined
Coq <
Coq < Definition value (t : tm) : Prop :=  bvalue t.
value is defined
Coq <
Coq < Inductive full eval : tm > tm > Prop :=
        | f value : forall t, value t > full eval t t
        | f iftrue : forall t1 t2 t3 t, full eval t1 tm true > full eval t2 t >  full
        eval (tm if t1 t2 t3) t
        | f iffalse : forall t1 t2 t3 t, full eval t1 tm false > full eval t3 t > full
        eval (tm if t1 t2 t3) t.
full eval is defined
full eval ind is defined
Coq <
Coq < Require Import Bool.
Coq <
Coq < Set Implicit Arguments.
Coq <
Coq < Inductive proc : Type := Charac : (A > bool) > proc.
proc is defined
proc rect is defined
proc ind is defined
proc rec is defined
Coq <
Coq < Definition charac (s:proc) (a:A) : bool := let (f) := s in fa.
```

```
charac is defined
Coq <
Coq < Parameter In : proc > tm > Prop.
In is assumed
Coq <
Coq < Definition Equal A A' := forall t:proc, In t A <> In t A'.
Equal is defined
Coq <
Coq < Definition Subset A A' := forall t:proc, In t A > In t A'.
Subset is defined
Coq <
Coq < Definition Empty A := forall t:proc, ~ In t A.
Empty is defined
```

We now add specific definitions for the notions of purpose and design in terms of the previously defined set-theoretic relations. In particular, we are interested in defining properties such as 'having no purpose' or 'having no design'.

```
Coq <
Coq < Definition purpose := exists t, exists A, In t A.
purpose is defined
Coq <
Coq < Definition no purpose := forall t, forall A, In t A.
no purpose is defined
Coq <
Coq < Definition design := exists t, forall t1, value t > Full eval t1 tm true.
design is defined
Coq <
Coq < Definition no design := exists t, forall t1, value t > full eval t1 tm false.
no design is defined
Coq <
```

The definitions of the notions of mistake, failure, malfunction and slip are inductively given, based on the original taxonomy provided in Primiero (2013). Here we define mistakes as missing, illdefined or wrongly termed types:

```
Coq <
Coq < Inductive mistake : Type > Type :=
        | missing type : mistake (forall A, Empty A)
        | type illdefined : mistake (forall t:proc, no purpose)
        | term retype : mistake (exists A, exists t:proc, In t A <> ~ In t A).
mistake is defined
mistake rect is defined
mistake ind is defined
mistake rec is defined
Coq <
```

failures as errors of rules (wrongly selected, badly defined, requiring wrong resources, missing resources):

```
Coq <
Coq < Inductive failure : Type > Type :=
        | wrong rule : failure (match A with match rule => ~ A end)
        | bad rule : failure (match A with context rule => ~ A end)
        | bad address : failure (match A with B => ~ B end)
        | no resources : failure (match A with t => ~ t end).
Warning: pattern B is understood as a pattern variable
failure is defined
failure rect is defined
failure ind is defined
failure rec is defined
Coq <
```

malfunctions as functional errors of unusable rule in context, unusable term in rule, unusable resource in term dependency and missing resource in term dependency;

```
Coq < Inductive malfunction : Type > Type :=
        | unusable wrong rule: malfunction (exists t, match t with match rule
        => ~t end)
        | unusable bad rule : malfunction (exists t, match t with context rule
        => ~ t end)
        | unusable bad address : malfunction (exists t, match t with B => ~t
        end)
        | unusable no resources : malfunction (exists t, match t with t' => ~t
        end).
Warning: pattern B is understood as a pattern variable
malfunction is defined
malfunction rect is defined
malfunction ind is defined
malfunction rec is defined
Coq <
```

Slips as material errors of repeated selection of rule, redundancy, recurrent use of data in inappropriate ways.

```
Coq <
Coq < Inductive slip : Type > Type :=
        | exception rule : slip (~forall t t', value t > full eval t t')
        | bad location : slip (~forall t1 t2 t3 t, full eval t1 tm true > full eval t2
        t) > full eval (tm if t1 t2 t3) t)
        | redundant process : slip (forall A, match A with match rule => ~A
        end)
        | recurrent data : slip (forall A, match A with t => ~t end).
slip is defined
slip rect is defined
slip ind is defined
slip rec is defined
Coq <
```

We then proceed by applying breach of conditions to each of purpose, design, routine, algorithm, algorithm implementation, execution. Validity conditions breaches at FSL:

```
Coq < Inductive invalid purpose : Type > Type :=
   | inconsistent req : invalid purpose (exists t:proc, no purpose).
invalid purpose is defined
invalid purpose rect is defined
invalid purpose ind is defined
invalid purpose rec is defined
Coq <
Coq < Inductive incorrect purpose : Type > Type :=
   | inaccurate req : incorrect purpose (match A with t => no purpose end)
   | incomplete req : incorrect purpose (exists A, match A with t => ~t end).
incorrect purpose is defined
incorrect purpose rect is defined
incorrect purpose ind is defined
incorrect purpose rec is defined
Coq <
```

Validity and correctness breaches at DSL:

```
Coq <
Coq < Inductive invalid design : Type > Type :=
        | inconsistent design : invalid design (exists A, exists t:proc, In t A <>
        ~ In t A).
invalid design is defined
invalid design rect is defined
invalid design ind is defined
invalid design rec is defined
Coq <
Coq < Inductive incorrect design : Type > Type :=
        | incomplete routine : incorrect design (match A with match rule => ~
        A end)
        | incomplete routine2 : incorrect design (match A with context rule
        => ~ A end)
        | inaccurate data : incorrect design (match A with B => ~ B end)
        | inaccessible data : incorrect design (match A with t => ~ t end).
Warning: pattern B is understood as a pattern variable
incorrect design is defined
incorrect design rect is defined
incorrect design ind is defined
incorrect design rec is defined
Coq <
```

Validity and correctness breaches at ADL:

```
Coq <
Coq < Inductive invalid algorithmdesign : Type > Type :=
    | inconsistent algorithm : invalid algorithmdesign (match invalid de-
    sign A with t => ~ t end)
    | incomplete algorithm : invalid algorithmdesign (match incomplete
    routine with t => ~A end)
    | incomplete algorithm2 : invalid algorithmdesign (match incomplete
    routine2 with t => ~A end)
    | irrelevant algorithm : invalid algorithmdesign
    (match A with t => no purpose end)
    | inaccurate algorithm : invalid algorithmdesign (match inaccurate
    req with t => ~A end).
invalid algorithmdesign is defined
invalid algorithmdesign rect is defined
invalid algorithmdesign ind is defined
invalid algorithmdesign rec is defined
Coq <
```

Validity and correctness breaches at AIL:

```
Coq <
Coq < Inductive incorrect algorithmimplementation : Type > Type :=
    | inaccessible data2 : incorrect algorithmimplementation(match inac-
    cessible data with t => ~A end)
    | insecure routine : incorrect algorithmimplementation
     (match bad rule with t => ~A end)| no flex data : incorrect algorith-
    mimplementation (match no resources with t => ~A end)
    | no precise IO1 : incorrect algorithmimplementation (match bad
    rule with t => ~A end)
    | no precise IO2 : incorrect algorithmimplementation (match bad ad-
    dress with t => ~A end)
    | no precise IO3 : incorrect algorithmimplementation (match excep-
    tion rule with t => ~A end)
    | no precise IO4 : incorrect algorithmimplementation (match bad lo-
    cation with t => ~A end)
    | no efficient task: incorrect algorithmimplementation (match redun-
    dant process with t => ~A end)
    | no efficient task2: incorrect algorithmimplementation (match re-
    current data with t => ~A end)
    | no reliable task: incorrect algorithmimplementation (match inaccu-
    rate req with t => ~A end)
    | no sufficient task: incorrect algorithmimplementation(match in-
    complete req with t => ~A end).
incorrect algorithmimplementation is defined
incorrect algorithmimplementation rect is defined
incorrect algorithmimplementation ind is defined
incorrect algorithmimplementation rec is defined
Coq <
```

Physical breaches at AEL:

```
Coq <
Coq < Inductive unusable execution : Type > Type :=
        | unusable system : unusable execution (match no purpose with t =>
        ~A end)
        | useless system : unusable execution (forall A, Empty A)
        | inaccessible data3 : unusable execution (match inaccessible data
        with useless system => ~A end).
unusable execution is defined
unusable execution rect is defined
unusable execution ind is defined
unusable execution rec is defined
Coq <
```

We now proceed to define the algorithm meant to check for the failure of IQ standard:

```
Coq <
Coq < Inductive Check invalid purpose : Type > Type :=
        | check inconsistent req : Check invalid purpose (exists A:Prop,
        match inconsistent req with A => no purpose end).
Warning: pattern A is understood as a pattern variable
Check invalid purpose is defined
Check invalid purpose rect is defined
Check invalid purpose ind is defined
Check invalid purpose rec is defined
Coq <
Coq < Inductive Check incorrect purpose : Type > Type :=
        | check inaccurate req : (forall A:Prop, forall t:proc, match A with t
        => purpose end) Coq < Check incorrect purpose (exists A:Prop,
        match inaccurate req with A => no purpose end)
        | check incomplete req : (exists A:Prop, match A with t => ~t end) >
        Check incorrect purpose (exists A:Prop, exists t, match in complete
        req with A => ~t end)
Warning: pattern A is understood as a pattern variable
Warning: pattern A is understood as a pattern variable
Check incorrect purpose is defined
Check incorrect purpose rect is defined
Check incorrect purpose ind is defined
Check incorrect purpose rec is defined
Coq <
Coq < Inductive Check invalid design : Type > Type :=
        | check inconsistent design : (forall A:Prop, forall t:proc, match A
        with t => purpose end)  > Check invalid design (exists A, exists
        t:proc, match inconsistent design with t => ~ A end).
Check invalid design is defined
Check invalid design rect is defined
Check invalid design ind is defined
Check invalid design rec is defined
```

```
Coq <
Coq < Inductive Check incorrect design : Type > Type :=
        | check incomplete routine : (forall A, exists t:proc, In t A) > Check
incorrect design (exists A:Prop, match A with match rule => ~ A end)
        | check incomplete routine2 : (forall A:Prop, exists t:proc, match A
with match rule => A end) > Check incorrect design (exists A:Prop,match A
with context rule => ~ A end)
        | check inaccurate data : (forall A:Prop, exists t:proc, match A with
        context rule => A end) > Check incorrect design (exists A:Prop, ex-
        ists B:Prop, match A with B => ~ B end)
        | check inaccessible data : (forall A:Prop, exists t:proc, match A with
        B => B end) > Check incorrect design (exists A:Prop, exists t:proc,
        match A with t => ~ t end).
Warning: pattern B is understood as a pattern variable
Warning: pattern B is understood as a pattern variable
Check incorrect design is defined
Check incorrect design rect is defined
Check incorrect design ind is defined
Check incorrect design rec is defined
Coq <
Coq < Inductive Check invalid algorithmdesign : Type > Type :=
        | check inconsistent algorithm : (match invalid design with t => ~ A
        end) > Check invalid algorithmdesign (exists A:Prop, exists t:proc,
        match inconsistent algorithm with t => ~ A end)
        | check incomplete algorithm : (match incomplete routine with t => ~
        A end) > Check invalid algorithmdesign (exists A:Prop, exists t:proc,
        match incomplete algorithm with t => ~A end)
        | check incomplete algorithm2 : (match incomplete routine2 with t =>
        ~ A end) > Check invalid algorithmdesign (match incomplete algo-
        rithm2 with t => ~A end)
        | check irrelevant algorithm : (match A with t => no purpose end) >
        Check invalid algorithmdesign (match irrelevant algorithm with t =>
        ~ A end)
        | check inaccurate algorithm : (match inaccurate req with t => ~ A
        end) > Check invalid algorithmdesign (match inaccurate algorithm
        with t => ~A end).
Check invalid algorithmdesign is defined
Check invalid algorithmdesign rect is defined
Check invalid algorithmdesign ind is defined
Check invalid algorithmdesign rec is defined
Coq <
Coq <
Coq < Inductive Check incorrect algorithmimplementation : Type > Type :=
        | check inaccessible data2 : (match inaccessible data with t => ~A
        end) > Check incorrect algorithmimplementation (exists A:Prop, ex-
        ists t:proc, match inaccessible data2 with t => ~A end)
        | check insecure routine : (match bad rule with t => ~A end)
```

```
        > Check incorrect algorithmimplementation (exists A:Prop, exists
        t:proc, match insecure routine with t => ~A end)
        | (check no flex data : (match no resources with t => ~A end) >
        Check incorrect algorithmimplementation (forall A:Prop, exists
        t:proc, match A with context rule => A end) > Check incorrect design
         (exists A:Prop, exists B:Prop, match A with B => ~ B end)
        | check inaccessible data : (forall A:Prop, exists t:proc, match A with
        B => B end) > Check incorrect design (exists A:Prop, exists t:proc,
        match A with t => ~ t end).
Warning: pattern B is understood as a pattern variable
Warning: pattern B is understood as a pattern variable
Check incorrect design is defined
Check incorrect design rect is defined
Check incorrect design ind is defined
Check incorrect design rec is defined
Coq <
Coq < Inductive Check invalid algorithmdesign : Type > Type :=
        | check inconsistent algorithm : (match invalid design with t => ~ A
        end) > Check invalid algorithmdesign (exists A:Prop, exists t:proc,
        match inconsistent algorithm with t => ~ A end)
        | check incomplete algorithm : (match incomplete routine with t => ~
        A end) > Check invalid algorithmdesign (exists A:Prop, exists t:proc,
        match incomplete algorithm with t => ~A end)
        | check incomplete algorithm2 : (match incomplete routine2 with t =>
        ~ A end) > Check invalid algorithmdesign (match incomplete algo-
        rithm2 with t => ~A end)
        | check irrelevant algorithm : (match A with t => no purpose end) >
        Check invalid algorithmdesign (match irrelevant algorithm with t =>
        ~ A end)
        | check inaccurate algorithm : (match inaccurate req with t => ~ A
        end) > Check invalid algorithmdesign (match inaccurate algorithm
        with t => ~A end).
Check invalid algorithmdesign is defined
Check invalid algorithmdesign rect is defined
Check invalid algorithmdesign ind is defined
Check invalid algorithmdesign rec is defined
Coq <
Coq <
Coq < Inductive Check incorrect algorithmimplementation : Type > Type :=
        | check inaccessible data2 : (match inaccessible data with t => ~A
        end) > Check incorrect algorithmimplementation (exists A:Prop, ex-
        ists t:proc, match inaccessible data2 with t => ~A end)
        | check insecure routine : (match bad rule with t => ~A end)
        > Check incorrect algorithmimplementation (exists A:Prop, exists
        t:proc, match insecure routine with t => ~A end)
        | check no flex data : (match no resources with t => ~A end) >
```

```
Check incorrect algorithmimplementation (exists t':proc, match no
flex data with t => ~A end)
| check no precise IO1 : (match bad rule with t => ~A end) > Check
incorrect algorithmimplementation (exists t':proc, match no precise
IO1 with t => ~A end)
| check no precise IO2 : (match bad address with t => ~A end) >
Check incorrect algorithmimplementation (exists t':proc, match no
precise IO2 with t => ~A end)
| check no precise IO3 : (match exception rule with t => ~A end) >
Check incorrect algorithmimplementation (exists t':proc, match no
precise IO3 with t => ~A end)
| check no precise IO4 : (match bad location with t => ~A end) >
Check incorrect algorithmimplementation (exists t:proc, match no
precise IO4 with t => ~A end)
| check no efficient task: (match redundant process with t => ~A end)
> Check incorrect algorithmimplementation (exists t:proc, match no
efficient task with t => ~A end)
| check no efficient task2: (match recurrent data with t => ~A end) >
Check incorrect algorithmimplementation (exists t:proc, match no
efficient task2 with t => ~A end)
| check no reliable task: (match inaccurate req with t => ~A end) >
Check incorrect algorithmimplementation (exists t:proc, match no
reliable task with t => ~A end)
| check no sufficient task: (match incomplete req with t => ~A end) >
Check incorrect algorithmimplementation (exists t:proc, match no
sufficient task with t => ~A end).
Check incorrect algorithmimplementation is defined
Check incorrect algorithmimplementation rect is defined
Check incorrect algorithmimplementation ind is defined
Check incorrect algorithmimplementation rec is defined
Coq <
Coq < Inductive Check unusable execution : Type >Type :=
          | check unusable system: (match no purpose with t => ~A end)
          > Check unusable execution (exists t:proc, match unusable system
          with t => ~A end)

          | check useless system : (forall A, Empty A) > Check unusable execu-
          tion (exists A':Prop, match useless system with A => ~A' end)
          | check inaccessible data3 :(match inaccessible data with incorrect
          design => ~A end) > Check unusable execution (forall t:proc, match
          inaccessible data3 with t => ~A end).
Warning: pattern A is understood as a pattern variable
Warning: pattern incorrect design is understood as a pattern variable
Check unusable execution is defined
Check unusable execution rect is defined
Check unusable execution ind is defined
Check unusable execution rec is defined
Coq <
```

It finally follows the resolution procedure which suggests how to repair the error at hand. Invalid purpose requires purpose re-assignment, while purpose incorrectness requires purpose term-rewriting.

```
Coq <
Coq < Inductive Resolve invalid purpose : Type > Type :=
        | resolve inconsistent req :(exists A:Prop, match inconsistent req with
        A => no purpose end) > Resolve invalid purpose (exists A':Prop,
        purpose).
Warning: pattern A is understood as a pattern variable
Resolve invalid purpose is defined
Resolve invalid purpose rect is defined
Resolve invalid purpose ind is defined
Resolve invalid purpose rec is defined
Coq <
Coq < Inductive Resolve incorrect purpose : Type >Type :=
        | resolve inaccurate req : (exists A:Prop, match inaccurate req with A
        => no purpose end) > (Resolve incorrect purpose (exists A':Prop,
        purpose))
        | resolve incomplete req : (exists A:Prop, exists t:proc, match A with t
        => ~t end) > Resolve incorrect purpose (exists t':proc, match A with
        t' => purpose end).
Warning: pattern A is understood as a pattern variable
Resolve incorrect purpose is defined
Resolve incorrect purpose rect is defined
Resolve incorrect purpose ind is defined
Resolve incorrect purpose rec is defined
Coq <
```

For design: invalidity requires purpose reassignment and term rewriting; incorrectness requires re-matching of one of: rule, context, data dependency and procedure.

```
Coq <
Coq < Inductive Resolve invalid design : Type > Type :=
        | resolve inconsistent design : (exists A, exists t:proc, match incon-
        sistent design with t => ~ A end)  > Resolve invalid design (exists
        t':proc, exists A', A' purpose).
Resolve invalid design is defined
Resolve invalid design rect is defined
Resolve invalid design ind is defined
Resolve invalid design rec is defined
Coq <
Coq <
Coq < Inductive Resolve incorrect design : Type > Type :=
        | resolve incomplete routine : Coq < (exists A:Prop, match A with
        match rule => ~ A end) > Coq < Resolve incorrect design (exists
        A':Prop, match A' with match rule => A' end)
        | resolve incomplete routine2 : (exists A:Prop,match A with context
        rule => ~ A end) > Resolve incorrect design (exists A':Prop, match A'
        with context rule => A' end)
        | resolve inaccurate data : (exists A:Prop, exists B:Prop, match A
        with B => ~ B end) > Resolve incorrect design (exists A':Prop, exists
        B:Prop, match A' with B => B end)
        | resolve inaccessible data : (exists A:Prop, exists t:proc, match A
        with t => ~ t end) > Resolve incorrect design (exists A:Prop, exists
        t':proc, match A with t' => t' end).
```

```
Warning: pattern B is understood as a pattern variable
Warning: pattern B is understood as a pattern variable
Resolve incorrect design is defined
Resolve incorrect design rect is defined
Resolve incorrect design ind is defined
Resolve incorrect design rec is defined
Coq <
```

For algorithm design: inconsistency requires routine re-definition; incompleteness rule re-matching or data fulfillment; irrelevancy requires both purpose and routine re-definition; inaccuracy require routine and purpose re-matching.

```
Coq <
Coq < Inductive Resolve invalid algorithmdesign : Type > Type :=
        | resolve inconsistent algorithm : (exists A:Prop, exists t:proc, match
        inconsistent algorithm with t => ~ A end)> Resolve invalid algo-
        rithmdesign (exists A, exists t':proc, match A with t => In t' A end)
        | resolve incomplete algorithm : (exists A:Prop, exists t:proc, match
        incomplete algorithm with t => ~A end) > Resolve invalid algo-
        rithmdesign  (exists A, match A with match rule => A end)
        | resolve incomplete algorithm2 : (match incomplete routine2 with t
        => ~ A end) > Resolve invalid algorithmdesign (exists A, match A
        with context rule => A end)
        | resolve irrelevant algorithm : (exists A:Prop, exists t:proc, match A
        with t => no purpose end) > Resolve invalid algorithmdesign (exists
        A', exists t':proc, match A with t' => A' end)
        | resolve inaccurate algorithm : (exists A:Prop, exists t:proc, match
        inaccurate req with t => ~ A end) > Resolve invalid algorithmdesign
        (exists A, exists t':proc, match A with t' => A end).
Resolve invalid algorithmdesign is defined
Resolve invalid algorithmdesign rect is defined
Resolve invalid algorithmdesign ind is defined
Resolve invalid algorithmdesign rec is defined
Coq <
```

For algorithm implementation: data inaccessibility requires data re-matching; security flaws context rule re-definition; imprecise I/O relations require term or data-dependency re-matching; efficiency requires purpose re-assessment while unreliability and insufficiency require procedure re-selection.

```
Coq <
Coq < Inductive Resolve incorrect algorithmimplementation : Type > Type: =
        | resolve inaccessible data2 : (exists A:Prop, exists t:proc, match inac-
        cessible data2 with t => ~A end) > Resolve incorrect algorith-
        mimplementation (exists A, exists t':proc, match A with t' => A end)
        | resolve insecure routine : (exists A:Prop, exists t:proc, match inse-
        cure routine with t => ~A end) > Resolve incorrect algorithmimple-
        mentation (exists A, exists t':proc, match A with context rule => A
        end)
        | resolve no flex data : (exists t':proc, match no flex data with t => ~A
        end) > Resolve incorrect algorithmimplementation (exists A, exists
        t'':proc, match A with t'' => A end)
        | resolve no precise IO1 : (exists t':proc, match no precise IO1 with t
        => ~A end) > Resolve incorrect algorithmimplementation (exists A,
        match A with context rule => A end)
        | resolve no precise IO2 : (exists A:Prop, exists t':proc, match no pre-
        cise IO2 with t => ~A end) > Resolve incorrect algorithmimplementa-
        tion (exists A:Prop, exists B':Prop, match A with B' => A end)
        | resolve no precise IO3 : (exists t':proc, match no precise IO3 with t
        => ~A end) > Resolve incorrect algorithmimplementation (exists
        A:Prop, exists t, exists t'', value t > full eval t t'')
        | resolve no precise IO4 : (exists A:Prop, exists t:proc, match no pre-
        cise IO4 with t => ~A end) > Resolve incorrect algorithmimplementa-
        tion (forall t1 t2 t3 t', full eval t1 tm true > full eval t2 t' > full eval
        (tm if t1 t2 t3) t')
        | resolve no efficient task: (exists t:proc, match no efficient task with
        t => ~A end) > Resolve incorrect algorithmimplementation (exists A',
        match A' with match rule => A' end)
        | resolve no efficient task2: (exists t:proc, match no efficient task2
                with t => ~A end) > Resolve incorrect algorithmimplementa-
        tion (exists t':proc, match A with t' => t' end)
        | resove no reliable task: (exists A:Prop, exists t:proc, match no relia-
        ble task with t => ~A end) > Resolve incorrect algorithmimplementa
        tion (exists t':proc, match A with t' => A end)
        | resolve no sufficient task: (exists t:proc, match no sufficient task
        with t => ~A end) > Resolve incorrect algorithmimplementation
        (match A with t' => forall t', exists t1, value t' > full eval t1 tm true
        end).
Resolve incorrect algorithmimplementation is defined
Resolve incorrect algorithmimplementation rect is defined
Resolve incorrect algorithmimplementation ind is defined
Resolve incorrect algorithmimplementation rec is defined
Coq <
```

Finally, for algorithm execution: un-usability requires both purpose and procedure re-assessment; uselessness wants purpose re-selection; for data inaccessibility, procedure re-definition.

```
Coq <
Coq < Inductive Resolve unusable execution : Type > Type :=
        | resolve unusable system: (exists A:Prop, exists t:proc, match unusa-
        ble system with t => ~A end) > Resolve unusable execution (exists
        A:Prop, exists t':proc, match A with t' => A end)
        | resolve useless system : (exists A':Prop, match useless system with A
        => ~A' end) > Resolve unusable execution (exists B:Prop, match B
        with A => B end)
        | resolve inaccessible data3 : (forall t:proc, match inaccessible data3
        with t => ~A end) > Resolve unusable execution (exists t':proc, match
        A with t' => A end).
Warning: pattern A is understood as a pattern variable
Warning: pattern A is understood as a pattern variable
Resolve unusable execution is defined
Resolve unusable execution rect is defined
Resolve unusable execution ind is defined
Resolve unusable execution rec is defined
  Coq <
```

References

AA.VV. The agile manifesto.

Abran, A., Al-Qutaish, R. E., Desharnais, J.-M., & Habra, N. (2008). Chapter 5: Iso-based models to measure software product quality. In B. Ravi Kumar Jain (Ed.), *Software quality measurement – Concepts and approaches* (pp. 61–96). Hyderabad: ICFAI University Press.

Batini, C., & Scannapieco, M. (2006). *Data quality: Concepts, methodologies and techniques.* Berlin/New York: Springer.

ISO/IEC 25010. (2011). Systems and software engineering – Systems and software quality requirements and evaluation (SQuaRE) – System and software quality models.

Kahn, B. K., Strong, D. M., & Wang, R. Y. (2002). Information quality benchmarks: Product and service performance. *Communications of the ACM, 45*(4), 184–192.

Primiero, G. (2013). A taxonomy of errors for information systems. *Minds & Machines.* doi:10.1007/s11023-0139307-5.

Royce, W. (1970). Managing the development of large software systems. *Proceedings of IEEE WESCON, 26*, 1–9.

Suryn, W., Abran, A., & April, A. (2003). Iso/iec square: The second generation of standards for software product quality. In *Proceedings of the 7th IASTED International Conference on Software Engineering and Applications* (ICSEA03), Marina del Rey, 3–5 Nov 2003, pp. 1–9.

Wand, Y., & Wang, R. Y. (1996). Anchoring data quality dimensions in ontological foundations. *Communications of the ACM, 39*, 87–95.

Chapter 8
The Varieties of Disinformation

Don Fallis

Abstract Intentionally misleading information (*aka* Disinformation) is ubiquitous and can be extremely dangerous. Emotional, financial, and even physical harm can easily result if people are misled by deceptive advertising, government propaganda, doctored photographs, forged documents, fake maps, internet frauds, fake websites, and manipulated Wikipedia entries. In order to deal with this serious threat to Information Quality, we need to improve our understanding of the nature and scope of disinformation. One way that work in philosophy can help with this task is by identifying and classifying the various types of disinformation, such as lies, spin, and even bullshit. If we are aware of the various ways that people might try to mislead us, we will be in a better position to avoid being duped by intentionally misleading information. Toward this end, this essay surveys and extends classification schemes that have been proposed by several noted philosophers—including Saint Augustine, Roderick Chisholm, and Paul Grice.

8.1 Introduction

Prototypical instances of *disinformation* include deceptive advertising (in business and in politics), government propaganda, doctored photographs, forged documents, fake maps, internet frauds, fake websites, and manipulated Wikipedia entries. Disinformation can be extremely dangerous. It can directly cause serious emotional, financial, and even physical harm if people are misled about important topics, such as investment opportunities, medical treatments, or political

D. Fallis (✉)
School of Information Resources, University of Arizona,
1515 East First Street, Tucson, AZ 85719, USA
e-mail: fallis@email.arizona.edu

L. Floridi and P. Illari (eds.), *The Philosophy of Information Quality*,
Synthese Library 358, DOI 10.1007/978-3-319-07121-3_8,

candidates. In addition, and possibly more importantly, it can cause harm indirectly by eroding trust and thereby inhibiting our ability to effectively share information with each other.

This threat to *information quality* has become much more prevalent in recent years. New information technologies are making it easier for people to create and disseminate inaccurate and misleading information (see Hancock 2007).[1] Investors have been duped by websites that "impersonate" the websites of reputable sources of information, such as *Bloomberg News* (see Fowler et al. 2001). Software now allows people to convincingly manipulate visual images (see Farid 2009). In addition, anyone with internet access can easily (and anonymously) insert inaccurate and misleading information into Wikipedia, "the free online encyclopedia that anyone can edit." Several individuals and organizations have been caught manipulating entries in this encyclopedia in self-serving ways (see Fallis 2008, 1665).

Inaccurate information (or *misinformation*) can mislead people whether it results from an honest mistake, negligence, unconscious bias, or (as in the case of disinformation) intentional deception. But disinformation is particularly dangerous because it is no accident that people are misled. Disinformation comes from someone who is actively engaged in an attempt to mislead. Thus, developing strategies for dealing with this threat to information quality is particularly pressing.

In order to develop such strategies, we first need to improve our understanding of the nature and scope of disinformation. For instance, we need to be able to distinguish disinformation from other forms of misinformation. After all, the clues that someone is lying to us are likely to be different from the clues that she just does not know what she is talking about. In addition, we need to be able to distinguish the various types of disinformation, such as lies, spin, and even bullshit (cf. Isaac and Bridewell 2014).

Work in philosophy can help with both of these tasks. This essay surveys the attempts by several noted philosophers—including Saint Augustine, Roderick Chisholm, and Paul Grice—to classify the different types of intentionally misleading information. In addition, this essay extends many of these classification schemes. Its goal is similar to the goal of books like Darrell Huff's (1954) *How to Lie with Statistics* and Mark Monmonier's (1991) *How to Lie with Maps*. Despite their titles, these books are not instruction manuals for liars. They are intended to help all of us to avoid being misled by showing us the various ways that people might try to mislead us. Before discussing the classification schemes, I begin in Sect. 8.2 by offering an analysis of what disinformation is.

---

[1] This problem arises with any new information technologies. For instance, when new printing technology first made books widely available, there was often a question of whether or not you held in your hands the authoritative version of a given text (see Johns 1998, 30–31). Techniques eventually developed for assuring ourselves of the authority and reliability of books. But such techniques are not always immediately available with new information technologies.

8.2 What Is Disinformation?

Disinformation is a type of *information*. That is, it is something that has *representational content* (see Floridi 2011, 80). For instance, the Wikipedia entry on the journalist John Seigenthaler was edited in 2005 to say that he was "directly involved in the Kennedy assassinations" (see Fallis 2008, 1665). This entry represented the world as being a certain way (as it happens, a way that it actually was not).[2] By contrast, in one of the *Sherlock Holmes* stories, Jonas Oldacre creates a fake thumbprint in order to frame John Hector McFarlane for murder.[3] Although it might cause someone to *infer* that the world is a certain way, the fake thumbprint does not *represent* the world as being that way. It is only an object (albeit an "informative" one).

More specifically, disinformation is information that is *intentionally misleading*. That is, it is information that—just as the source of the information intended—is likely to cause people to hold false beliefs. For instance, the person who edited the aforementioned Wikipedia entry intended readers to believe that Seigenthaler was directly involved in the Kennedy assassinations. By contrast, although the *Chicago Tribune*'s famous headline in 1948 that "Dewey defeats Truman" probably misled many readers, it was not intended to be misleading. It was an "honest mistake" (Table 8.1).

The most notable type of disinformation is the *lie*. Indeed, philosophers have had much more to say about lying than about disinformation in general. According to the traditional philosophical analysis, a lie is a false statement the speaker believes to be false and that is intended to mislead (see Mahon 2008, §1).[4] For instance, in Shakespeare's *Othello*, Iago famously says to Othello, "such a handkerchief—I am

Table 8.1 Major types of disinformation

Lies
Visual disinformation
Examples – doctored photographs, fake maps
True disinformation
Examples – false implicature
Side effect disinformation
Examples – inaccuracies inserted to test Wikipedia

[2] Some philosophers (e.g., Floridi 2011, 80) claim that representational content only counts as information if it is *true*. This paper will use the term *information* more broadly, to refer to representational content that is false as well as representational content that is true (cf. Fallis 2011, 202–203).

[3] In "The Adventure of the Norwood Builder," Oldacre uses a little of his own blood, and McFarlane's thumbprint from a wax seal on an envelope, to place a bloody thumbprint on the wall.

[4] Many philosophers (e.g., Chisholm and Feehan 1977, 152; Fallis 2009, 34) claim that lies do not have to be false. Also, some philosophers (e.g., Fallis 2009, 34; Carson 2010, 30) claim that lies do not even have to be intended to mislead. However, if there are such lies, they would clearly not count as disinformation. So, they can safely be set aside for purposes of this paper.

sure it was your wife's—did I today see Cassio wipe his beard with." This is a false statement and Iago is aware that it is false. (Although Iago ultimately plants the handkerchief in Cassio's lodgings, he has not yet done so at the time of his conversation with Othello. So, he could not possibly have seen it in Cassio's possession.) In addition, Iago makes this statement in order to convince Othello that Desdemona has been unfaithful.

Some philosophers simply equate disinformation with lying. For instance, James Fetzer (2004, 231) claims that disinformation "should be viewed more or less on a par with acts of lying. Indeed, the parallel with lying appears to be fairly precise."[5] However, it is important to note that such an analysis of disinformation is too narrow. Lies are not the only type of disinformation.

First, unlike lies, disinformation does not have to be a statement. Fetzer's analysis incorrectly rules out what we might call *visual disinformation*. Doctored photographs and fake maps are clearly examples of disinformation. For instance, during the 2004 Presidential campaign, a photograph was circulated which appeared to show John Kerry and Jane Fonda sharing the stage at an anti-Vietnam war rally. But it was really a composite of two separate photographs taken at two separate events (see Farid 2009, 98). Also, during the Cold War, the Soviets deliberately falsified maps in an attempt to fool their enemies about where important sites were located (see Monmonier 1991, 115–118).

Second, unlike lies, disinformation does not have to be false. Fetzer's analysis incorrectly rules out what we might call *true disinformation*. As several philosophers (e.g., Vincent and Castelfranchi 1981; Adler 1997; Fallis 2011, 209; Manson 2012) have pointed out, even accurate information can be intentionally misleading. For instance, if a villain who means to harm my friend asks me where he is, I might truthfully reply, "He's been hanging around the *Nevada* a lot" intending the villain to draw the false conclusion that my friend could be at this diner now (see Adler 1997, 437–438). Similarly, even if Iago had waited until he had seen Cassio wiping his beard with the handkerchief, his statement to Othello would still have been misleading about Desdemona having been unfaithful.[6] These are both examples of "false implicature" (Adler 1997, 452).

Third, unlike lies, disinformation does not have to be intended to mislead. Fetzer's analysis incorrectly rules out what we might call *side effect disinformation*. For instance, researchers have inserted inaccurate information into Wikipedia to see how long it takes to get corrected (see Fallis 2008, 1666). These researchers do not

[5] Floridi's (2011, 260) analysis of disinformation is very similar to Fetzer's. He claims that "misinformation is 'well-formed and meaningful data (i.e. semantic content) that is false.' 'Disinformation' is simply misinformation purposefully conveyed to mislead the receiver into believing that it is information."

[6] Some people (e.g., O'Neill 2003, §4; Ekman 2009, 28) have a broader notion of lying that counts such statements as lies. But the traditional analysis of lying requires that the statement is believed by the speaker to be false (see Mahon 2008, §1.2).

intend to mislead anyone. Doing so would not be a means to their ends. But they do intend to create information that is inaccurate and misleading.[7]

So, it is better to analyze disinformation as *intentionally misleading information*.[8] In addition to capturing the prototypical instances of disinformation, this analysis captures visual disinformation, true disinformation, and side effect disinformation. But even though disinformation is a broader category than lying, much of what philosophers have said about lying can be applied to disinformation in general.

8.3 Augustine on the Purpose of Lying

Saint Augustine (1952, 86–88) was the first philosopher to explicitly classify different types of lying and deception. He identified eight kinds of lies (Table 8.2).

Augustine's scheme is based largely on the *purpose* for which the lie is told. And it is certainly useful to know about the different reasons *why* people might want to mislead us. This can make us more aware that a person might have a motivation to mislead us. For instance, it would have benefited Othello to reflect on whether or not Iago might have a motivation to mislead him about Desdemona and Cassio.

Moreover, this classification scheme can clearly be extended to disinformation in general. For instance, we might consider disinformation used in the teaching of religion and disinformation created from a desire to please others.

Even so, this particular scheme is probably not the most useful for dealing with the threat to information quality that disinformation poses. After all, almost all dis-

Table 8.2 Augustine's classification of lies

A lie which is "uttered in the teaching of religion."
A lie which "helps no one and harms someone."
A lie which is "beneficial to one person while it harms another."
A lie which is "told solely for the pleasure of lying."
A lie which is "told from a desire to please others."
A lie which "harms no one and benefits some person" materially (as when the lie keeps her money from being "taken away unjustly").
A lie which "harms no one and benefits some person" spiritually (as when the lie gives her the "opportunity for repentance").
A lie which is "harmful to no one and beneficial to the extent that it protects someone from physical defilement."

[7] Although Floridi's analysis does not rule out visual disinformation, it does rule out true disinformation and side effect disinformation.

[8] It may be no accident that a piece of information is misleading even if it is not intended to be misleading (see Skyrms 2010, 80; Fallis 2011, 211–212). For instance, false rumors can spread even when everybody passing them along believes what they are saying. But this paper will focus exclusively on *intentionally* misleading information.

information falls into just the one category of being beneficial to one person while it harms another. Fortunately, other philosophers have subsequently offered what promise to be more useful classification schemes.

8.4 Chisholm and Feehan on the Epistemic Goal of Deception

In more recent work, Chisholm and Thomas Feehan (1977) offer another classification of types of deception (Table 8.3). Much like Augustine, Chisholm and Feehan focus primarily on the purpose of the deception. But they focus on the immediate *epistemic* goals that a deceiver might have.

The epistemic goal of deception is usually to cause someone to acquire a new false belief. For instance, Iago made his statement about the handkerchief in order to get Othello to believe falsely that Desdemona had been unfaithful. Chisholm and Feehan call this "positive deception *simpliciter*." But there are at least three other epistemic goals that a deceiver might have.

First, instead of causing someone to acquire a *new* false belief, the goal might just be to cause someone to continue to hold an *existing* false belief. For instance, it might have been that Othello already believed that Desdemona had been unfaithful and that Iago made his statement about the handkerchief in order to keep Othello thinking that. (Perhaps, her protestations of innocence would have raised doubt in his mind otherwise.) Chisholm and Feehan call this "positive deception *secundum quid*."

Second, instead of causing someone to acquire, or to continue to hold, a false belief, the goal might be to cause someone give up a true belief. In other words, the goal might be to make someone ignorant on some topic. For instance, it might have been that Othello started out with the true belief that Desdemona had been faithful and that Iago made his statement about the handkerchief, not in order to convince him that she had been unfaithful, but simply to make him uncertain. Chisholm and Feehan call this "negative deception *simpliciter*."

Finally, the goal might just be to cause someone *not* to acquire a *new* true belief. In other words, the goal might be to *keep* someone ignorant on some topic. For instance, it might have been that Othello started out uncertain about Desdemona's faithfulness and that Iago made his statement about the handkerchief, not in order to convince him that she had been unfaithful, but simply to keep him from coming to believe that she *had been* faithful. (Perhaps, her protestations of innocence would

Table 8.3 Chisholm and Feehan's classification of deception

Positive deception (i.e., causing a false belief)
Positive deception *simpliciter* (i.e., creating a new false belief)
Positive deception *secundum quid* (i.e., maintaining an existing false belief)
Negative deception, *aka* keeping someone in the dark (i.e., causing ignorance)
Negative deception *simpliciter* (i.e., causing the loss of a true belief)
Negative deception *secundum quid* (i.e., preventing the acquisition of a true belief)

have been successful otherwise.) Chisholm and Feehan call this "negative deception *secundum quid*." As J. Bowyer Bell and Barton Whaley (1991, 48–49) and Paul Ekman (2009, 28–29) also point out, a deceiver can *show the false* or she can just do things to *hide the truth*.

Admittedly, many philosophers only count "positive deception" as deception (see Mahon 2008, §2.2). On their view, if you do not cause someone to acquire, or at least to continue to hold, a false belief, you are merely *keeping her in the dark* (see Carson 2010, 53–55). But some philosophers (e.g., Skyrms 2010, 81–82; Lackey 2013) do concur with Chisholm and Feehan that failing to make someone as epistemically well off as she could have been counts as deception.[9]

It is worth noting that the term *deception* is sometimes used in the broad sense that Chisholm and Feehan have in mind. For instance, we say that a magician is being deceptive even if he simply conceals from the audience how the trick was done. He does not have to create a false belief in the audience that he has actually sawn his assistant in half. Moreover, it is not immediately clear why such "negative deception" should not count as deception. It involves the same sort of manipulation of someone's epistemic state as does "positive deception." Why should it matter so much that the suboptimal epistemic state that she ends up in is ignorance rather than false belief?

While Chisholm and Feehan were interested in deception in general, disinformation in particular clearly comes in these same four varieties. But the possibility of "negative deception" does indicate that our analysis of disinformation as intentionally *misleading* information is too narrow. Strictly speaking, disinformation should be analyzed as information that intentionally causes someone to be epistemically worse off than she could have been. However, in the remainder of this paper, I will focus primarily on how information can be used to create false beliefs. Even so, analogous techniques can often be used to create ignorance as well.

Chisholm and Feehan also point out that we might try to accomplish any of these four epistemic goals through acts of omission as well as through acts of commission. It is certainly possible to keep someone ignorant on some topic by (*passively*) failing to give her information. But it is even possible to create false beliefs just by failing to give someone information. For instance, a tax advisor who maliciously fails to mention a legitimate and lucrative exemption "intentionally causes her client to believe falsely that there is no way for him to save more money on his taxes" (Carson 2010, 56). However, in this paper, I will focus on the *active* manipulation of information.

8.5 Lying About What?

Some useful classifications of disinformation can be derived from debates between philosophers about lying. While most philosophers agree that a lie must be intended to create a false belief, they disagree about what that false belief must be about. For

[9] Skyrms does not say this explicitly. But it is a consequence of his analysis of deception (see Fallis forthcoming).

Table 8.4 A classification of deceptive goals

Mislead about the content being accurate
Examples – lies
Mislead about the source believing the content
Mislead about the identity of the source
Mislead about an implication of the content being accurate
Examples – false implicature

instance, many philosophers claim that a liar must intend to mislead someone about the accuracy of what he actually says (see Mahon 2008, §1). Although Iago's ultimate goal is to convince Othello that Desdemona has been unfaithful (which is not what he actually said), he first has to convince Othello that he saw Cassio wiping his beard with Desdemona's handkerchief (which is what he said).

In contrast, some philosophers (e.g., Mahon 2008, §1.6) claim that a liar only needs to intend to mislead about *his believing* what he says. For instance, suppose that a crime boss, Tony, has discovered that one of his henchmen, Sal, has become an FBI informant. But Tony does not want Sal to find out that his treachery has been uncovered. So, to keep his disloyal henchman at ease, Tony says with pride to Sal one day, "I have a really good organization here. There are no rats in my organization." Although Tony certainly seems to be lying here, he does not intend Sal to believe what he says. It is not as if Tony's statement is going to lead Sal to think to himself, "Well, I guess that I am not a rat after all." Tony only intends Sal to believe that he (Tony) believes what he says.

When you say something to someone, you usually intend for her to believe it and for her to believe that you believe it. In fact, the latter is typically the means to the former. For instance, Iago's statement only suggests that he actually saw Cassio wiping his beard with Desdemona's handkerchief because it suggests that *he believes* that he saw this. But the crime boss case shows that these two goals can come apart.

A few philosophers (e.g., Newey 1997, 100–102) go even further and claim that a liar just has to intend to mislead someone about something. So, in addition to (a) the accuracy of what he says and (b) his believing what he says, there may be other things that a liar might intend to mislead about. Instead of trying to resolve this debate about exactly what is required for lying, we can treat it as a classification of different types of misleading utterances. In fact, as I discuss below, it suggests an important classification of intentionally misleading information in general (Table 8.4).

8.5.1 Misleading About the Accuracy of the Content

In order to apply this classification scheme to disinformation in general, we cannot just talk about whether or not information is intended to be misleading about the accuracy of *what is said*. With disinformation in general, there may be nothing that is literally said. For instance, a map does not express a particular proposition. Nevertheless, as noted above, information always has some sort of representational content. So, we can talk about whether or not information is intended to be misleading about the *accuracy of the content*. Iago's statement to Othello falls under this category because Iago intended Othello to believe that its content was accurate. But in addition, a map might be intended to mislead people about the geography of the area depicted on the map. This was the intention behind the aforementioned Soviet maps (see Sect. 8.2 above).

As Soo Young Rieh (2002, 152) points out, information quality is a "multidimensional concept." The standard dimensions include accuracy, completeness, currency, cognitive authority, accessibility, etc. Of course, accuracy is arguably the *sine qua non*.[10] Thus, information that is intentionally misleading about the accuracy of the content lacks a critically important dimension of information quality.[11]

8.5.2 Misleading About the Source Believing the Content

In the standard case, disinformation is intended to be misleading about the accuracy of the content. But as suggested by the philosophical discussion of lying, there are several other possibilities. Most notably, a piece of information might be intended to be misleading about the source of the information *believing* the content. As in the case of Iago, misleading about the source believing the content is often the means to misleading about the accuracy of the content. But a piece of information can be misleading about the former without being misleading about the latter. For instance, much like Tony's statement that there is no rat in his organization, a map might just be intended to mislead people about what the person who drew the map *believed* about the geography of the area depicted on the map.

A real life case of this sort is the *Vinland Map* housed at the Beinecke Library of Yale University. It is a modern forgery that was made to look like a map of the world drawn in the Middle Ages (see Monmonier 1995, 72–104).[12] The map depicts a large island ("Vinland") to the southwest of Greenland with a large inland lake (reminiscent of Hudson Bay) connected to the ocean by a narrow inlet and a larger inlet to the south (reminiscent of the Gulf of St. Lawrence). But the map was not

[10] The fact that accuracy is critical does not mean that the other dimensions are not also critical. For instance, in addition to misleading people by disseminating inaccurate information, governments often keep people in the dark by making accurate information inaccessible.

[11] Even if a piece of information is *intended* to be misleading about the accuracy of the content, it might—unbeknownst to the source—actually be accurate. (As the protagonist of Oscar Wilde's *The Importance of Being Earnest* laments when he learns that his name really is Ernest, "it is a terrible thing for a man to find out suddenly that all his life he has been speaking nothing but the truth.") But since it is not misleading, such accurate information would clearly not count as disinformation.

[12] A little bit of controversy actually remains about the genuineness of the map (see Monmonier 1995, 102).

created in order to fool anybody about the geography of North America. Instead, it was apparently created in order to fool people in the twentieth-century about what people in the fifteenth-century knew about the geography of North America. If the map were genuine, it would mean that there must have been extensive pre-Columbian exploration of North America by the Norse (e.g., by Leif Ericson).

8.5.3 Misleading About the Identity of the Source

Another important category of disinformation is when a piece of information is intended to be misleading about *who* is the source of the information. For instance, confidence tricksters often try to mislead people about their true identity. Someone putting on the *big con* known as "The Wire" (which was portrayed in the movie *The Sting*) needs to convince his "mark" that he can make sure bets on horse races because he can get the results before they are sent to bookmakers (see Maurer 1999, 34). So, he might claim that he is the "manager of the central office of the Western Union here in New York."[13]

While this sort of disinformation is nothing new, new information technologies have made it much more prevalent. In fact, Jeffery Hancock (2007, 290) distinguishes two main types of digital deception. In addition to "message-based digital deception" (which is misleading about the accuracy of the content), there is "identity-based digital deception." One example is when someone creates a website that "impersonates" or "mimics" another website. But an even more common practice is the creation of fake online identities. Such "sock puppets" are usually manipulated one at time in order to post fake reviews of books, restaurants, hotels, etc. But as Peter Ludlow (2013) points out, private intelligence agencies have developed systems for the government "that allowed one user to control multiple online identities ("sock puppets") for commenting in social media spaces, thus giving the appearance of grass roots support."

It is worth noting that there are a few different ways in which such "mimicking," as Barton and Whaley (1991, 56) call it, can be carried out. First, you might impersonate a specific individual. For instance, people have created fake Twitter accounts for Brittany Spears and for the Pope. Second, you might just impersonate a type of person.[14] For instance, websites offering medical advice and selling medical products have been posted by people who falsely claimed to have medical qualifications (see Detwiler 2002, 28–31).

Disinformation in this category tends to lack another important dimension of information quality. For instance, a medical website posted by someone falsely claiming to have medical expertise does not have what Rieh (2002, 146) calls "cognitive

---

[13] The Vinland Map is another example where a piece of information was intended to be misleading about the identity of the source.

[14] If the type of person that you impersonate is one that never existed before, Bell and Whaley (1991, 58) call the technique "inventing" rather than "mimicking."

authority." It only appears to come from someone who is "credible and worthy of belief." As a result, misleading people about the identity of the source is often a means to misleading people about the accuracy of the content.

Disinformation that is intended to mislead about the identity of the source typically is intended to mislead about the accuracy of the content as well (see Hancock 2007, 290). In fact, a statement about the identity of the source might even be part of the content as when the conman says that he works for Western Union. But these two goals can come apart. For instance, while the Vinland Map was intended to mislead people about who created it, it was not intended to mislead about the accuracy of the content. We can even imagine cases where the disinformation is *only* intended to mislead about the identity of the source. For instance, someone might adopt a French accent and talk a lot about Paris in order to give the false impression that he is from France. He might not intend to mislead anyone about what he says or about his believing it. Everything that he literally says might be true.

Finally, it should be noted that, instead of intending to mislead about *who* sent a piece of information, you might just intend to mislead about *where* you were when you sent the information or about *when* you sent it. Twitter was used extensively by protesters of the 2009 presidential election in Iran to communicate with each other and with the outside world. So, a campaign was quickly started to have Twitter users from around the world change their location to Tehran. This was intended to support the protesters by making it more difficult for the Iranian government to identify which Twitter users were actually in Iran (see Morozov 2011, 15). In a similar vein, you might alter the time stamp on an email message to suggest that you actually sent it a day earlier.[15]

8.5.4 Misleading About an Implication of the Accuracy of the Content

Another important category of disinformation is when a piece of information is intended to be misleading about something that is not part of the content, but that is implied by the content being accurate. For instance, Iago's statement falls under this category. Iago did not actually say that Desdemona had been unfaithful. But if it were true that Iago had seen Cassio wiping his beard with the handkerchief, that would suggest that Desdemona had been unfaithful. As noted above, while his statement was misleading about the former as well, it is the latter that Iago really wanted to mislead Othello about.

As in the case of Iago, misleading about the accuracy of the content is often the means to misleading about some implication of the accuracy of the content. But it is important to note that a piece of information can be misleading about the latter without being misleading about the former. For instance, the examples of false implicature

[15] This is different from the feature (announced on April Fools' Day) that supposedly allowed Gmail users to actually send messages into the past (see Google 2008).

(see Sect. 8.2 above) fall into this category. If I truthfully say that my friend has been hanging around the *Nevada* a lot, I do not intend to mislead the villain about how much my friend hangs out at this diner. I only intend to mislead him about the possibility of my friend being there now. Similarly, if Iago had waited until he had seen Cassio wiping his beard with the handkerchief before making his statement, he would not have intended to mislead Othello about what he said. He would only have intended to mislead him about Desdemona having been unfaithful.

It should also be noted that it does not have to be *open* that the false thing that you want someone to believe is an implication of the accuracy of the content. When Iago made his statement, he and Othello were discussing whether or not Desdemona had been unfaithful. Thus, it was clear to both of them that Iago having seen Cassio wiping his beard with the handkerchief suggested that Desdemona had been unfaithful. However, we can easily imagine an alternative scenario in which Desdemona's fidelity was not the topic under discussion. Iago might just have mentioned (out of the blue and without giving any indication that he was even aware that the handkerchief belonged to Desdemona) that he saw Cassio wiping his beard with "a handkerchief spotted with strawberries." Othello would still have been likely to draw the conclusion that Desdemona had been unfaithful (as Iago intended). But Othello would not have been aware that Iago was aware of this implication of his statement. (The issue of *openness* is discussed in more detail in Sect. 8.6 below.)

But this category does not (at least in general) include disinformation that is intended to be misleading about the source believing the content.[16] The content being accurate does not imply that the source believes the content. Instead, it is the fact that the source is disseminating the content that suggests that the source believes it. For instance, it is the fact that Tony *says* that there is no rat in his organization, rather than the nonexistence of the rat, which implies that Tony believes that there is no rat.

8.6 Grice on Showing and Telling

Yet another classification of disinformation can be derived from Grice's work in the philosophy of language (Table 8.5). In his influential article "Meaning," Grice (1989, 217–218) considered three ways of getting someone to believe something. First, there is "telling" someone that *p*. For instance, Herod might get Salome to believe that John the Baptist is dead by telling her so. The distinctive feature of tellings is that you intend someone to believe something by virtue of her recognizing your intention that she believe it. Basically, she believes what you tell her because you have offered your *assurance* that it is true (see Moran 2005, 6; Faulkner 2007, 543).

[16] There are special cases where the content being accurate might imply that the source believes the content. For instance, the claim that the source believes the content might be part of the content. Alternatively, it might be that this particular source is likely to have a lot of true beliefs on this particular topic. So, if the content is accurate it means that the source is likely to know it.

Table 8.5 A Gricean classification of deception based on showing and telling

- Tell **X** that *p*
  - *Examples* – lies, false implicature
- Show **X** that *p*
  - *Examples* – doctored photographs
- Mislead **X** without showing or telling **X** that *p*
  - That involve telling
    - Tell **X** that *q*
    - Pretend to tell **X** that *q*
      - *Examples* – double bluffs
    - Pretend to tell **Y** that *p*
      - *Examples* – tricking eavesdroppers
  - That involve showing
    - Show **X** that *q*
    - Pretend to show **X** that *q*
    - Pretend to show **Y** that *p*
  - That do not involve showing or telling at all
    - *Examples* – fake diaries

Second, there is "deliberately and openly letting someone know" that *p*. In other words, you might *show* someone that *p*. For instance, Herod might get Salome to believe that John the Baptist is dead by bringing her the preacher's decapitated head. With showings as well as tellings, it is *open* that you intend someone to believe a certain thing. However, the distinctive feature of showings is that this person's recognition of your intention is not what you intend to cause her to hold this belief. As Richard Moran (2005, 14) notes with respect to Herod bringing Salome the head, "while his intention regarding her belief is indeed manifest … it isn't doing any epistemological work of its own … the relevant belief could be expected to be produced whether or not the intention behind the action were recognized."

Third, there is "getting someone to think" that *p* without showing or telling her that *p*. For instance, Herod might get Salome to believe that John the Baptist is dead by leaving the preacher's head lying around somewhere that he expects Salome will run across it. As with showings and tellings, you make it appear to someone that *p* is true. But the distinctive feature here is that you do so without revealing your intention that she believe that *p*.

As Collin O'Neil (2012, 325–331) points out, this gives us three ways to get someone to believe something *false*. First, you can tell someone something false. For instance, Iago gets Othello to believe that he has seen Cassio wiping his beard with Desdemona's handkerchief by saying that he has. Second, you can show someone something false.[17] For instance, in George R. R. Martin's (2005) *A Feast for Crows*, in an attempt to earn the reward that Queen Cersei has offered, several peo-

[17] It might be suggested that you can only *show* someone something if that thing is *true*. But the sense of showing (or "deliberately and openly letting someone know") that Grice has in mind is not factive.

ple bring her what appears to be Tyrion Lannister's head in order to convince her that the Imp is dead.[18] Third, you can get someone to think something false without showing or telling her that thing. For instance, the fake thumbprint from the *Sherlock Holmes* story got the police to believe that McFarlane had committed murder. In fact, this sort of covert manipulation is most commonly associated with attempts to mislead.[19]

8.6.1 Telling Them with Information

We want a classification scheme for types of disinformation rather than for all types of deception. As I argue below, all three of Grice's categories can involve the dissemination of information. This is clearly the case when you tell someone the false thing that you want her to believe. A liar, such as Iago, literally says that *p* is the case.

But saying that *p* is not the only way to tell someone that *p* through the dissemination of information.[20] Examples of false implicature can also be acts of telling someone that *p*. For instance, if the villain asks me where my friend is and I truthfully reply, "He's been hanging around the *Nevada* a lot," I am *telling* him that my friend could be there now even though I do not literally say so.[21] I still intend to get him to believe that my friend could be at this diner by virtue of his recognizing my intention that he believe it.

In fact, it is even possible to tell someone that *p* by *saying* the exact opposite. For instance, if a teenager says *sarcastically* to a friend, "Yeah, right. I want *Jimmy* to be my boyfriend" (even though she really does like Jimmy), she intends her friend to believe that she does not like Jimmy by virtue of her friend recognizing her intention that her friend believe it. As Jocelyne Vincent and Cristiano Castelfranchi (1981, 766) would put it, the teenager is "pretending to joke."

Finally, it should be noted that telling (or showing) someone that *p* does not preclude your intending to mislead her about your identity. For instance, the conman

---

[18] As we learn in Martin's (2011) *A Dance with Dragons*, the Imp is actually alive and well across the Narrow Sea.

[19] But this technique can also be used to get someone to believe something true. Regardless of whether *p* is true or false, you might create fake evidence for *p* or you might just arrange it so that someone will run across existing evidence for *p*.

[20] In *House of Cards*, Francis Urquhart regularly tells people something without actually saying it by using his catchphrase, "You might very well think that; I couldn't possibly comment."

[21] When I say that my friend has been hanging around the *Nevada* a lot, I am not telling the villain that my friend *is there* now. In fact, my reply suggests that I do not know for sure where my friend is. However, there are examples of false implicature where the speaker does tell his audience precisely what they ask for. For instance, suppose that François is a Frenchman who does not know how to cook. When someone asks him whether he knows how to cook, François might haughtily reply, "I am French!" In that case, he has *told* his audience (falsely) that he knows how to cook (see Recanati 2004, 5).

clearly tells his mark that he works for Western Union. That is, he intends his mark to believe that he works for Western Union by virtue of her recognizing his intention that she believe it. This is so even though the mark is seriously mistaken about exactly who he is.

8.6.2 Showing Them with Information

Several people tried to show Queen Cersei that Tyrion was dead using an object (viz., a head). But it is also possible to show someone something false with information. Most notably, you can show someone something false with a doctored photograph. For instance, you might show someone that John Kerry and Jane Fonda attended the same anti-Vietnam rally by producing the aforementioned photograph (see Sect. 8.2 above). Your audience's recognition that you intend her to believe that Kerry and Fonda were together at the rally plays no epistemological role. The photograph "speaks for itself."

But it is important to note that not all displays of visual information are acts of showing. For instance, if you draw someone a treasure map, you have not "shown" her where the treasure is. You have only "told" her where the treasure is (cf. Moran 2005, 11). She only believes that the treasure is in that location because of your assurance.[22] Thus, tellings sometimes involve visual information instead of propositional information.

8.6.3 Telling Them Something Else

In addition to showing or telling someone the false thing that you want her to believe, there are several other ways to use information to bring about this result. For instance, if you want someone to falsely believe that *p*, you might tell her that *q* with the intention that she infer that *p* is also the case. When pursuers—who did not recognize him—asked, "Where is the traitor Athanasius?," Saint Athanasius replied, "Not far away," which misled them into thinking that he was not Athanasius (see Faulkner 2007, 535). Vincent and Castelfranchi (1981, 760) give an example of another "indirect lie" of this sort. A child truthfully says to a new acquaintance, "My Dad works at the B. B. C." intending his audience to draw the false conclusion that his dad is "something glamorous or important like a reporter or a cameraman" when he is really "a cleaner at the studios."

[22]The Three Stooges fall victim to a fake treasure map in the 1937 film *Cash and Carry*. Such disinformation can be quite convincing because of "the alluring believability of cartographic images. Maps have an authority that even scholars are reluctant to question" (Monmonier 1995, 103).

Admittedly, this strategy (as well as the following two strategies) does involve a telling. For instance, Athanasius did offer his assurance that Athanasius was not far away. However, he did not assure his pursuers about the false thing that he intended them to believe. He did not intend them to believe that he was not Athanasius by virtue of their recognizing his intention that they believe it. Thus, unlike with the examples of false implicature, he did not tell them the false thing that he wanted them to believe. This is so even though both Athanasius and his pursuers would have been aware that it was reasonable to infer from what he said that he was not Athanasius.

8.6.4 Pretending to Tell Them Something Else

You can also get someone to believe something false by *pretending* to tell her something else (sometimes even the exact opposite of what you want her to believe). One strategy that falls under this category is the *double bluff* (see Augustine 1952, 57; Vincent and Castelfranchi 1981, 764–766; Moran 2005, 17). For instance, Sigmund Freud (1960, 137–138) tells the following joke:

> Two Jews met in a railway carriage at a station in Galicia. "Where are you going?" asked one. "To Cracow", was the answer. "What a liar you are!" broke out the other. "If you say you're going to Cracow, you want me to believe you're going to Lemberg. But I know that in fact you're going to Cracow. So, why are you lying to me?"

Although one of the men (call him **A**) is going to Cracow, he wants the other man (call him **B**) to falsely believe that he is going to Lemberg. But **A** knows that **B** does not trust him. As a result, if **A** says that he is going to one destination (Cracow), he expects **B** to conclude that **A** is lying and, thus, that **A** must be going to the other possible destination (Lemberg).[23] As Vincent and Castelfranchi (1981, 764) would put it, **A** is "pretending to lie" to **B**.

Such double bluffs fall under the category of *pretending* to tell someone something else, rather than under the previous category of *actually* telling someone something else. Admittedly, there is a weak sense of telling in which **A** does tell **B** that he is going to Cracow. After all, this is what he literally says to **B**. But this is not the sense of telling that Grice had in mind. **A** does not intend **B** to believe that he is going to Cracow at all, much less by virtue of recognizing **A**'s intention that **B** believe it (see Moran 2005, 17).

This category of disinformation is not just a hypothetical possibility. For instance, Malcolm Gladwell (2010) describes a real life example of a double bluff:

> At one point, the British discovered that a French officer in Algiers was spying for the Germans. They "turned" him, keeping him in place but feeding him a steady diet of false and misleading information. Then, before D Day—when the Allies were desperate to convince Germany that they would be invading the Calais sector in July—they used the French

[23] Let us assume that it is common knowledge that Cracow and Lemberg are the only two possible destinations. A more common version of this joke uses the Russian cities of Minsk and Pinsk.

officer to tell the Germans that the real invasion would be in Normandy on June 5th, 6th, or 7th. The British theory was that using someone the Germans strongly suspected was a double agent to tell the truth was preferable to using someone the Germans didn't realize was a double agent to tell a lie.

It should be noted that you can also get someone to believe something false by pretending to tell her the very thing that you want her to believe (see Newey 1997, 98; Faulkner 2007, 536–537). In fact, there are at least two ways that this might happen. First, continue to suppose that, while **A** is going to Cracow, he wants **B** to believe that he is going to Lemberg. But in this case, **A** knows that **B** believes that he is *incompetent* (in particular, that he believes that he is going one place when he is really going somewhere else), as well as insincere. Thus, if **A** says that he is going to Lemberg, he expects **B** to conclude that **A** believes that he is going to Cracow (because **A** is insincere). Thus, **A** expects **B** to conclude that he is actually going to Lemberg (because **A** is incompetent).

Second, continue to suppose that, while **A** is going to Cracow, he wants **B** to believe that he is going to Lemberg. But in this case, **B** knows that **A** knows that **B** does not trust him, and **A** knows that **B** knows. Thus, if **A** says that he is going to Lemberg, he expects **B** to conclude, not that **A** is lying, but that **A** is *pretending* to lie (i.e., that he is trying a double bluff). Thus, **A** expects **B** to conclude he is actually going to Lemberg. We might call this a *triple bluff*.[24]

Although **A** intends **B** to believe what he says in both of these variations on the double bluff, he does not intend **B** to believe what he says by virtue of recognizing **A**'s intention that **B** believe it. If **B** ends up believing this false thing, it is not because **A** has assured him that it is true. So again, it is not an act of telling in the sense that Grice had in mind.

Admittedly, these variations on the double bluff are somewhat complicated. However, as Glen Newey (1997, 98) points out, while such cases "may be thought farfetched … it is relatively straightforward compared with some forms of confidence trick." Newey and Christian Plunze (2001, 185–187) both give more detailed and somewhat more realistic versions of the first variation.

8.6.5 *Pretending to Tell Someone Else*

You can also get someone to believe something false by pretending to tell it to *someone else*. You just need to arrange it so that the person that you want to mislead overhears your conversation with this other person. For instance, in Shakespeare's *Much Ado About Nothing*, this is how Benedick's friends trick him into believing that Beatrice is in love with him. When Leonato says to Don Pedro and Claudio that it is "most wonderful that she should so dote on Signior Benedick, whom she hath

[24]The obvious reading of Freud's joke is that it was an unsuccessful double bluff. But it might have been a successful triple bluff. As is suggested by the "battle of wits" between Vizzini and the Dread Pirate Roberts in *The Princess Bride*, even higher-order bluffs are possible.

in all outward behaviors seemed ever to abhor," he intends to mislead Benedick who is hiding in the bushes eavesdropping on the conversation. But he does not assure *Benedick* that Beatrice is in love with him.[25]

A famous real life example of this technique was part of Operation Fortitude South, the successful operation to mislead the Germans about the intended location of the D-Day invasion (see Rankin 2008, 399). Among other deceits, the Allies sent out fake radio transmissions, which they expected the Germans to intercept. These transmissions suggested that a large force in East Anglia was preparing to attack Calais rather than Normandy (the actual site of the invasion).

It should be noted that, as long as they are not in on your scheme, it is possible to *actually* tell these other people the false thing. But if you are trying to mislead an eavesdropper, you are typically going to be speaking to a confederate that you do not intend to mislead. For instance, since Don Pedro and Claudio are in on the scheme, Leonato is only pretending to tell them that Beatrice is in love with Benedick.

Finally, it should be noted that letting someone overhear misleading information may actually be more effective than providing him with the information directly. He will not be as worried that the information has been intentionally crafted for his benefit.[26]

8.6.6 Further Variations

As we have now seen, there are three different ways to get someone to believe a false thing that involve telling, but that do not involve telling her that very thing. First, you can tell her something else (from which she infers the false thing that you want her to believe). Second, you can pretend to tell her something else (as in double bluffing cases). Third, you can pretend to tell someone else that false thing (as in eavesdropping cases). But in addition, these three strategies can be combined. For instance, you can get someone to believe that p by telling (or pretending to tell) someone else something else. You just need to make sure that your intended victim will overhear what you say and is likely to infer that p is the case from what you say.

In addition, for each of the strategies involving telling, there is an analogous strategy involving showing. First, if you want someone to falsely believe that p, you might show her something that you expect will lead her to infer that p is the case, even though it is not open that this thing implies that p is the case. (In that case, you

[25] It is not common knowledge between Leonato and Benedick that Benedick is eavesdropping. But it could be completely open that the person that you want to mislead is listening in. For instance, you might make (or just pretend to make) a phone call in her presence. Since you are not addressing her, you are still not assuring *her* that what you say is true.

[26] As a result, this technique of allowing yourself to be overheard can also be effective at getting someone to believe something true. For instance, Adele Faber and Elaine Mazlish (1980, 205) recommend that, in order to promote self-esteem, you should let your children overhear you saying something positive about them.

would be "openly letting them know" something. But you would not be "openly letting them know" the false thing that you want them to believe.) Second, along the lines of a double bluff, you might (pretend to) show your intended victim something that you expect will lead her to infer that *p* is the case based on her distrust of you. Third, you might (pretend to) show someone else something with the expectation that your intended victim will be observing and will be led to infer that *p* is the case.

Finally, there are also ways to use information to mislead someone that do not involve showing or telling at all. For instance, you might write a bunch of false claims about your sexual exploits in a private diary with the intention that someone will steal the diary, read it, and be misled (cf. Allen 1988, 21). Thus, not all informational deception involves the observation of an act (real or apparent) of showing or telling.

8.7 Grice on Norm Violations

Grice's (1989, 26–30) work on *norms of communication* in his William James lectures at Harvard suggests another way to classify disinformation (Table 8.6). A norm of communication is a type of *social norm*. Social norms (e.g., "Do not speak with your mouth full") are rules of behavior (**a**) that people usually obey and (**b**) that people think that people *ought* to obey (see Pettit 1990). Norms of communication are those social norms that are especially adapted to promote the effective exchange of information. For instance, as Grice pointed out, in normal conversations, you should "not say what you believe to be false," you should "not say that for which you lack adequate evidence," you should "make your contribution as informative as is required," you should "avoid ambiguity," you should "avoid obscurity of expression," etc.

It is pretty clear how obedience to these norms facilitates communication. But Grice was also interested in what happens when we *disobey* them. Most notably, he studied how a speaker might "blatantly fail to fulfill" these norms in order to communicate things beyond what he literally says. For instance, when the heroes of *Star Wars* use a garbage chute to escape from the detention block of the Death Star, they land in a huge garbage compactor and Han Solo says *sarcastically*, "The garbage chute was a really wonderful idea. What an incredible smell you've discovered!" By

Table 8.6 A Gricean classification of deception based on norm violations

Say what you believe to be false
Examples – lies
Say that for which you lack adequate evidence
Examples – bullshit
Be less informative than is required
Examples – half-truths, spin
Be ambiguous

openly disobeying the norm against saying what he believes to be false, Solo is able to communicate that he really thinks that the garbage chute was a *bad* idea.

But as Grice (1989, 30) was aware, a speaker may also "quietly and unostentatiously *violate* a maxim; if so, in some cases he will be liable to mislead."[27] Moreover, violations of these different norms represent different ways to mislead people. As several philosophers (e.g., Wilson and Sperber 2002, 586; Fallis 2009, 34; Dynel 2011, 143) have pointed out, lying clearly involves *violating* the norm against saying what you believe to be false. But there are several other possibilities. For instance, you can certainly mislead someone by violating the norm against saying that for which you lack adequate evidence.[28]

In fact, this particular phenomenon may be an especially important type of disinformation. Several prominent philosophers (e.g., Black 1983; Frankfurt 2005) have noted the serious problem of *humbug* and/or *bullshit* in modern society. They describe this phenomenon as a sort of misrepresentation that falls "short of lying" and involves a "lack of connection to a concern with truth." You certainly lack concern for the truth if you are willing to say something without having enough evidence to believe that it is true. Thus, a few of philosophers (e.g., Fallis 2009, 30–31; Dynel 2011, 152–153) have suggested that, whereas lying involves a violation of Grice's first maxim of quality ("Do not say what you believe to be false"), bullshitting involves a violation of Grice's second maxim of quality ("Do not say that for which you lack adequate evidence").

You can also mislead someone simply by failing to "make your contribution as informative as is required." For instance, with a reply that was not fully forthcoming, Athansius misled his pursuers about his identity.[29] Disinformation in this category is often referred to as a "half-truth" or "spin" (see Vincent and Castelfranchi 1981, 762; Carson 2010, 57–58; Manson 2012). In order to mislead people into adopting a certain view, you are selective about the information that you provide.[30] You hold back any evidence that argues against the view that you want people to adopt. You only present evidence that supports it. The important dimension of information quality that this sort of disinformation lacks is *completeness*. Incomplete information can often be as misleading as inaccurate information.

Finally, yet another possibility is to mislead someone by failing to "avoid ambiguity." In most cases, disinformation (such as Iago's statement to Othello) is only

[27] Violations of Gricean norms of conversation are only misleading if the audience assumes that the speaker is being cooperative. But it is also possible (e.g., with a double bluff) to mislead someone even if she assumes that you are not being cooperative.

[28] In addition to misleading them about what you say, you are likely to mislead your audience about your having good evidence for what you say.

[29] Athanasius also misled them into thinking that he did not know exactly where Athanasius was. If you provide fewer details than your audience clearly would like, it is legitimate for them to conclude that you do not know any more details (see Grice 1989, 33; Fallis forthcoming, §3.2).

[30] This technique of being selective about the information that you provide can also be used to get someone to believe something true. This seems to be what happened during the "Climategate" controversy. In public reports, scientists left out data that might have suggested to people that temperatures were not increasing (see Tierney 2009).

misleading when it is understood by the person who is to be misled. But disinformation is sometimes intended to be *misunderstood.* For instance, when President Bill Clinton stated during the Lewinsky Scandal, "there is no improper relationship," he meant that there *currently* was no such relationship, but he wanted people to conclude that there never had been such a relationship. As Vincent and Castelfranchi (1981, 763) would put it, Clinton engaged in "deliberate ambiguity."

8.8 Manipulating the Flow of Information

So far, I have focused on how people can create misleading information. However, as Luciano Floridi (1996, 509) points out, "the process of information is defective" in many other ways.[31] We might want to reserve the term *disinformation* for actual pieces of misleading information rather than for activities that interfere with the flow of information. But such activities are, nevertheless, an important threat to information quality. Thus, it would be useful to include them in our classification schemes (Table 8.7).

8.8.1 Restricting What Information They Have Access To

The most obvious example of manipulating the flow of information is *censorship* (see Floridi 1996, 511; Morozov 2011; Fallis 2011, 204). This sort of manipulation can take place at various stages of the communication process. People can be stopped from sending certain pieces of information. People can be stopped from

Table 8.7 Major ways to manipulate the flow of information

Disseminate misleading information
Examples – disinformation
Restrict information access
Examples – censorship
Bias information access
Examples – search engine personalization
Hide information
Mask
Examples – steganography
Repackage
Dazzle
Decoy
Make information access difficult

[31] Floridi (1996, 510) is also interested in cases where the process is accidentally defective. However, the focus here is on activities that are *intentionally* misleading.

accessing certain pieces of information. Or certain pieces of information can simply be diverted or destroyed.

Censorship is often just intended to keep people in the dark. For instance, several repressive regimes have recently restricted access to the internet in order to keep their citizens ignorant of protests going on in their own or other countries (see Richtel 2011). However, it is also possible to actually mislead people by restricting access to information (cf. Sect. 8.4 above). For instance, it can serve to maintain false beliefs that would be overturned if people had access to more information. In fact, restricting access to information can even be used to create new false beliefs under some circumstances. For instance, in the *Attack of the Clones*, Obi-wan Kenobi searches for the planet Kamino in the Jedi Archive. However, Count Dooku had previously erased the files about Kamino.[32] And this is a problem because the Jedi Archive is supposed to be complete. As the archivist tells Obi-wan, "if an item does not appear in our records, it does not exist." Thus, Obi-wan might easily have concluded that Kamino does not exist when he failed to find it in the archive.

8.8.2 Biasing What Information They Have Access To

Of course, censorship is not the only way to interfere with the flow of information in order to mislead people. We can simply make some information less accessible and/or make some information more accessible. One way that this might happen is with the "personalization" of search engine results. If two people do the very same search using the very same search engine (e.g., Google, Yahoo, or Bing), they will *not* get the very same results (see Simpson 2012). Utilizing the huge amount data that they have collected about us, these search engines are able to return results that are tailored specifically to our personal profile. The epistemic drawback of such personalization is that we are less likely to be exposed to viewpoints that differ from our own. As a result, we are more likely to form beliefs on the basis of information that is biased and incomplete.[33]

At the moment, such biasing of beliefs may only be an unintended consequence of search engines trying to make their products more attractive to internet users. But this sort of manipulation of search results could also be done malevolently in order

[32] Depending on your ontology of databases, it might be argued that this actually is a case of creating misleading information rather than just a case of removing information. Dooku arguably created a new database that lacks certain information (cf. Renear and Wickett 2009).

[33] During the 2012 Presidential election, Bing created a special website for news about the election (see Bing 2012). This website put some of the personalization back in the hands of internet users. It had a "thermometer" that allowed the user to change the list of stories that would be featured by selecting *strongly left leaning*, *moderately left leaning*, *center*, *moderately right leaning*, or *strongly right leaning*. This feature probably increases biasing since left (right) leaning internet users probably ended up selecting left (right) leaning stories. But this feature could also be used to fight biasing since epistemologically savvy users could select right leaning stories if they were left leaning (or vice versa).

to mislead people. Search engines could simply put those websites that support the view that they want people to adopt at the top of the results. This technique would be very similar to telling a half-truth and, unlike censorship, it would not require actually blocking any results. Internet users tend to check only the first few search results (see Simpson 2012, 434). So, search engines would just need to put those websites that support the contrary view way down in the results.

In addition to the search engines themselves, website owners can also try to manipulate search results. For instance, as Clifford Lynch (2001, 13–14) points out, website owners have attempted to fool the automated "crawlers" sent out by search engines to index the internet. Suppose that you have just started selling a product that competes with another product *Y*. When an automated crawler asks for your webpage to add to its index, you might send it a copy of the webpage for product *Y*. That way, when someone uses the search engine to search for product *Y*, your webpage will appear at the top of the search results.

8.8.3 Hiding Information from Them

In addition to manipulating the flow of information between other parties, you can hide your own information from others in order to keep them in the dark. Bell and Whaley (1991, 47–61) identify several different techniques for "hiding the real." With *masking* (or *camouflage*), the person or the thing to be hidden is not intended to be seen at all. A prime example is a chameleon changing its color to blend in with the surrounding environment. By contrast, with *repackaging*, the person or the thing to be hidden is made to look like something else. For instance, several species of insects have evolved to look like sticks or leaves. Unlike with masking, this is not an attempt to keep people from seeing the disguised item, but just to keep them from recognizing it for what it is.

When pursuers know that a particular person or thing is in a particular location, masking and repackaging are not going to be effective techniques. However, it is still possible to confound the pursuers with *dazzling*. For instance, an octopus might shoot out ink to confuse a predator and escape. Finally, *decoying* is yet another way to hide the real. For instance, a bird will sometimes lure predators away from its nest by pretending that it has a broken wing.[34]

It is also worth noting that these techniques can sometimes be combined. For instance, something is often disguised with the hope that no one will even notice it (i.e., masking). However, the disguise may be such that, if someone does notice it, she will not recognize it for what it really is (i.e., repackaging). For instance, in another story ("The Adventure of the Empty House"), Holmes disguises himself as

[34] Bell and Whaley actually classify decoying as way to show the false rather than as a way to hide the real. But while it might involve showing the false (e.g., a bird *faking* a broken wing), it might not (e.g., a bird with a real broken wing might also attempt to distract predators from her nest). In any event, the ultimate goal of decoying is clearly to hide the real.

an "elderly deformed" book collector so that Professor Moriarty's gang will not notice him at all. But if they do notice him, as they probably do when Watson bumps into him and upsets his books, they are still unlikely to recognize him as the famous consulting detective.

Several of these techniques can be used to hide information as well as to hide people and things. Steganography is essentially the study of how to *mask* information (see Cole 2003). It is one step beyond cryptography. Not only does it keep other people from deciphering a message. It keeps other people from even knowing that there is a message.

A famous example of the *repackaging* of information is Edgar Allan Poe's *The Purloined Letter*. A letter containing compromising information is stolen. Instead of finding an elaborate hiding place for it (as the Parisian police expected him to do), the thief makes it appear to be a different letter and then hides it "in plain sight" in his rooms. (While the ruse fools the police, the letter is discovered by the private detective C. Auguste Dupin.)

Finally, *dazzling* is actually quite common in the context of information. For instance, much like placing "a needle in a haystack," law firms often provide boxes and boxes of documents so that the opposition will not be able to find the one incriminating document. In a similar vein, a credit card company might try to hide its various usurious fees from customers in a mass of fine print (cf. Carson 2010, 53–54). Vincent and Castelfranchi (1981, 764) refer to this particular technique as "obfuscation."

8.8.4 Making It More Difficult for Them to Access Information

In addition to hiding accurate information that you do not want others to know about, you might actually mislead people by making it more difficult for them to access your own inaccurate and misleading information. This seems counter-intuitive. After all, people ultimately need to access such information if they are going to be misled by it. However, making it more difficult for people to access it can make the information more convincing once they succeed in acquiring it. For instance, at the beginning of their conversation, Othello really has to work in order to get Iago to reveal his suspicions about Desdemona and Cassio.

A notable real life example of this strategy is Operation Mincemeat. In 1943, the British military got a body from a London morgue, dressed it up to look like a military courier, gave it fake ID, and dumped it in the ocean off the coast of Spain. In addition, an attaché case chained to the body contained documents indicating that "American and British forces planned to cross the Mediterranean from their positions in North Africa, and launch an attack on German-held Greece and Sardinia." When the body was recovered by a fisherman, the local British vice-consul worked to keep the Spanish authorities from simply releasing the body and the attaché case to the Germans. As Gladwell (2010) points out, it looks like "one of the reasons the Germans fell so hard for the Mincemeat ruse is that they really had to struggle to gain access to the documents."[35]

8.9 Conclusion

This essay has shown how real life examples of disinformation—historical cases as well as cases involving the latest information technologies—can be located within multiple classification schemes. For instance, Athanasius misled his pursuers about an *implication* of the information that he provided (rather than about the accuracy of the information itself). He provided this information to his pursuers by *telling* them. Thus, he misled them by *telling them something else* other than the false thing that he wanted them to believe. In particular, he did so by *violating the norm of conversation against providing less information than is required.* In contrast, several Allied disinformation campaigns during World War II did mislead the Germans about the *accuracy of the information itself.* However, since the communications in question were not addressed to the Germans, the Allies misled them by *pretending to tell someone else* the false thing that the Germans were intended to believe.

Awareness of the diverse ways in which people might try to mislead us can potentially help us to avoid being misled by disinformation. For instance, the various classifications schemes presented here might be used as checklists to insure that, when we receive a piece of information, we consider all of the different possible ways in which we might actually be the target of disinformation. In addition, a better understanding of the nature and scope of disinformation can potentially facilitate research on techniques for dealing with this serious threat to information quality. For instance, different types of disinformation are likely to be susceptible to different methods of detection. With such goals in mind, this essay has surveyed and extended classification schemes for disinformation that have been proposed in the philosophical literature.[36]

---

[35] Since the Germans had to intercept the documents, the Mincemeat ruse is also an example of misleading by pretending to tell someone else. In fact, it might even be an example of a triple bluff. Despite many flaws in the Mincemeat ruse (e.g., the body was much more decomposed than it should have been), the Germans did not notice them and were simply fooled by the documents. But as Gladwell (2010) points out, if the Germans had noticed these flaws …

> maybe they would have found the flaws in Mincemeat a little *too* obvious, and concluded that the British were trying to deceive Germany into thinking that they were trying to deceive Germany into thinking that Greece and Sardinia were the real targets—in order to mask the fact that Greece and Sardinia *were* the real targets.

[36] I would like to thank Tony Doyle, Phyllis Illari, and Kay Mathiesen for extremely helpful feedback. This research was supported by a Research Professorship from the Social and Behavioral Sciences Research Institute at the University of Arizona.

References

Adler, J. E. (1997). Lying, deceiving, or falsely implicating. *Journal of Philosophy, 94*, 435–452.
Allen, A. L. (1988). *Uneasy access*. Totowa: Rowman & Littlefield.
Augustine, St. (1952). *Treatises on various subjects*, ed. R. J. Deferrari. New York: Fathers of the Church.
Bell, J. B., & Whaley, B. (1991). *Cheating and deception*. New Brunswick: Transaction Publishers.
Bing. (2012). *Election 2012*. http://www.bing.com/politics/elections
Black, M. (1983). *The prevalence of humbug and other essays*. Ithaca: Cornell University Press.
Carson, T. L. (2010). *Lying and deception*. New York: Oxford University Press.
Chisholm, R. M., & Feehan, T. D. (1977). The intent to deceive. *Journal of Philosophy, 74*, 143–159.
Cole, E. (2003). *Hiding in plain sight*. Indianapolis: Wiley.
Detwiler, S. M. (2002). Charlatans, leeches, and old wives: Medical misinformation. In A. P. Mintz (Ed.), *Web of deception* (pp. 23–49). Medford: Information Today.
Dynel, M. (2011). A web of deceit: A neo-Gricean view on types of verbal deception. *International Review of Pragmatics, 3*, 139–167.
Ekman, P. (2009). *Telling lies*. New York: W. W. Norton.
Faber, A., & Mazlish, E. (1980). *How to talk so kids will listen & listen so kids will talk*. New York: Avon Books.
Fallis, D. (2008). Toward an epistemology of Wikipedia. *Journal of the American Society for Information Science and Technology, 59*, 1662–1674.
Fallis, D. (2009). What is lying? *Journal of Philosophy, 106*, 29–56.
Fallis, D. (2011). Floridi on disinformation. *Etica & Politica, 13*, 201–214.
Fallis, D. (forthcoming). Skyrms on the possibility of universal deception. *Philosophical Studies*. http://dx.doi.org/10.1007/s11098-014-0308-x
Farid, H. (2009). Digital doctoring: Can we trust photographs? In B. Harrington (Ed.), *Deception* (pp. 95–108). Stanford: Stanford University Press.
Faulkner, P. (2007). What is wrong with lying? *Philosophy and Phenomenological Research, 75*, 535–557.
Fetzer, J. H. (2004). Disinformation: The use of false information. *Minds and Machines, 14*, 231–240.
Floridi, L. (1996). Brave.net.world: The internet as a disinformation superhighway? *Electronic Library, 14*, 509–514.
Floridi, L. (2011). *The philosophy of information*. Oxford: Oxford University Press.
Fowler, B., Franklin, C., & Hyde, R. (2001). Internet securities fraud: Old trick, new medium. *Duke Law and Technology Review*. http://dltr.law.duke.edu/2001/02/28/internet-securities-fraud-old-trick-new-medium/
Frankfurt, H. G. (2005). *On bullshit*. Princeton: Princeton University Press.
Freud, S. (1960). *Jokes and their relation to the unconscious*. ed. James Strachey. New York: W. W. Norton.
Gladwell, M. (2010). Pandora's briefcase. *The New Yorker*. http://www.newyorker.com/arts/critics/atlarge/2010/05/10/100510crat_atlarge_gladwell
Google. (2008). *Introducing Gmail custom time*. http://mail.google.com/mail/help/customtime/
Grice, P. (1989). *Studies in the way of words*. Cambridge: Harvard University Press.
Hancock, J. T. (2007). Digital deception: When, where and how people lie online. In A. N. Joinson, K. Y. A. McKenna, T. Postmes, & U.-D. Reips (Eds.), *Oxford handbook of internet psychology* (pp. 289–301). Oxford: Oxford University Press.
Huff, D. (1954). *How to lie with statistics*. New York: Norton.
Isaac, A. M. C., & Bridewell, W. (2014). Mindreading deception in dialog. *Cognitive Systems Research, 28*, 12–19.
Johns, A. (1998). *The nature of the book*. Chicago: University of Chicago Press.
Lackey, J. (2013). Lies and deception: An unhappy divorce. *Analysis, 73*, 236–248.

Ludlow, P. (2013). The real war on reality. *New York Times*. http://opinionator.blogs.nytimes.com/2013/06/14/the-real-war-on-reality/
Lynch, C. A. (2001). When documents deceive: Trust and provenance as new factors for information retrieval in a tangled web. *Journal of the American Society for Information Science and Technology, 52*, 12–17.
Mahon, J. E. (2008). The definition of lying and deception. *Stanford Encyclopedia of Philosophy*. http://plato.stanford.edu/entries/lying-definition/
Manson, N. C. (2012). Making sense of spin. *Journal of Applied Philosophy, 29*, 200–213.
Martin, G. R. R. (2005). *A feast for crows*. New York: Random House.
Martin, G. R. R. (2011). *A dance with dragons*. New York: Random House.
Maurer, D. W. (1999). *The big con*. New York: Doubleday.
Monmonier, M. (1991). *How to lie with maps*. Chicago: University of Chicago Press.
Monmonier, M. (1995). *Drawing the line*. New York: Henry Holt.
Moran, R. (2005). Getting told and being believed. *Philosophers' Imprint*. http://www.philosophersimprint.org/005005/
Morozov, E. (2011). *The net delusion*. New York: PublicAffairs.
Newey, G. (1997). Political lying: A defense. *Public Affairs Quarterly, 11*, 93–116.
O'Neil, C. (2012). Lying, trust, and gratitude. *Philosophy & Public Affairs, 40*, 301–333.
O'Neill, B. (2003). *A formal system for understanding lies and deceit*. http://www.sscnet.ucla.edu/polisci/faculty/boneill/bibjer5.pdf
Pettit, P. (1990). *Virtus Normativa*: Rational choice perspectives. *Ethics, 100*, 725–755.
Plunze, C. (2001). Try to make your contribution one that is true. *Acta Philosophica Fennica, 69*, 177–189.
Rankin, N. (2008). *A genius for deception*. New York: Oxford University Press.
Recanati, F. (2004). *Literal meaning*. Cambridge: Cambridge University Press.
Renear, A. H. & Wickett, K. M. (2009). Documents cannot be edited. *Proceedings of Balisage: The Markup Conference*. http://www.balisage.net/Proceedings/vol3/html/Renear01/BalisageVol3-Renear01.html
Richtel, M. (2011). Egypt cuts off most internet and cell service. *New York Times*. http://www.nytimes.com/2011/01/29/technology/internet/29cutoff.html
Rieh, S. Y. (2002). Judgment of information quality and cognitive authority in the web. *Journal of the American Society for Information Science and Technology, 53*, 145–161.
Simpson, T. W. (2012). Evaluating Google as an epistemic tool. *Metaphilosophy, 43*, 426–445.
Skyrms, B. (2010). *Signals*. New York: Oxford University Press.
Tierney, J. (2009). E-mail fracas shows peril of trying to spin science. *New York Times*. http://www.nytimes.com/2009/12/01/science/01tier.html
Vincent, J. M., & Castelfranchi, C. (1981). On the art of deception: How to lie while saying the truth. In H. Parret, M. Sbisa, & J. Verschueren (Eds.), *Possibilities and limitations of pragmatics* (pp. 749–777). Amsterdam: John Benjamins B. V.
Wilson, D., & Sperber, D. (2002). Truthfulness and relevance. *Mind, 111*, 583–632.

Chapter 9
Information Quality in Clinical Research

Jacob Stegenga

Abstract Clinical studies are designed to provide information regarding the effectiveness of medical interventions. Such information is of varying quality. 'Quality assessment tools' (QATs) are designed to measure the quality of information from clinical studies. These tools are designed to take into account various methodological details of clinical studies, including randomization, blinding, and other features of studies deemed relevant to minimizing bias and systematic error in the generated information. There are now dozens of these tools available. The various QATs on offer differ widely from each other, and second-order empirical studies show that QATs have low inter-rater reliability and low inter-tool reliability. This is an instance of a more general problem I call the underdetermination of evidential significance. Disagreements about the quality of information can be due to different—but in principle equally good—weightings of the fine-grained methodological features which constitute QATs.

9.1 Introduction

We want to make well-informed judgements about what medical interventions will work for us. Clinical studies are designed to provide information which is relevant to assessing causal hypotheses regarding the effectiveness of medical interventions. The diversity of such studies is impressive, including experiments on cell and tissue cultures, experiments on laboratory animals, mathematical models, epidemiological

J. Stegenga (✉)
Department of Philosophy, University of Utah, 215 South Central Campus Drive, Carolyn Tanner Irish Humanities Building, 4th Floor, Salt Lake City, UT 84112, USA

Department of Philosophy, University of Johannesburg, Johannesburg, South Africa
e-mail: jacob.stegenga@utoronto.ca; http://individual.utoronto.ca/jstegenga

L. Floridi and P. Illari (eds.), *The Philosophy of Information Quality*,
Synthese Library 358, DOI 10.1007/978-3-319-07121-3_9,

studies of human populations, randomized controlled trials (RCTs), and meta-level summaries based on techniques such as meta-analysis and social processes such as consensus conferences. Moreover, each of these kinds of methods has many variations. Epidemiological studies on humans, for instance, include case-control studies, retrospective cohort studies, and prospective cohort studies. As a result of this great diversity of kinds of methods in medical science, the information generated by such studies has varying degrees of credibility and relevance for the hypotheses of interest. Different kinds of methods are susceptible to different kinds of inductive errors. To properly assess causal hypotheses regarding the effectiveness of medical interventions, one must assess not only the extent to which the available information confirms the hypothesis, but how *good* the available information is. In other words, to assess such hypotheses, one must consider information quality.

To do this, one must take into account substantive details of the methods that generated the information, and how the information was analyzed and used. Information quality in medical research is constituted by the extent to which the design, conduct, analysis, and report of a clinical trial minimizes potential biases and systematic errors in the generated information. Biases and systematic errors are, roughly, the various ways that a research method can fail to provide truth-conducive information about its target subject. Medical scientists attempt to account for the various dimensions of information quality in a number of ways.

Information quality in clinical research is a complex multi-dimensional property that one cannot simply intuit, and so formalized tools have been developed to aid in the assessment of the quality of information from clinical studies. Information from clinical studies is often assessed rather crudely by rank-ordering the types of methods according to an ‘evidence hierarchy’. Systematic reviews and specifically meta-analyses are typically at the top of such hierarchies, RCTs are near the top, non-randomized cohort and case-control studies are lower, and near the bottom are laboratory studies and anecdotal case reports.[1] Evidence from methods at the top of this hierarchy, especially evidence from RCTs, is often assessed with more fine-grained tools that I call quality assessment tools (QATs). There are now many such tools on offer. The most important use to which they are put is to estimate the quality of information generated by clinical studies, especially when such information is amalgamated in a systematic review.

A systematic review is a summary of the available literature on a particular hypothesis, which often takes the form of a quantitative estimate of the strength of a particular causal relation as estimated by the various individual studies that have

[1] I discuss evidence hierarchies in more detail in §6. Such evidence hierarchies are commonly employed in evidence-based medicine. Examples include those of the Oxford Centre for Evidence-Based Medicine, the Scottish Intercollegiate Guidelines Network (SIGN), and The Grading of Recommendations Assessment, Development and Evaluation (GRADE) Working Group. These evidence hierarchies have recently received much criticism. See, for example, Bluhm (2005), Upshur (2005), Borgerson (2008), and La Caze (2011), and for a specific critique of placing meta-analysis at the top of such hierarchies, see Stegenga (2011). In footnote 4 below I cite several recent criticisms of the assumption that RCTs ought to be necessarily near the top of such hierarchies.

already been performed. These quantitative systematic reviews are referred to as meta-analyses. Meta-analysis is a tool employed by medical scientists, epidemiologists, and policy-makers, in an attempt to gather as much of the best available evidence as possible which is relevant to a particular causal hypothesis. The output of a meta-analysis is a weighted average of estimates of causal relation: such estimates are generated by primary-level clinical studies, and the 'weights' can be derived by scores from a QAT. Thus, the output of a meta-analysis can be thought of as second-order information. Such second-order information is usually thought to be the best available information for estimating causal relations in clinical research (though see Stegenga 2011 for a critical evaluation of meta-analysis). Since most causal hypotheses in medicine are assessed with systematic reviews and specifically meta-analysis, and since meta-analysis often involves the use of QATs, much of what we think we know about causal hypotheses in medicine is influenced by QATs. A QAT score is, fundamentally, a measure of the quality of information generated by clinical studies.

The purpose of using a QAT is to evaluate the quality of information from clinical studies in a fine-grained way. Their domain of application is relatively focused, therefore, since they do not apply to other kinds of information that is typically available for causal hypotheses in medicine (such as information about mechanisms generated by basic science research, or information from experiments on animals), but for better or worse it is usually only RCTs, meta-analyses of RCTs, and observational studies that are considered when assessing causal hypotheses in medicine, and it is these types of methods that QATs are typically designed to assess.

By 'information', I mean roughly what Floridi (2004) calls the 'semantic' account of information: information is semantic content which affords and constrains certain inferences. This is commonly how the term 'evidence' is employed in scientific contexts. An assessment of the quality of information aids in determining precisely what inferences are afforded and constrained by the given information. Data from clinical studies are information about the effectiveness of particular medical interventions, and the quality of such information influences the extent to which certain inferences are afforded and constrained.

In what follows I examine the use of QATs as codified tools for assessing the quality of information in clinical research. Although there has been some criticism of QATs in the medical literature, they have received little philosophical critique.[2] I begin by describing general properties of QATs, including the methodological features that many QATs share and how QATs are typically employed (Sect. 9.2). I then turn to a discussion of empirical studies which test the inter-rater reliability (Sect. 9.3) and inter-tool reliability (Sect. 9.4) of QATs: most QATs are not very

[2]Although one only needs to consider the prominence of randomization in QATs to see that QATs have, in fact, been indirectly criticized by the recent literature criticizing the assumed 'gold standard' status of RCTs (see footnote 4). In the present paper I do not attempt a thorough normative evaluation of any particular QAT. Considering the role of randomization suggests what a large task a thorough normative evaluation of a particular QAT would be. But for a systematic survey of the most prominent QATs, see West et al. (2002). See also Olivo et al. (2007) for an empirical critique of QATs.

good at constraining intersubjective assessments of hypotheses, and more worrying, the use of different QATs to assess the same primary-level information leads to widely divergent quality assessments of that information. This is an instance of a more general problem I call the underdetermination of evidential significance, which holds that in a rich enough empirical situation, the strength of the evidence (or quality of information) is underdetermined (Sect. 9.5). Despite this problem, I defend the use of QATs in clinical research. I end by comparing QATs to the widely employed evidence hierarchies, and argue that despite the problems with QATs, they are better than evidence hierarchies for assessing the quality of information in clinical research (Sect. 9.6).

9.2 Quality Assessment Tools

A quality assessment tool (QAT) for information from clinical studies can be either a scale with elements that receive a quantitative score representing the degree to which each element is satisfied by a clinical study, or else a QAT can be simply a checklist with elements that are marked as either present or absent in a clinical study. Given the emphasis on randomized controlled trials (RCTs) in medical research, most QATs are designed for the evaluation of RCTs, although there are several for observational studies and systematic reviews.[3] Most QATs share several elements, including questions about how subjects were assigned to experimental groups in a trial, whether or not the subjects and experimenters were concealed to the subjects' treatment protocol, whether or not there was a sufficient description of subject withdrawal from the trial groups, whether or not particular statistical analyses were performed, and whether or not a report of a trial disclosed financial relationships between investigators and companies.[4] Most QATs provide instructions on how to score the individual components of the QAT and how to determine an overall quality-of-information score of a trial.

A comprehensive list of QATs developed by the mid-1990s was described by Moher et al. (1995). The first scale type to be developed, known as the Chalmers scale, was published in 1981. By the mid-1990s there were over two dozen QATs, and by 2002 West et al. were able to identify 68 for RCTs or observational studies.

[3] The view that RCTs are the 'gold standard' of evidence has recently been subjected to much philosophical criticism. See, for example, Worrall (2002), Worrall (2007), Cartwright (2007), and Cartwright (2012); for an assessment of the arguments for and against the gold standard status of RCTs, see Howick (2011). Observational studies also have QATs, such as QATSO (Quality Assessment Checklist for Observational Studies) and NOQAT (Newcastle-Ottawa Quality Assessment Scale – Case Control Studies).

[4] Sometimes the term 'trial' in the medical literature refers specifically to an experimental design (such as a randomized controlled trial) while the term 'study' refers to an observational design (such as a case control study), but this use is inconsistent. I will use both terms freely to refer to any method of generating information relevant to causal hypotheses in clinical research, including both experimental and observational designs.

Some are designed for the evaluation of any medical trial, while others are designed to assess specific kinds of trials, or trials from a particular medical sub-discipline, or even particular token trials. Some are designed to assess the quality of a trial itself (as described in trial design protocols and methods sections of publications), while others are designed to assess the quality of a report of a trial (as a published article), and some assess both.

QATs are now widely used for several purposes. As described above, a QAT score is, at bottom, a measure of the quality of information. When performing a systematic review of the available evidence for a particular hypothesis, then, QATs help reviewers take the quality of information into account. This is typically done in one of two ways. First, QAT scores can be used to generate a weighting factor for the technique known as meta-analysis. Meta-analysis usually involves calculating a weighted average of so-called effect sizes from individual medical studies, and the weighting of effect sizes can be determined by the score of the respective trial on a QAT.[5] Second, QAT scores can be used as an inclusion criterion for a systematic review, in which any primary-level clinical study that achieves a QAT score above a certain threshold would be included in the systematic review (and conversely, any trial that achieves a QAT score below such a threshold would be excluded). This application of QATs is perhaps the most common use to which they are put, and is perhaps motivated by the view that only information of a certain minimum quality ought to be relied on when performing a systematic review. Finally, QATs can be used for purposes not directly associated with a particular systematic review or meta-analysis, but rather to investigate relationships between information quality and other properties of clinical trials. For instance, several findings suggest that there is an inverse correlation between information quality (as measured by a QAT score) and effect size (in other words, information from higher quality trials tends to support lower estimates of the efficacy of medical interventions).[6]

Why should medical scientists bother assessing the quality of information from clinical studies? Consider the following argument, similar to an argument for following the principle of total evidence, based on a concern to take into account any possible 'defeating' properties of one's information. Suppose your available information seems to provide support for some hypothesis, H_1. But then you learn that there was a systematic error in the method which generated your information. Taking into account this systematic error, the information no longer supports H_1 (perhaps instead the information supports a competitor hypothesis, H_2). Had you not taken into account the fine-grained methodological details regarding the systematic error, you would have unwarranted belief in H_1. You do not want to have unwarranted

[5] There are several commonly employed measures of effect size, including mean difference (for continuous variables), or odds ratio, risk ratio, or risk difference (for dichotomous variables). The weighting factor is sometimes determined by the QAT score, but a common method of determining the weight of a trial is simply based on the size of the trial (Egger, Smith, and Phillips 1997), often by using the inverse variability of the data from a trial to measure that trial's weight (because inverse variability is correlated with trial size).

[6] See, for example, Moher et al. (1998), Balk et al. (2002), and Hempel et al. (2011).

belief in a hypothesis, so you ought to take into account fine-grained methodological details relevant to assessing the quality of information.

Here is a related argument: if one does not take into account all of one's evidence, including one's old evidence, then one is liable to commit the base-rate fallacy. In terms of Bayes' Theorem—$p(H|e) = p(e|H)p(H)/p(e)$—one commits the base-rate fallacy if one attempts to determine $p(H|e)$ without taking into account $p(H)$. Similarly, if one wants to determine $p(H|e)$ then one ought to take into account the detailed methodological features which determine $p(e|H)$ and $p(e)$. For a Bayesian, these quantities are constituted by one's available information and the quality of such information.

One need not be a Bayesian to see the importance of assessing information at a fine-grain with QATs. For instance, Mayo's notion of 'severe testing', broadly based on aspects of frequentist statistics, also requires taking into account fine-grained methodological details. The Severity Principle, to use Mayo's term, claims that "passing a test *T* (with *e*) counts as a good test of or good evidence for *H* just to the extent that *H* fits *e* and *T* is a *severe test* of *H*" (Mayo 1996). Attending to fine-grained methodological details to ensure that one has minimized the probability of committing an error is central to ensuring that the test in question is severe, and thus that the Severity Principle is satisfied. So, regardless of one's doctrinal commitment to Bayesianism or frequentism, the employment of tools like QATs to take into account detailed features about the methods used to generate the available information ought to seem reasonable.

One of the simplest QATs is the Jadad scale, first developed in the 1990s to assess clinical studies in pain research. Here it is, in full:

1. Was the study described as randomized?
2. Was the study described as double blind?
3. Was there a description of withdrawals and dropouts?

A 'yes' to question 1 and question 2 is given one point each. A 'yes' to question 3, in addition to a description of the number of withdrawals and dropouts in each of the trial sub-groups, and an explanation for the withdrawals or dropouts, receives one point. An additional point is given if the method of randomization is described in the paper, and the method is deemed appropriate. A final point is awarded if the method of blinding is described, and the method is deemed appropriate. Thus, a trial can receive between zero and five points on the Jadad scale.

The Jadad scale has been praised by some as being easy to use—it takes about ten minutes to complete for each study—which is an obvious virtue when a reviewer must assess hundreds of studies for a particular hypothesis. On the other hand, others complain that it is too simple, and that it has low inter-rater reliability (discussed in Sect. 9.3). I describe the tool here not to assess it but merely to provide an example of a QAT for illustration.

In contrast to the simplicity of the Jadad scale, the Chalmers scale has 30 questions in several categories, which include the trial protocol, the statistical analysis, and the presentation of results. Similarly, the QAT developed by Cho and Bero (1994) has 24 questions. At a coarse grain some of the features on the Chalmers QAT and the Cho and Bero QAT are similar to the basic elements of the Jadad QAT: these scales

Table 9.1 Number of methodological features used in six QATs, and weight assigned to three widely shared methodological features

Scale	Number of items	Weight of randomization	Weight of blinding	Weight of withdrawal
Chalmers et al. (1981)	30	13.0	26.0	7.0
Jadad et al. (1996)	3	40.0	40.0	20.0
Cho and Bero (1994)	24	14.3	8.2	8.2
Reisch et al. (1989)	34	5.9	5.9	2.9
Spitzer et al. (1990)	32	3.1	3.1	9.4
Linde et al. (1997)	7	28.6	28.6	28.6

Adapted from Jüni et al. (1999)

both include questions about randomization, blinding, and subject withdrawal. (In Sect. 9.5 I briefly describe how Cho and Bero developed their QAT, as an illustration of the no-best-weighting argument). In addition, these more detailed QATs include questions about statistical analyses, control subjects, and other methodological features deemed relevant to minimizing systematic error. These QATs usually take around 30–40 min to complete for each study. Despite the added complexity of these more detailed QATs, their scoring systems are kept as simple as possible. For instance, most of the questions on the Cho and Bero QAT allow only the following answers: 'yes' (2 points), 'partial' (1 point), 'no' (0 points), and 'not applicable' (0 points). This is meant to constrain the amount of subjective judgment required when generating a QAT score.

Although most QATs share at least several similar features, the relative weight of the overall score given to the various features differs widely between QATs. Table 9.1 lists the relative weight of three central methodological features—subject randomization, subject allocation concealment (or 'blinding'), and description of subject withdrawal—for the above QATs, in addition to three other QATs.

Note two aspects of Table 9.1. First, the number of items on a QAT is highly variable, from 3 to 34. Second, the weight given to particular methodological features is also highly variable. Randomization, for instance, constitutes 3.1 % of the overall information quality score on the QAT designed by Spitzer et al. (1990), whereas it constitutes 40 % of the overall information quality score on the QAT designed by Jadad et al. (1996). The differences between QATs explains the low inter-tool reliability, which I describe in Sect. 9.4. But first I describe the low inter-rater reliability of QATs.

9.3 Inter-rater Reliability

The extent to which multiple users of the same rating system achieve similar ratings is usually referred to as 'inter-rater reliability'. Empirical evaluations of the inter-rater reliability of QATs have shown a wide disparity in the outcomes of a QAT when applied to the same primary-level study by multiple reviewers; that is, when assessing information quality, the inter-rater reliability of QATs is usually low.

The typical set-up of evaluations of inter-rater reliability of a QAT is simple: give a set of manuscripts to multiple reviewers who have been trained to use the QAT, and compare the information quality scores assigned by these reviewers to each other. A statistic called kappa (κ) is typically computed which provides a measure of agreement between the information quality scores produced by the QAT from the multiple reviewers (although other statistics measuring agreement are also used, such as Kendall's coefficient of concordance and the intraclass correlation coefficient).[7] Sometimes the manuscripts are blinded as to who the authors were and what journals the manuscripts were published in, but sometimes the manuscripts are not blinded, and sometimes both blinded and non-blinded manuscripts are assessed to evaluate the effect of blinding. In some cases the manuscripts all pertain to the same hypothesis, while in other cases the manuscripts pertain to various subjects within a particular medical sub-discipline.

For example, Clark et al. (1999) assessed the inter-rater reliability of the Jadad scale, using four reviewers to evaluate the quality of information from 76 manuscripts of RCTs. Inter-rater reliability was found to be "poor", but it increased substantially when the third item of the scale (explanation of withdrawal from study) was removed and only the remaining two questions were employed.

A QAT known as the 'risk of bias tool' was devised by the Cochrane Collaboration (a prominent organization in the so-called evidence-based medicine movement) to assess the degree to which the results of a study "should be believed." A group of medical scientists subsequently assessed the inter-rater reliability of the risk of bias tool. They distributed 163 manuscripts of RCTs among five reviewers, who assessed the RCTs with this tool, and they found the inter-rater reliability of the quality assessments to be very low (Hartling et al. 2009).

Similarly, Hartling et al. (2011) used three QATs (Risk of Bias tool, Jadad scale, Schulz allocation concealment) to assess 107 studies on a medical intervention (the use of inhaled corticosteriods for adults with persistent asthma). This group employed two independent reviewers who scored the 107 studies using the three QATs. They found that inter-rater reliability was 'moderate'. However, the claim that inter-rater reliability was moderate was based on a standard scale in which a κ measure between 0.41 and 0.6 is deemed moderate. The κ measure in this paper was 0.41, so it was just barely within the range deemed moderate. The next lower category,

[7] For simplicity I will describe Cohen's Kappa, which measures the agreement of two reviewers who classify items into discrete categories, and is computed as follows:

$$\kappa = [p(a) - p(e)] / [1 - p(e)]$$

where p(a) is the probability of agreement (based on the observed frequency of agreement) and p(e) is the probability of chance agreement (also calculated using observed frequency data). Kappa was first introduced as a statistical measure by Cohen (1960). For more than two reviewers, a measure called Fleiss' Kappa can be used. I give an example of a calculation of κ below.

with a κ measure between 0.21 and 0.4, is deemed 'fair' by this standard scale. But at least in the context of measuring inter-rater reliability of QATs, a κ of 0.4 represents wide disagreement between reviewers.

Here is a toy example to illustrate the disagreement that a κ measure of 0.4 represents. Suppose two teaching assistants, Beth and Sara, are grading the same class of 100 students, and must decide whether or not each student passes or fails. Their joint distribution of grades is:

		Sara	
		Pass	Fail
Beth	Pass	40	10
	Fail	20	30

Of the 100 students, they agree on passing 40 students and failing 30 others, thus their frequency of agreement is 0.7. But the probability of random agreement is 0.5, because Beth passes 50 % of the students and Sara passes 60 % of the students, so the probability that Beth and Sara would agree on passing a randomly chosen student is 0.5×0.6 (=0.3), and similarly the probability that Beth and Sara would agree on failing a randomly chosen student is 0.5×0.4 (=0.2) (and so the overall probability of agreeing on passing or failing a randomly chosen student is $0.3 + 0.2 = 0.5$). Applying the kappa formula gives:

$$(0.7 - 0.5)/(1 - 0.5) = 0.4$$

Importantly, Beth and Sara disagree about 30 students regarding a relatively simple property (passing). It is natural to suppose that they disagree most about 'borderline' students, and their disagreement is made stark because Beth and Sara have a blunt evaluative tool (pass/fail grades rather than, say, letter grades). But a finer-grained evaluative tool would not necessarily mitigate such disagreement, since there would be more categories about which they could disagree for each student; a finer-grained evaluative tool would increase, rather than decrease, the number of borderline cases (because there are borderline cases between each letter grade). This example is meant to illustrate that a κ measure of 0.4 represents poor agreement between two reviewers.[8] A κ score is fundamentally an arbitrary measure of disagreement, and the significance of the disagreement that a particular κ score represents presumably varies with context. This example, I nevertheless hope, helps to

[8] I owe Jonah Schupbach thanks for noting that a κ measure can not only seem inappropriately low, as in the above cases of poor inter-rater reliability, but can seem inappropriately high as well. If a κ measure approaches 1, this might suggest agreement which is 'too good to be true'. Returning to my toy example, if Beth and Sara had a very high a κ measure, then one might wonder if they colluded in their grading. Thus when using a κ statistic to assess inter-rater reliability, we should hope for a κ measure above some minimal threshold (below which indicates too much disagreement) but below some maximum threshold (above which indicates too much agreement). What exactly these thresholds should be are beyond the scope of this paper (and are, I suppose, context sensitive).

illustrate the extent of disagreement found in empirical assessments of the inter-rater reliability of QATs.

In short, different users of the same QAT, when assessing the same information, generate diverging assessments of the quality of that information. In most tests of the inter-rater reliability of QATs, the information being assessed comes from a narrow range of study designs (usually all the studies are RCTs), and the information is about a narrow range of subject matter (usually all the studies are about the same causal hypothesis regarding a particular medical intervention). The poor inter-rater reliability is even more striking considering the narrow range of study designs and subject matter from which the information is generated.

9.4 Inter-tool Reliability

The extent to which multiple instruments have correlated measurements when applied to the same property being measured is referred to as inter-tool reliability. One QAT has inter-tool reliability with respect to another if its measurement of the quality of information correlates with measurements of the quality of information by the other QAT. A QAT score is a measure on a relatively arbitrary scale, and the scales between multiple QATs are incommensurable, so constructs such as 'high quality' and 'low quality' are developed for each QAT which allow the results from different QATs to be compared. That is, when testing the inter-tool reliability of multiple QATs, what is usually being compared is the extent of their agreement regarding the categorization of information from particular clinical studies into pre-defined bins of quality. Similar to assessments of inter-rater reliability, empirical evaluations of the inter-tool reliability have shown a wide disparity in the outcomes of multiple QATs when applied to the same primary-level studies; that is, the inter-tool reliability of QATs is poor. I should note, however, that there are few such assessments available, and those published thus far have varied with respect to the particular QATs assessed, the design of the reliability assessment, and the statistical analyses employed.[9]

An extensive investigation of inter-tool reliability was performed by Jüni and colleagues (1999). They amalgamated data from 17 studies which had tested a particular medical intervention (the use of low molecular weight heparin to prevent post-operative thrombosis), and they used 25 QATs to assess the quality of information from these 17 studies (thereby effectively performing 25 meta-analyses). The QATs that this group used were the same that Moher et al. (1995) had earlier described,

[9] For this latter reason I refrain from describing or illustrating the particular statistical analyses employed in tests of the inter-tool reliability of QATs, as I did in §3 on tests of the inter-rater reliability of QATs. Nearly every published test of inter-rater reliability uses a different statistic to measure agreement of quality assessment between tools. Analyses include Kendall's rank correlation coefficient (τ), Kendall's coefficient of concordance (W), and Spearman's rank correlation coefficient (ρ).

which varied in the number of assessed study attributes, from a low of three attributes to a high of 34, and varied in the weight given to the various study attributes. Jüni and his colleagues noted that "most of these scoring systems lack a focused theoretical basis." Their results were troubling: the amalgamated effect sizes between these 25 meta-analyses differed by up to 117 %—*using exactly the same primary evidence*. They found that information deemed high quality according to one QAT could be deemed low quality according to another. The authors concluded that "the type of scale used to assess trial [information] quality can dramatically influence the interpretation of meta-analytic studies."

Perhaps the most recent evaluation of inter-tool reliability is Hartling et al. (2011), discussed above in Sect. 9.3. Recall that this group used three QATs (Risk of Bias tool, Jadad scale, Schulz allocation concealment) to assess 107 trials on a particular medical intervention. They also found that the inter-tool reliability was very low.

Yet another example of a test of inter-tool reliability of QATs was reported by Moher et al. (1996). This group used six QATs to evaluate 12 trials of a medical intervention. Again, the inter-tool reliability was found to be low.

Low inter-tool reliability of QATs is troubling: it is a quantitative empirical demonstration that the determination of the quality of information from a clinical study depends on the choice of QAT. Moreover, in Sect. 9.2 I noted that there are many QATs available, and between them there are substantial differences in their design. Thus the *best* tools that medical scientists have to determine the quality of information generated by what are typically deemed the *best* study designs (RCTs) are relatively unconstraining and liable to produce conflicting assessments. Such low inter-tool reliability of QATs has important practical consequences. Elsewhere I show that multiple meta-analyses of the same primary evidence can reach contradictory conclusions regarding particular causal hypotheses, and one of the conditions which permits such malleability of meta-analysis is the choice of QAT (Stegenga 2011).[10] The discordant results from the 25 meta-analyses performed by Moher et al. (1995) are a case in point. Moreover, this low inter-tool reliability has philosophical consequences, which I explore in Sect. 9.5.

Such low inter-tool reliability might be less troubling if the various QATs had distinct domains of application. The many biases present in medical research are pertinent to varying degrees depending on the details of the particular circumstances at hand, and so one might think that it is a mistake to expect that one QAT ought to apply to all circumstances. For some causal hypotheses, for instance, it is difficult or impossible to conceal the treatment from the experimental subjects and/or the investigators (that is, 'blinding' is sometimes impossible)—hypotheses regarding chiropractic spinal manipulation are a case in point. Thus, no study relevant to such a hypothesis will score well on a QAT that gives a large weight to allocation concealment.

[10] Low inter-tool reliability of QATs is only one of several problems with meta-analysis. Other parameters of meta-analysis that render this method malleable include the choice of primary-level studies to include in the analysis, the choice of outcome measure to employ, the choice of kind of data to amalgamate (patient-level or study-level), and the choice of averaging technique to employ. See Stegenga (2011) for a critical account of meta-analysis.

Such a QAT would be less sensitive to the presence or absence of sources of bias other than lack of allocation concealment, relative to QATs that give little or no weight to allocation concealment. In such a case one might argue that since the absence of allocation concealment is fixed among the relevant studies, an appropriate QAT to use in this case should not give any weight to allocation concealment, and would only ask about the presence of those properties of a study that might vary among the relevant studies.

On the other hand, one might argue that since we have principled reasons for thinking that the absence of allocation concealment can bias the information generated from a study, even among those studies that cannot possibly conceal subject allocation, an appropriate QAT to use in this case *should* evaluate the presence of allocation concealment (in which case all of the relevant studies would simply receive a zero score on allocation concealment), just as a QAT ought to evaluate the presence of allocation concealment in a scenario in which the studies in fact can conceal subject allocation. The former consideration is an appeal to determining the *relative* quality between studies, and the latter consideration is an appeal to determining the *absolute* quality of studies. The latter consideration should be more compelling in most cases, since, as discussed above, the typical use of QATs is to help estimate the true efficacy of a medical intervention, and such estimates ought to take into account the full extent of the potential for biases in the relevant information, regardless of whether or not it was possible for the respective studies to avoid such biases.

There are scenarios, though, in which we might have reasons to think that a property of a study that causes bias in other scenarios does not cause bias (or perhaps causes less bias) in these scenarios. For example, the placebo effect might be stronger in studies that are designed to assess the *benefits* of pharmaceuticals compared with studies that are designed to assess the *harms* of pharmaceuticals. Such a difference could be independently and empirically tested. If this were true, then the different scenarios would indeed warrant different QATs, suitable for the particularities of each scenario at hand. If the low inter-tool reliability of QATs were merely the result of employing multiple QATs to different kinds of empirical scenarios (different kinds of studies, say, or studies of different kinds of hypotheses, such as benefits versus harms of pharmaceuticals), then such low inter-tool reliability would hardly be troubling. Indiscriminate use of QATs might lead to low inter-tool reliability, such thinking would go, but discriminate use would not.

Similarly, low inter-tool reliability of QATs would be less troubling if one could show that in principle there is only one good QAT for a given domain, or at least a small set of good ones which are similar to each other in important respects, because then one could dismiss the observed low inter-tool reliability as an artefact caused by the inclusion of poor QATs in addition to the good ones.

Unfortunately, on the whole, these considerations do not mitigate the problem of low inter-tool reliability of QATs. There are, in fact, a plurality of equally fine QATs, designed for the same kinds of scenarios (namely, assessing the quality of information generated by RCTs regarding the efficacy of pharmaceuticals). A systematic review by medical scientists concluded that there were numerous QATs that “represent acceptable approaches that could be used today without major modifications” (West et al. 2002). Moreover, all of the empirical demonstrations of their low inter-tool reliability involve the assessment of the quality of studies from a very narrow domain: for instance, the low inter-tool reliability of QATs shown in Jüni et al. (1999) involved assessing studies of a *single* design (RCTs) about a *single* causal hypothesis, and these QATs had been developed with the purpose of assessing the quality of information from that very type of study design. Although there are some QATs which are arguably inferior to others, at least among the reasonably good ones I argue below that we lack a theoretical basis for distinguishing among them, and so we are stuck with a panoply of acceptable QATs which disagree widely about the quality of information from particular medical studies.

One might agree with the view that there is no uniquely best QAT, but be tempted to think that this is due only to the fact that the quality of information from a study depends on particularities of the context (e.g. the particular kind of study in question and the form of the hypothesis being tested by that study). Different QATs might, according to this thought, be optimally suited to different contexts. While this latter point is no doubt true—above I noted that some QATs are designed for assessing particular kinds of *studies*, and others are designed for assessing studies in particular *domains* of medicine—it does not explain the low inter-tool reliability of QATs. That is because, as above, the low inter-tool reliability of QATs is demonstrated in narrowly specified contexts. Moreover, the research groups that design QATs usually claim (explicitly) that their QATs are meant to be applicable to a given study design (usually RCTs) in most domains of medical research. In short, QATs are intended to apply to a broad range of contexts, but regardless, the empirical demonstrations of their low inter-tool reliability are almost always constrained to a single particular context (though as described above, such demonstrations have been repeated in multiple contexts).

Despite their widespread and growing use, among medical scientists there is some debate about whether or not QATs ought to be employed at all (see, for example, Herbison et al. (2006)). Their low inter-rater and inter-tool reliability might suggest that resistance to their use is warranted. There are three reasons, however, that justify the continuing improvement and application of QATs to assessing the quality of information from clinical studies. First, when performing a meta-analysis, a decision to not use an instrument to differentially weight the quality of the primary-level information is equivalent to weighting all the primary-level information to an equal degree. So whether one wishes to or not, when performing a meta-analysis one is forced, in principle, to weight the primary-level information, and the remaining question then is simply how arbitrary one’s method of weighting is. Assigning equal weights regardless of information quality is maximally arbitrary. The use of QATs to differentially weight primary-level information is an attempt to minimize such arbitrariness. Second, as argued in Sect. 9.2 above, one must account for fine-grained methodological features in order to guarantee that one avoids potential defeating properties of information, and QATs can help with this. Third—but closely related to the second point—there is some empirical evidence which suggests that information of lower quality has a tendency to over-estimate the efficacy of medical interventions (see footnote 7), and thus the use of QATs helps to accurately estimate the efficacy of medical interventions. In short, despite their low inter-rater and inter-tool reliability, QATs are an important component of medical research, and should be employed when performing a systematic review or meta-analysis.

9.5 Underdetermination of Evidential Significance

The primary use of QATs is to estimate the quality of information from particular medical studies, and the primary use of such information is to estimate the strength (if any) of causal relations. The relata in these purported causal relations are, of course, the medical intervention under investigation and the change in value of one or more parameters of a group of subjects. The best available QATs appropriate to a given domain differ substantially in the weight assigned to various methodological properties (Sect. 9.2), and thus generate discordant estimates of information quality when applied to the same information (Sect. 9.4). The differences between the best available QATs are fundamentally arbitrary. Although I assume that there must be a unique value (if at all) to the strength of purported causal relations in the domains in which these tools are employed, the low inter-tool reliability of QATs—together with the fundamentally arbitrary differences of their content—suggests that, in such domains and for such relations, there is no uniquely correct estimate of the quality of information. This is an instance of the general problem I call the underdetermination of evidential significance.

Disagreement regarding the quality of information in particular scientific domains has been frequently documented with historical case studies. One virtue of examining the disagreement generated by the use of QATs is that such disagreements occur in highly controlled settings, are quantifiable using measures such as the κ statistic, and are about subjects of great importance. Such disagreements do not necessarily represent shortcomings on the part of the disagreeing scientists, and nor do such disagreements necessarily suggest a crude relativism. Two scientists who disagree about the quality of information from a particular study can both be rational because their differing assessments of the quality of that information can be due to their different weightings of fine-grained features of the methods which generated the information. This explains (at least in part) the low inter-rater and inter-tool reliability of QATs.

Concluding that there is no uniquely correct determination of the quality of information by appealing to the poor inter-rater and inter-tool reliability of QATs is not merely an argument from disagreement. If it were, then the standard objection would simply note that the mere fact of disagreement about a particular subject does not imply that there is no correct or uniquely best view on the subject. Although different QATs disagree about the quality of information from a particular study, this does not imply that there is no true or best view regarding the quality of information from this particular trial—goes the standard objection—since the best QATs might agree with each other about the quality of information from this trial, and even more ambitiously, agreement or disagreement among QATs would be irrelevant if we just took into account the quality of information from this particular trial by the uniquely best QAT. The burden that this objection faces is the identification of the single best QAT or at least the set of good ones (and then hope that multiple users of the best QAT will have high inter-rater reliability, or that the set of good QATs will have high inter-tool reliability). As noted in Sect. 9.4, medical scientists involved in the development and assessment of QATs claim that there are simply a plurality of decent QATs that differ from one another in arbitrary respects. More fundamentally, we lack a theory of scientific inference that would allow us to referee between the most sophisticated QATs. Recall the different weightings of the particular methodological features assessed in QATs, noted in Table 9.1. Another way to state the burden of the 'mere argument by disagreement' objection is that to identify the best QATs, one would have to possess a principled method of determining the optimal weights for the methodological features included on a QAT. That we do not presently have such a principled method is an understatement.

Consider this compelling illustration of the arbitrariness involved in the assignment of weights to methodological features in QATs. Cho and Bero (1994) employed three different algorithms for weighting the methodological features of their QAT (discussed in Sect. 9.2). Then they tested the three weighting algorithms for their effect on information quality scores from medical trials, and their effect on the inter-rater reliability of such scores. They selected for further use—*with no principled basis*—the weighting algorithm that had the highest inter-rater reliability. Cho and Bero explicitly admitted that nothing beyond the higher inter-rater reliability warranted the choice of this weighting algorithm, and they rightfully claimed that such arbitrariness was justified because "there is little empiric [sic] evidence on the relative importance of the individual quality criteria to the control of systematic bias."[11] Medical scientists have no principled foundation for developing a uniquely good QAT, and so resort to a relatively arbitrary basis for their development.

One could press the standard objection by noting that while it is true that we *presently* lack an inductive theory that could provide warrant for a unique system for weighting the various methodological features relevant to information quality, it is overly pessimistic to think that we will *never* have a principled basis for identifying a uniquely best weighting system. It is plausible, this objection goes, to think that someday we will have a uniquely best QAT, or perhaps uniquely best QATs for particular kinds of epistemic scenarios, and we could thereby achieve agreement regarding the quality of information from clinical studies. To this one would have to forgive those medical scientists, dissatisfied with this response, who are concerned with assessing quality of information today. But there is another, deeper reason why such a response is not compelling.

[11] There is a tendency among medical scientists to suppose that the relative importance of various methodological features is merely an empirical matter. One need not entirely sympathize with such methodological naturalism to agree with the point expressed by Cho and Bero here: we lack reasons to prefer one weighting of methodological features over another, regardless of whether one thinks of these reasons as empirical or principled.

It is not a mere argument from present disagreement—I reiterate—to claim that the poor inter-tool reliability of QATs implies that the quality of information from particular clinical studies is underdetermined. That is because, as the example of the Cho and Bero QAT suggests, the disagreements between QATs are due to arbitrary differences in how the particular methodological features are weighed in the various QATs. There are, to be sure, better and worse QATs. But that is about as good as one can do when it comes to distinguishing between QATs. Of those that account for the majority of relevant methodological features, some weight those features in a slightly different manner than others, and we have no principled grounds for preferring one weighting over another. We do not possess a theory of scientific inference that could help determine the weights of the methodological features in QATs. If one really wanted to, one could sustain the objection by claiming that it is possible that in the future we will develop a theory of inference which would allow us to identify a uniquely best QAT. There is a point at which one can no longer argue against philosophical optimism. The underdetermination of evidential significance is a hard problem; like other hard philosophical problems, it does not preclude optimism.

One could put aside the aim of finding a *principled* basis for selecting among the available QATs, and instead perform a selection based on their *historical* performance. Call this a 'naturalist' selection of QATs.[12] Since QATs are employed to estimate the quality of information from clinical studies, and such information is used to estimate the strength of causal relations, the naturalist approach would involve selecting QATs based on a parameter determined by the 'fit' between (i) the strength of presently known causal relations and (ii) the quality of information for such causal relations available at a particular time, as determined in retrospect by currently available QATs. The best QAT would be the one with the best average fit between (i) and (ii). Such an assessment of QATs would be of some value. It would be limited, though, by a fundamental epistemic circularity. In the domains in which QATs are employed, the best epistemic access to the strength of causal relations is the total evidence from all the available medical studies, summarized by a careful systematic review (which, in this domain, usually takes the form of a meta-analysis), appropriately weighted to take into account the quality of information from those studies. But of course, those very weightings are simply generated by QATs. The naturalist approach to assessing QATs, then, itself requires the employment of QATs.

The underdetermination of evidential significance is *not* the same problem that is often associated with Duhem and Quine. One formulation of the standard underdetermination problem—underdetermination of theory by evidence—holds that there are multiple theories compatible with one's available information. The underdetermination of evidential significance is the prior problem of settling on the quality of information in the first place. Indeed, one may wish to say that an appropriate name for the present problem is just the inverse of the Quinean locution: *underdetermination of evidence by theory*. Our best theories of inference underdetermine the strength of evidence, exemplified by tools such as QATs.

[12] Such an approach was first suggested to me by Jim Tabery.

9.6 QATs and Hierarchies

The most frequently used tools for assessing the quality of information from of clinical studies are not QATs, but rather evidence hierarchies. An evidence hierarchy is a rank-ordering of kinds of methods according to the potential for bias in that kind of method. The potential for bias is usually based on one or very few parameters of study designs, most prominently randomization. QATs and evidence hierarchies are not mutually exclusive, since an evidence hierarchy can be employed to generate a rank-ordering of types of methods, and then QATs can be employed to evaluate the quality of tokens of those methods. However, judicious use of QATs should replace evidence hierarchies altogether. The best defense of evidence hierarchies that I know of is given by Howick (2011), who promotes a sophisticated version of hierarchies in which the rank-ordering of a particular study can increase or decrease depending on parameters distinct from the parameter first used to generate the ranking. Howick's suggestion, and any evidence hierarchy consistent with his suggestion (such as that of GRADE), ultimately amounts to an outright abandonment of evidence hierarchies. Howick gives conditions for when mechanistic evidence and evidence from non-randomized studies should be considered, and also suggests that sometimes evidence from RCTs should be doubted. If one takes into account methodological nuances of medical research, in the ways that Howick suggests or otherwise, then the metaphor of a hierarchy of evidence and its utility in assessing quality of information seem less compelling than more quantitative tools like QATs, because QATs can take into account more parameters of information, and in more detail, than can evidence hierarchies.

For instance, the GRADE evidence hierarchy employs more than one property to rank methods. GRADE starts with a quality assignment based on one property and takes other properties into account by subsequent modifications of the quality assignment (shifting the assignment up or down). Formally, the use of n properties to rank methods is equivalent to a scoring system based on n properties which discards any information that exceeds what is required to generate a ranking. QATs generate scores that are measured on scales more informative than ordinal scales (such as interval, ratio, or absolute scales). From any measure on one of these supra-ordinal scales, a ranking can be inferred on an ordinal scale, but not vice versa (from a ranking on an ordinal scale it is impossible to infer measures on supra-ordinal scales). Thus hierarchies (including the more sophisticated ones such as GRADE) provide evaluations of evidence which are *necessarily less informative* than evaluations provided by QATs.

Moreover, because these sophisticated hierarchies begin with a quality assignment based on one methodological property and then shift the quality assignment by taking other properties into account, the weights that can be assigned to various methodological properties are constrained. With QATs, on the other hand, the weight assigned to any methodological property is completely open, and can be determined based on rational arguments regarding the respective importance of the various properties, without arbitrary constraints imposed by the structure of the scoring system. In short, despite the widespread use of evidence hierarchies and the defense of such use by Howick (2011), and despite the problems that I raise for QATs above, QATs are superior to evidence hierarchies for assessing the great volume of evidence in contemporary medical research.

9.7 Conclusion

An examination of QATs suggests that coarse-grained features of information in medical research, like freedom from systematic error, are themselves amalgams of a complex set of considerations; that is why QATs take into account a plurality of methodological features such as randomization and blinding. The various aspects of a specific empirical situation which can influence an assessment of the quality of information are numerous, often difficult to identify and articulate, and if they can be identified and articulated (as one attempts to do with QATs), they can be evaluated by different scientists to varying degrees and by different quality assessment tools to various degrees. In short, there are a variety of features of information that must be considered when assessing the quality of that information, and there are numerous and potentially contradictory ways to do so. Our best theories of scientific inference provide little guidance on how to weigh the relevant methodological features included in tools like QATs.

A group of medical scientists prominent in the literature on QATs notes that "the quality of controlled trials is of obvious relevance to systematic reviews" but that "the methodology for both the assessment of quality and its incorporation into systematic reviews are a matter of ongoing debate" (Jüni, Altman, and Egger 2001). I have argued that the use of QATs are important to minimize arbitrariness when assessing information in medical research. However, available QATs vary in their constitutions, and when information in medical research is assessed using QATs their inter-rater reliability and inter-tool reliability is low. This, in turn, is a compelling illustration of a more general problem: the underdetermination of evidential significance. Disagreements about the quality of information are, of course, ubiquitous in science. Such disagreement is especially striking, however, when it results from the employment of carefully codified tools designed to quantitatively assess the quality of information. QATs are currently the *best* instruments available to medical scientists to assess the quality of information, yet when applied to what is purported to be the *best* kind of information in medical research (namely, evidence from RCTs), different users of the same QAT, and different QATs applied to the same information, lead to widely discordant assessments of the quality of that information.

Acknowledgements This paper has benefited from discussion with audiences at the University of Utah, University of Toronto, the Canadian Society for the History and Philosophy of Science, and a workshop on Information Quality organized by Phyllis Illari and Luciano Floridi. I am grateful for financial support from the Banting Fellowships Program administered by the Social Sciences and Humanities Research Council of Canada.

References

Balk, E. M., Bonis, P. A., Moskowitz, H., Schmid, C. H., Ioannidis, J. P., Wang, C., & Lau, J. (2002). Correlation of quality measures with estimates of treatment effect in meta-analyses of randomized controlled trials. *JAMA: The Journal of the American Medical Association, 287*(22), 2973–2982.

Bluhm, R. (2005). From hierarchy to network: A richer view of evidence for evidence-based medicine. *Perspectives in Biology and Medicine, 48*(4), 535–547.

Borgerson, K. (2008). *Valuing and evaluating evidence in medicine.* PhD dissertation, University of Toronto.

Cartwright, N. (2007). Are RCTs the gold standard? *Biosocieties, 2*, 11–20.

Cartwright, N. (2012). Presidential address: Will this policy work for you? Predicting effectiveness better: How philosophy helps. *Philosophy of Science, 79*(5), 973–989.

Chalmers, T. C., Smith, H., Blackburn, B., et al. (1981). A method for assessing the quality of a randomized control trial. *Controlled Clinical Trials, 2*, 31–49.

Cho, M. K., & Bero, L. A. (1994). Instruments for assessing the quality of drug studies published in the medical literature. *JAMA: The Journal of the American Medical Association, 272*, 101–104.

Clark, H. D., Wells, G. A., Huët, C., McAlister, F. A., Salmi, L. R., Fergusson, D., & Laupacis, A. (1999). Assessing the quality of randomized trials: Reliability of the Jadad scale. *Controlled Clinical Trials, 20*, 448–452.

Cohen, J. (1960). A coefficient of agreement for nominal scales. *Educational and Psychological Measurement, 20*(1), 37–46.

Egger, M., Smith, G. D., & Phillips, A. N. (1997). Meta-analysis: Principles and procedures. *British Medical Journal, 315*, 1533–1537.

Floridi, L. (2004). Outline of a theory of strongly semantic information. *Minds and Machines, 14*, 197–222.

Hartling, L., Ospina, M., Liang, Y., Dryden, D., Hooten, N., Seida, J., & Klassen, T. (2009). Risk of bias versus quality assessment of randomised controlled trials: Cross sectional study. *British Medical Journal, 339*, b4012.

Hartling, L., Bond, K., Vandermeer, B., Seida, J., Dryen, D. M., & Rowe, B. H. (2011). Applying the risk of bias tool in a systematic review of combination long-acting beta-agonists and inhaled corticosteroids for persistent asthma. *PLoS One, 6*(2), 1–6. e17242.

Hempel, S., Suttorp, M. J., Miles, J. N. V., Wang, Z., Maglione, M., Morton, S., Johnsen, B., Valentine, D., & Shekelle, P. G. (2011). Empirical evidence of associations between trial quality and effect sizes. *Methods Research Report* (AHRQ Publication No. 11-EHC045-EF). Available at: http://effectivehealthcare.ahrq.gov

Herbison, P., Hay-Smith, J., & Gillespie, W. J. (2006). Adjustment of meta-analyses on the basis of quality scores should be abandoned. *Journal of Clinical Epidemiology, 59*, 1249–1256.

Howick, J. (2011). *The philosophy of evidence-based medicine*. Chichester: Wiley-Blackwell.

Jadad, A. R., Moore, R. A., Carroll, D., et al. (1996). Assessing the quality of reports of randomized clinical trials: Is blinding necessary? *Controlled Clinical Trials, 17*, 1–12.

Jüni, P., Witschi, A., Bloch, R., & Egger, M. (1999). The hazards of scoring the quality of clinical trials for meta-analysis. *JAMA: The Journal of the American Medical Association, 282*(11), 1054–1060.

Jüni, P., Altman, D. G., & Egger, M. (2001). Assessing the quality of randomised controlled trials. In M. Egger, G. D. Smith, & D. G. Altman (Eds.), *Systematic reviews in health care: Meta-analysis in context.* London: BMJ Publishing Group.

La Caze, A. (2011). The role of basic science in evidence-based medicine. *Biology and Philosophy, 26*(1), 81–98.

Linde, K., Clausius, N., Ramirez, G., et al. (1997). Are the clinical effects of homoeopathy placebo effects? *Lancet, 350*, 834–843.

Mayo, D. (1996). *Error and the growth of experimental knowledge*. Chicago: University of Chicago Press.

Moher, D., Jadad, A. R., Nichol, G., Penman, M., Tugwell, P., & Walsh, S. (1995). Assessing the quality of randomized controlled trials: An annotated bibliography of scales and checklists. *Controlled Clinical Trials, 16*, 62–73.

Moher, D., Jadad, A. R., & Tugwell, P. (1996). Assessing the quality of randomized controlled trials. Current issues and future directions. *International Journal of Technology Assessment in Health Care, 12*(2), 195–208.

Moher, D., Pham, B., Jones, A., Cook, D. J., Jadad, A. R., Moher, M., Tugwell, P., & Klassen, T. P. (1998). Does quality of reports of randomised trials affect estimates of intervention efficacy reported in meta-analyses? *Lancet, 352*(9128), 609–613.

Olivo, S. A., Macedo, L. G., Gadotti, I. C., Fuentes, J., Stanton, T., & Magee, D. J. (2007). Scales to assess the quality of randomized controlled trials: A systematic review. *Physcial Therapy, 88*(2), 156–175.

Reisch, J. S., Tyson, J. E., & Mize, S. G. (1989). Aid to the evaluation of therapeutic studies. *Pediatrics, 84*, 815–827.

Spitzer, W. O., Lawrence, V., Dales, R., et al. (1990). Links between passive smoking and disease: A best-evidence synthesis. A report of the Working Group on Passive Smoking. *Clinical and Investigative Medicine, 13*, 17–42.

Stegenga, J. (2011). Is meta-analysis the platinum standard of evidence? *Studies in History and Philosophy of Biological and Biomedical Sciences, 42*, 497–507.

Upshur, R. (2005). Looking for rules in a world of exceptions: Reflections on evidence-based practice. *Perspectives in Biology and Medicine, 48*(4), 477–489.

West, S., King, V., Carey, T. S., Lohr, K. N., McKoy, N., Sutton, S. F., & Lux, L. (2002). Systems to rate the strength of scientific evidence. *Evidence Report/Technology Assessment Number* 47 (AHRQ Publication No. 02-E016).

Worrall, J. (2002). *What* evidence in evidence-based medicine? *Philosophy of Science, 69*, S316–S330.

Worrall, J. (2007). Why there's no cause to randomize. *The British Journal for the Philosophy of Science, 58*, 451–488.

Chapter 10
Educating Medical Students to Evaluate the Quality of Health Information on the Web

Pietro Ghezzi, Sundeep Chumber, and Tara Brabazon

Abstract Google and googling pose an array of challenges for information professionals. The Google search engine deskills information literacy, so that many people can find some information. Yet the great challenge is knowing what we do not know. We cannot put words into Google that we do not know. Therefore the instruments for diagnosis are blunt and brutal. The field of e-health has great possibilities, yet the lack of information literacy undermines the expertise of professionals and creates misinformation and confusion. This chapter analyzes the means of assessing the quality of health information and describes an approach to improve the ability of medical students to navigate through the various health information available and to critically evaluate a research publication. Improving Internet literacy is required not only to meet the standards for medical education but also to prepare future doctors to deal with patients exposed to an information overload.

P. Ghezzi (✉) • S. Chumber
Division of Clinical and Laboratory Investigation,
Brighton & Sussex Medical School,
Falmer BN19RY, UK
e-mail: p.ghezzi@bsms.ac.uk

T. Brabazon
School of Teacher Education,
Charles Sturt University, Bathurst,
NSW 2795, Australia

L. Floridi and P. Illari (eds.), *The Philosophy of Information Quality*,
Synthese Library 358, DOI 10.1007/978-3-319-07121-3_10,

10.1 Quality of Health Information on the Internet

10.1.1 The Internet as a Source of Health Information

According to statistics up to 61 % of the general American and European adult population seek medical and health related information online (Andreassen et al. 2007; Fox and Jones 2009). Furthermore, a large proportion of patients have become dependent on health information on the Internet, with roughly 70 % of those who seek health advice online, using the information obtained to make significant health decisions (Berland et al. 2001). Hence, medical information supplied on the Internet has encouraged patients to be more proactive in managing their own health and disease (Lee et al. 2010).

There are several additional benefits of using the Internet as a means of attaining health-related information. Firstly the Internet is a cost-effective and convenient resource, allowing easy and immediate access to multiple sources (Wagner et al. 2004; Hanauer et al. 2003; Lustria 2007). These sources can also be searched anonymously, thereby facilitating patient confidentiality (Hanauer et al. 2003; Lustria 2007; McKay et al. 1998). Furthermore, the Internet is an interactive medium, which enables individuals to communicate with each other and attain information specific to their needs, thus providing an interpersonal dimension (Lustria 2007). Individuals can further use online sources to seek emotional and psychological support, hence improving health outcomes (Jadad and Gagliardi 1998; Weed 1997; Gawande and Bates 2000). It is therefore evident why the Internet has developed as one of the most popular, rapidly expanding health information tools.

Despite this, individuals who seek health information online must do so with some caution. The sheer breadth of information can prove to be overwhelming and confusing for both patients and medical professionals alike (Berland et al. 2001; Wagner et al. 2004; Price and Hersh 1999; Lindberg and Humphreys 1998). Moreover, a large portion of Internet health sites remain unregulated, thus increasing the risk of poor quality, unreliable, inaccurate, and unsuitable information (Berland et al. 2001; Harland and Bath 2007; Maloney et al. 2005). This is a concern given the increasing reliance both physicians and patients have on the Internet, as misleading information could potentially have serious consequences (Price and Hersh 1999; McKay et al. 2001; Winker et al. 2000; Lopez-Jornet and Camacho-Alonso 2009; Sajid et al. 2008; Weiss and Moore 2003).

10.1.2 Methods of Assessing the Quality of Online Health Information

There are various tools available to help evaluate the quality of online health information. These can be categorised into five broad groups: codes of conduct (e.g., American Medical Association), quality labels [e.g., Health On the Net Foundation

Table 10.1 HON code principles

Authoritativeness	Indicate the qualifications of the authors
Complementarity	Information should support, not replace, the doctor-patient relationship
Privacy	Respect the privacy and confidentiality of personal data submitted to the site by the visitor
Attribution	Cite the source(s) of published information, date medical and health pages
Justifiability	Site must back up claims relating to benefits and performance
Transparency	Accessible presentation, accurate email contact
Financial disclosure	Identify funding sources
Advertising policy	Clearly distinguish advertising from editorial content

Source: Health on the net foundation (2013)

Table 10.2 JAMA benchmarks

Authorship	Authors and contributors, their affiliations, and relevant credentials should be provided
Attribution	References and sources for all content should be listed clearly, and all relevant copyright information noted
Disclosure	Web site "ownership" should be prominently and fully disclosed, as should any sponsorship, advertising, underwriting, commercial funding arrangements or support, or potential conflicts of interest
Currency	Dates that content was posted and updated should be indicated

Source: Silberg et al. (1997)

(HON) code], user guidance systems (e.g., DISCERN), filtering tools [e.g., OMNI (intute.ac.uk)], and third-party quality and accreditation labels [e.g., Utilization Review Accreditation Commission (URAC)] (Wilson 2002).

The HON code is one of the most well-known and widely used quality labels (Wilson 2002; Health on the net foundation 2009, 2013). Currently, over 5,000 health and medical sites have received HON code certification (Health on the net foundation 2009). The HON foundation is a non-profit, non-governmental organisation, created in 1996, with a purpose of highlighting reliable, comprehensible, relevant, and trustworthy sources of online health and medical information (Health on the net foundation 2009). Web sites accredited with the HON seal must adhere to eight ethical principles, set out by the HON code (Table 10.1) (Health on the net foundation 2009, 2013). HON code accreditation has been found to be a reliable indicator of web site quality, correlating with higher scores obtained from other quality assessment instruments (Bruce-Brand et al. 2013). Similarly, The Journal of the American Medical Association (JAMA) set up its own benchmarks, consisting of four core standards, which can help users assess whether a web site is trustworthy or credible (Table 10.2) (Silberg et al. 1997). Some of these core standards are also referred to in the HON code (Health on the net foundation. The HON code of conduct for medical and health web sites (HONcode) [updated Feb 4th 2013). The author's affirm that health and medical websites should be considered suspect if they fail to meet at least three of the JAMA benchmarks (Silberg et al. 1997).

Examples of other quality evaluation tools include the Health-Related Website Evaluation Form (HRWEF) and Quality Component Scoring System (QCSS) (Martins and Morse 2005; Peterlin et al. 2008). Patients and, primarily, health professionals can use the HRWEF to assess the appropriateness of web sites. The form highlights multiple criteria: content, accuracy, author, currency, audience, navigation, external links and structure. Each criterion is listed with several statements, which the assessor can disagree with (1), agree with (2), or find non-applicable (0). Web sites are deemed to have poor quality, i.e. questionable validity and reliability, if they receive an overall score which is less than 75 % of that of the possible total. The QCSS is a comparable quality assessment, which offers scores on: ownership, purpose, authorship, author qualification, attribution, interactivity, and currency (Martins and Morse 2005; Peterlin et al. 2008). These quality assessment tools draw parallels with the HON code and JAMA benchmark, yet none of them provide complete coverage of all the criteria. Thus, a combination of multiple tools is required to provide a comprehensive evaluation of the quality of information available on the Internet (Harland and Bath 2007; Bernstam et al. 2005; Hanif et al. 2009).

The DISCERN instrument is another popular assessment tool, designed to enable patients and information providers to critically appraise the quality of written information related to treatment choices, thereby facilitating the production of novel, high quality and validated consumer health information (Charnock et al. 1999). DISCERN consists of 15 key questions, each representing an individual quality criterion, plus an overall quality rating (Table 10.3). Each question is given a score from 1 to 5 depending on whether the publication does (5), does not (1), or only partially adheres to the criterion in question (2–4) (Charnock et al. 1999). Although the DISCERN instrument has been found to be a reliable and valid instrument for judging the quality of written consumer health information, it does require a degree of subjectivity and may not be appropriate for all publications. It may also be best suited for experienced users, health care professionals, and information providers, thus limiting its use by patients (Charnock et al. 1999; Rees et al. 2002).

10.1.3 Quality of Health Information on the Internet

Numerous studies have been conducted to assess the quality of health information online, in relation to specific conditions. A large proportion of these studies have focused on cancer, where generally the quality of online health information has been shown to be poor (Lopez-Jornet and Camacho-Alonso 2009; Ni Riordain and McCreary 2009; Meric et al. 2002; Bichakjian et al. 2002; Friedman et al. 2004; Nasser et al. 2012). For instance, one study assessed the quality of online information related to oral cancer, using the two search engines Google and Yahoo. The websites were analysed using the validated DISCERN rating instrument and the

Table 10.3 DISCERN rating instrument

Section 1: Is the publication reliable?	**Score[a]**
1. Are the aims clear?	1–5
2. Does it achieve its aims?	1–5
3. Is it relevant?	1–5
4. Is it clear what sources of information were used to compile the publication (other than the author or producer)?	1–5
5. Is it clear when the information used or reported in the publication was produced?	1–5
6. Is it balanced and unbiased?	1–5
7. Does it provide details of additional sources of support and information?	1–5
8. Does it refer to areas of uncertainty?	1–5
Section 2: How good is the quality of information on treatment choices?	**Score[a]**
9. Does it describe how each treatment works?	1–5
10. Does it describe the benefits of each treatment?	1–5
11. Does it describe the risks of each treatment?	1–5
12. Does it describe what would happen if no treatment is used?	1–5
13. Does it describe how the treatment choices affect overall quality of life?	1–5
14. Is it clear that there may be more than one possible treatment choice?	1–5
15. Does it provide support for shared decision-making?	1–5
Section 3. Overall rating of the publication	**Score[a]**
16. Based on the answers to all of the above questions, rate the overall quality of the publication as a source of information about treatment choices	1–5

Sections 1 and 2		**Section 3**	
Score	**Description**	**Score**	**Description**
1	**Low:** *serious or extensive shortcomings*	**1**	*No*
2		**2**	
3	**Moderate:** *potentially important but not serious shortcomings*	**3**	*Partially*
4		**4**	
5	**High**: *minimal shortcomings*	**5**	*Yes*

From Charnock et al. (1999)
[a]Scoring

JAMA benchmarks. Of the 29 Google web sites visited, just two (6.9 %) met the four criteria set out by the JAMA benchmarks, in comparison to only one (4.5 %) of the 22 Yahoo websites. Furthermore, none of the sites obtained a maximum score for the DISCERN instrument, whilst only eight Google sites (27.6 %) and four Yahoo sites (18.2 %) were HON code certified. In addition, serious deficiencies were noted in the vast majority of Google (72.5 %) and Yahoo (68.2 %) sites. The researchers therefore reported a lack in the quality of online health information related to oral cancer (Lopez-Jornet and Camacho-Alonso 2009). Similarly, a study of online information regarding head and neck cancer, also utilised the DISCERN instrument and the JAMA benchmarks. A total of 33 sites were analysed from Google, of which 45 % obtained all four JAMA benchmarks, whilst 18 % achieved only one benchmark. No websites received the maximum overall DISCERN instrument score (Ni Riordain and McCreary 2009).

Another study investigated the possible link between website popularity and quality of information, in relation to breast cancer. The first 200 sites from Google were divided into 'more popular' and 'least popular', based on link popularity (the number of links to a site from other sites). The accessible 184 sites were subsequently assessed using the JAMA benchmarks. Results showed that only 16 (9 %) of sites had all 4 JAMA benchmarks, whilst only 27 (15 %) sites displayed the HON code seal. Interestingly none of the sites, displaying the HON code seal, complied with all eight HON code criteria or JAMA benchmarks, although commercial sites were more likely to display the HON code seal than professional sites. In addition, 12 (7 %) sites contained inaccurate medical statements, the majority of which (Gawande and Bates 2000) were commercial sites. The researchers concluded that there was no correlation between site popularity and quality and accuracy of information, although there was a relationship with type of content (Meric et al. 2002). Therefore, popularity and reliability are distinct ideologies and – in the case of health – can exhibit some concerning consequences.

In contrast, a 35-point checklist rating system was developed by researchers to assess the accuracy and extensiveness of online information related to melanoma. The rating system consisted of several specified factors, derived from a 'gold standard' (The University of Michigan Multidisciplinary Melanoma Clinic (MDMC) multispecialty consensus and the National Comprehensive Cancer Network melanoma guidelines). The review evaluated a total of 74 websites, retrieved from eight search engines. The majority of websites failed to meet each rating system factor, whilst ten websites were noted to have a total of 13 inaccuracies (Bichakjian et al. 2002).

As well as assessing the accuracy and quality of online information, numerous studies have measured the readability of health information using specific, validated readability formulae (Friedman et al. 2004; Nasser et al. 2012). In the majority of cases, online health information was found to be written at a school grade level 12, which is much higher than the recommended level of 6–8 (Friedman et al. 2004; Nasser et al. 2012; van der Marel et al. 2009; Croft and Peterson 2002). For example, researchers assessing the readability level of 55 web sites on breast (n = 20), colorectal (n = 18), and prostate (n = 17) cancers found that 63.6 % were written at a school grade level of 13 or greater, according to the SMOG formula (Friedman et al. 2004). Likewise, in a study on online warfarin information, only 27.3 % of the websites had a reading grade level thought to be representative of an adult patient population with low literacy rates (Nasser et al. 2012). This is a concern given that individuals with lower literacy rates generally have poorer health and, as the Internet is becoming an increasing popular source of health information, are likely to use online information (Friedman et al. 2004). Furthermore, approximately 16 % of 60–65 year olds and 58 % of individuals aged 85 and above are thought to have poor literacy skills (Estrada et al. 2000). This age group is also associated with declining health, co-morbidities, multiple medications and cognitive deficits (Kunst and Khan 2002). Online health information should therefore be written at a level that is suitable and comprehensible for all members of the general adult patient population, including those with poor literacy skills (Friedman et al. 2004; Nasser et al. 2012;

Estrada et al. 2000; Kunst and Khan 2002). The relationship between literacy, numeracy and information management skills should not be assumed.

Nevertheless, in some instances the quality of online health information has been found to be satisfactory. For instance, despite poor readability, Nasser et al. rated the overall quality and suitability of Internet derived warfarin information as adequate. The quality of 11 web sites was assessed using the HRWEF and the QCSS, whilst suitability was based on the Suitability Assessment of Materials (SAM) score (Nasser et al. 2012). In addition, an alternative study used the WebMedQual scale to analyse the quality of for-profit and non-profit sites related to schizophrenia. The majority of sites were found to offer comprehensive and useful information, with for-profit sites scoring more highly than non-profit sites (Guada and Venable 2011).

The quality of online health information is therefore highly varied. The general consensus is that there is a lack of quality health information available on the internet, as supported by multiple studies on a variety of health related topics, as well as cancer (Lopez-Jornet and Camacho-Alonso 2009; Sajid et al. 2008; Weiss and Moore 2003; Ni Riordain and McCreary 2009; Meric et al. 2002; Bichakjian et al. 2002; Friedman et al. 2004; Nasser et al. 2012; van der Marel et al. 2009; Croft and Peterson 2002; Estrada et al. 2000; Kunst and Khan 2002). However, it is evident that there is no distinct means of assessing the quality of information, with research differing in methods, quality criteria, definitions, study population, and selected topic (Eysenbach et al. 2002). A major limitation of multiple studies is that they use subjective methods, such as evaluating accuracy, to assess the quality of information (Meric et al. 2002; Bichakjian et al. 2002; Nasser et al. 2012; Eysenbach et al. 2002). Furthermore, some tools such as the SAM instrument are only suitable for specific patient populations, i.e. the general adult population with limited literacy skills, and do not take into account other patient groups i.e. older patients (Nasser et al. 2012). Assessing a greater number of web sites could also strengthen the reliability of results (Lopez-Jornet and Camacho-Alonso 2009; Meric et al. 2002; Nasser et al. 2012). Moreover, different tools may provide conflicting scores for the same site, as reported in Nasser et al., when using the HRWEF and the QCSS evaluation tools (Nasser et al. 2012). It is therefore difficult to ascertain the usability, validity and reliability of such quality assessment tools (Deshpande and Jadad 2009). Hence, a validated, objective, and universally employable tool is required to assess the quality of all online health information.

10.2 Internet Literacy and Medical Education

The availability of large amounts of unverified, non-peer-reviewed information freely available on the internet is a well known and debated issue not only in primary and secondary education but also in higher education, with the student copying and pasting from the internet rather than going to the library (online or not) being a challenge for many teachers. There is a confusion between having the ability to use hardware, software and applications and holding the knowledge, vocabulary and information literacy to deploy the technology with care and rigour. However, this is less of a problem in life sciences. In this field, most students are knowledgeable of what are the databases for finding scientific information. In general, students are well aware of where to look for health information on the internet and even the use of non-specialized search engines, including Google and not only Pubmed, may be useful to medical students and young doctors in making a diagnosis (Falagas et al. 2009; Nalliah et al. 2010). It should be noted that Google is often the first choice when searching for health information also by doctors, not just students (Hider et al. 2009; Sim et al. 2008).

The ability of students to navigate through the various medical information available and to critically evaluate a research publication is also required to meet the standards for medical education in the UK, set by the General Medical Council (GMC) and published in the guidance document "Tomorrow's doctor" (GMC 2009). The 2009 edition lists among the "outcomes for graduates" the following: "Critically appraise the results of relevant diagnostic, prognostic and treatment trials and other qualitative and quantitative studies as reported in the medical and scientific literature" and "Access information sources and use the information in relation to patient care, health promotion, giving advice and information to patients, and research and education" (GMC 2009).

Most of the attention on IT in medical education is based on providing the students with online systems that allow them to take their textbooks, lessons at home and everywhere using the Internet on computers or mobile phones. The existing teaching around Internet literacy is focused on explaining how to look up scientific evidence using proper databases. These include Pubmed, from the US National Library of Medicine (http://www.ncbi.nlm.nih.gov/pubmed); the US Food and Drug Administration clinical trial database (http://www.clinicaltrials.gov/); Sciencedirect (http://www.sciencedirect.com); the Cochrane collaboration (http://www.cochrane.org/); and the UK National Health Service (http://www.evidence.nhs.uk/) and National Institute of Clinical Excellence (http://www.nice.org.uk). The problem remains on how to educate students to critically evaluate information obtained using popular search engines. Once more, the challenge is that students do not know what they do not know.

On the other hand, future doctors may encounter other problems arising from Internet literacy in other ways. In fact, it is more and more common for patients, particularly those with chronic diseases, to look for information on the Internet. Most of them will not be aware of the databases used by scientists, like Pubmed, but will rely on Google and, if we are lucky, on the websites of patients' advocacy associations (such as the MS Society in the UK or the National MS Society in the USA, or the Stroke Association). One reason for this is that a search in Pubmed will normally retrieve a large number of publications, too specialist, with no special ranking order other than the date of publication, and on scientific journals that, in most cases, are not "Open Access" and therefore require expensive subscriptions to access the full text.

As a research immunologist, one of the authors experienced phone calls from patients with chronic and often incurable or fatal diseases, or their relatives, asking about a new therapy they had found on the internet or read about in a popular magazine, often triumphantly described as a real breakthrough and a magic bullet. In a few cases, the caller was a GP to whom the patient had brought the information. It is often regarded as not acceptable anymore to just respond in an authoritarian way to the patients that they should just take what is offered by their respective national health system (also because often patients will know that what is on the reimbursement list of the national health systems varies from one country to another, the USA often being regarded as the gold standard).

Internet literacy and information literacy are thus important for future doctors, as it would enable them to interpret the information available on Google and other popular search engines that are accessed by patients. While Internet can be useful to patients in self-management programs for chronic diseases (Lorig et al. 2008, 2010), it can also lead to hazardous behaviour by patients in terms of self-medication or change or suspend the therapy recommended by the GP (Siliquini et al. 2011).

The problem of identifying scientifically valid information in the ocean of information available on the Internet is a challenge at every level, even for postgraduate students, and even limiting to scientific journals and recognized databases. In fact, Pubmed/MEDLINE include over 20 million citations (National Center for Biotechnology Information USNLoM 2005) and there are about 100,000 worldwide, with a core of 2,000 journals publishing most of the works (Ioannidis 2006), making it very difficult to identify the reliability of the published information and requiring the development of sophisticated methods of analysis (Evans and Foster 2011; Renear and Palmer 2009).

Therefore, it is important to teach students what are the criteria by which to identify important and reliable information. While this can be rather easily done in the scientific literature, based on criteria such as the impact factor of the journal or the list of previous publication by the same authors, one can imagine the problems encountered when the information are searched among ALL the sources available on the internet.

We will describe here a course for teaching medical students to critically appraise the medical information available on the Internet, which is based on the preparation of an annotated bibliography described in "University of Google" (Brabazon 2007). It is both significant and important to evaluate the applicability of this study beyond the humanities and the social sciences, and through to the health professionals.

10.3 Designing a Course of Medical Internet Literacy

The course is designed as a Student Selected Component (SSC) for 2nd-year medical students. These are optional modules in medical schools in the UK. The General Medical Council requires that SSCs represent at least 10 % of course time (GMC 2009). A typical SSC is made of seven to eight 1-hour weekly sessions and is taught to groups of 10–12 students.

Each student is asked to choose a health-related topic. A list of topics was offered but students could propose a topic of their own interest. Because the SSC was part of a module that included immunology, the topics were broadly related to the field of infective, inflammatory and autoimmune diseases. Students would then compile an annotated bibliography from a variety of sources (the instructions prescribed the exact number of citations for each source) and would be evaluated according to the quality of the annotation or critique.

10.3.1 Topics

Topics are chosen from those that have received media coverage and for which there was enough material available on the Internet or in printed form. These included either new drugs approved or in development (e.g. the antiobesity drug tetrahydrolipstatin, the multiple sclerosis drugs fingolimod, dimethyl fumarate), new treatments still to be approved (e.g. cannabis, nanoparticles or stem cell in the treatment of multiple sclerosis) or unorthodox or alternative treatments (e.g. chronic cerebrospinal venous insufficiency for MS, vitamin C or vitamin E for viral infections; arginine/citrulline, resveratrol or red wine for cardiovascular disease prevention; plant products such as allicin, curcumin, cranberry juice, fermented papaya preparation for a variety of other indications).

10.3.2 Structure of the Course

All sessions are held in a room with several PCs, one for each student. In the first session the students are told the learning outcomes, the structure of the course and the modality of assessment.

10.3.2.1 Learning Outcome

Develop the ability to critically evaluate the health information available (to you as well to your future patients) on the Internet

They are first suggested to get a general idea of the biomedical basis for the chosen topic (for instance, a student who had chosen vitamin C, an antioxidant, and viral infections would be asked to find 2–3 review articles on oxidative stress and inflammation in infection).

Once the student has an idea of the scientific basis of a topic, they are asked to find ten sources they will have to comment on. These have to be chosen as in Table 10.4.

The type of sources is specified as follows:

Table 10.4 Composition of the bibliography. Find ten sources of the following type

No. of sources required	Source type
1	Book (monograph)
2–3	Journal, print-based refereed articles
1	Journal, web-based refereed articles
1–2	Websites or blogs
1	Social networking site
1–2	Articles in the press
1–2	Video, podcast, audio etc.

- Books. When possible should be a monograph, either published by a scholarly publisher (e.g. a University Press) or by a general publisher (e.g. Penguin Books).
- Journal, print-based refereed articles. (Refereeing is the process whereby a journal sends out an article to scholars in the field to assess if it is of international quality and rigour). Students know that articles are refereed because the journal lists an editorial board, and the "instructions to authors" explain the review process. Examples of refereed journals include the *Journal of Immunology, The Lancet, Science, Nature* etc. Refereed journals will be normally found searching biomedical databases such as Pubmed (http://www.ncbi.nlm.nih.gov/pubmed), Science Direct (http://www.sciencedirect.com/), Scopus (http://www.scopus.com/search/form.url), Google Scholar (scholar.google.com). Print-based means that the journal exists in a printed form. However, almost all journals are also available online and this is how the students will normally access them. Refereed articles are also accessible through full-text, refereed databases of the Library. The search databases above will often give you automatically the external link to the full-text (available if our library has a paid subscription or if the journal is "Open Access").
- Journal-web-based refereed articles. Students must ensure that the articles they use are in a refereed online journal (e.g. *PLoS Medicine, Frontiers in Inflammation, BMC Microbiology* etc.). For a list of biomedical journals: http://www.biomedcentral.com/browse/journals/ OR http://www.plos.org OR www.frontiersin.org. The student will have to check the home page of the journal to ensure that it does not have a print version.
- Articles in the press or in the news. For instance BBC News, The Times, The Daily Mail, The Guardian, Hallo, The Sun etc. We realize that in most cases you will access these magazines online.

After each source is listed – students must then write 100 (min)–300 (max) words about the source, for each source, in which the student should: (1) summarize the message (or the conclusions of the source) (2) identify the authors and their expertise/qualifications and therefore reliability; (3) identify any commercial/financial interest of the source; (4) comment on reputation, relevance, reliability of the source.

The assignment is then marked based on quality and calibre of annotation. The lower level of literacy will normally be the ability to summarize the content of the sources, higher levels will require doing some research about the authors, the publisher, and the relationship to the main message, such as commercial conflict of interest. The students also gain expertise in multimodality: how the form of information (podcast, vodcast, journalism or a scholarly monograph) impact on the tenor, level and literacy of the content.

The students will work weekly for 1 h under the teacher's supervision. The teacher will make sure that the students encounter no problems in locating all the sources, that they keep track of their research using a notebook or online note-taking applications, and that they are not behind in the project. Sometimes students will be asked to explain to their colleagues what stage of the work they are and what are the difficulties encountered. Students will continue the work at home or in the IT suite.

10.3.3 Results. Detecting Snake Oil, Press Releases and Scientific Misconduct

While a few students do not go much further than just summarising the source, most of them do very good research on the source. Of note, this course is not to check for the validity of the information (that is, its scientific basis). If a source claims that a new drug cures arthritis, the student is not asked to say whether the claim is true, but only if the claim makes reference to scientific evidence or clinical trials, rather than just showing the picture of someone in a white coat.

All students easily recognized that many websites ranking top ten in a web search on topics such as "vitamin C AND influenza" or "resveratrol and cardiovascular disease" contain pseudo-scientific information and are basically commercial sites selling vitamins or herbal supplements.

However, the student will learn to recognize many typical occurrences that they were not aware of, two of them, particularly, becoming more and more frequent: press releases and scientific misconduct.

10.3.3.1 Press Releases

A few students notice that most articles reporting new discoveries or new drugs in the media or the news are not original articles say, by a journalist of the BBC or the Guardian but just a press release. For instance, one article on the beneficial effects of a component of red wine, resveratrol ("New drug being developed using compound found in red wine 'could help humans live until they are 150'", The Daily Mail 10 March 2013) (Crossley 2013) appears to be signed by a journalist and we may think that it is the scientific journalist in that newspaper reporting in quotes what seems to be an interview with a scientist: "Genetics professor David Sinclair, based at Harvard University, said: 'Ultimately, these drugs would treat one disease, but unlike drugs of today, they would prevent 20 others. In effect, they would slow ageing'" (Crossley 2013). However, a simple Google search shows the same quote in other websites and magazines and was originally a press release (Anti-ageing drug breakthrough 2013). The Daily Mail and other newsmagazines were clearly just echoing a press release by the University where the scientist worked or by the scientific journal where the work was published (all scientific journals now release to the press some key articles a few days in advance of the publication). Interestingly, the "disclosure of financial interest", required in the academic world, is only found in the original press release on the University website ("Professor Sinclair formed a start-up company Sirtris to develop the anti-ageing technology. This was subsequently sold to GlaxoSmithKline (GSK). Professor Sinclair is now a scientific advisor to GSK. Several other authors on the paper work for GSK or an affiliated company.") (Anti-ageing drug breakthrough 2013) or, with a different wording, in the scientific article that prompted the press release (Hubbard et al. 2013), but this information is always lost in the news. The first rule that scientists need to follow to get a press release published is to make the information attractive and easy to read, there is no room for conflict of interests.

The case of press releases is something that is very common, not just in websites, but also in established newspapers and magazines that, being often understaffed, are more than willing to act as megaphones. In fact, most universities now make generous use of press releases to make "impact". Likewise, biotech companies sponsoring new drugs use press releases to increase their stock's value. This does not diminish the importance or even the reliability of the information, but it should not be overlooked either.

10.3.3.2 Scientific Misconduct

Searching the topic on the health effects of resveratrol, a student has discovered a case of data fabrication that, if known to the specialists in the field, would not have been identified. Analysing a paper published in an academic journal (Mukherjee et al. 2009), second-year medical student Nikoo Aziminia wrote: "Whether the authors of this paper declared no conflict of interests is no longer relevant; this is because this paper has been since retracted. The University of Connecticut notified the journal about the possibility of data manipulation and fabrication and possible scientific misconduct by the fifth author, Professor Dipak Das. Professor Das has since been found guilty of 145 counts of data falsification, and several of his papers have been consequentially retracted, including the title above. Although the majority of the lay public seeking information on resveratrol and its cardioprotective effects might not seek information from scientific journals, those who would read this prior to retraction would be potentially misguided by Professor Das' credentials; he is a professor of surgery at University of Connecticut Health Centre, an institution known for the quality of the researchers it recruits. This might serve as reassurance for the reader, who might measure the reliability of a source by the rank and credentials of its author, hence rendering them likely to place their faith in the findings of the paper. Whether these are true or false, Professor Das' scientific misconduct and the ensuing scandal would remove all faith in this."

10.4 Conclusions

First year medical students are taught early in their curriculum where and how to look at health information using online databases. At BSMS this is done from the first few months in the academic tutorials and as part of the teaching of academic skills. Luckily enough, we do not seem to risk that when we will go to see our doctor. He or she will type our symptoms and some keywords in our medical history in the Google search bar to make a diagnosis and prescribe a therapy.

On the other hand, students will have to face patients that will have collected large amounts of unreliable and unverified information. There is confusion between the ability to use a search engine and the capacity to interpret the returned results. To build a good relationship they must be prepared to discuss them with the patient rather than just dismissing all of them en bloc. Teaching Internet literacy to medical students will also help in educating future researchers in this field. This model can be further developed to carry out research projects on health information quality.

Acknowledgements We thank Nikoo Aziminia for giving her permission to quote part of her essay.

References

Andreassen, H. K., Bujnowska-Fedak, M. M., Chronaki, C. E., Dumitru, R. C., Pudule, I., Santana, S., et al. (2007). European citizens' use of E-health services: A study of seven countries. *BMC Public Health, 7*, 53. PubMed PMID: 17425798. Pubmed Central PMCID: 1855923.

Anti-ageing drug breakthrough (2013, August 24). Available from: http://newsroom.unsw.edu.au/news/health/anti-ageing-drug-breakthrough

Berland, G. K., Elliott, M. N., Morales, L. S., Algazy, J. I., Kravitz, R. L., Broder, M. S., et al. (2001). Health information on the Internet: Accessibility, quality, and readability in English and Spanish. *JAMA: The Journal of the American Medical Association, 285*(20), 2612–2621. PubMed PMID: 11368735.

Bernstam, E. V., Shelton, D. M., Walji, M., & Meric-Bernstam, F. (2005). Instruments to assess the quality of health information on the World Wide Web: What can our patients actually use? *International Journal of Medical Informatics, 74*(1), 13–19. PubMed PMID: 15626632.

Bichakjian, C. K., Schwartz, J. L., Wang, T. S., Hall, J. M., Johnson, T. M., & Biermann, J. S. (2002). Melanoma information on the Internet: Often incomplete – A public health opportunity? *Journal of Clinical Oncology, 20*(1), 134–141. PubMed PMID: 11773162.

Brabazon, T. (2007). *The University of Google: Education in the (post) information age*. Aldershot: Ashgate.

Bruce-Brand, R. A., Baker, J. F., Byrne, D. P., Hogan, N. A., & McCarthy, T. (2013). Assessment of the quality and content of information on anterior cruciate ligament reconstruction on the internet. *Arthroscopy, 29*(6), 1095–1100. PubMed PMID: 23582738.

Charnock, D., Shepperd, S., Needham, G., & Gann, R. (1999). DISCERN: An instrument for judging the quality of written consumer health information on treatment choices. *Journal of Epidemiology and Community Health, 53*(2), 105–111. PubMed PMID: 10396471. Pubmed Central PMCID: 1756830.

Croft, D. R., & Peterson, M. W. (2002). An evaluation of the quality and contents of asthma education on the World Wide Web. *Chest, 121*(4), 1301–1307. PubMed PMID: 11948066.

Crossley, L. (2013). *New drug being developed using compound found in red wine 'could help humans live until they are 150'* [cited 2013, August 24]. Available from: http://www.dailymail.co.uk/health/article-2291254/New-drug-developed-using-compound-red-wine-help-humans-live-150.html

Deshpande, A., & Jadad, A. R. (2009). Trying to measure the quality of health information on the internet: Is it time to move on? *The Journal of Rheumatology, 36*(1), 1–3. PubMed PMID: 19208527.

Estrada, C. A., Hryniewicz, M. M., Higgs, V. B., Collins, C., & Byrd, J. C. (2000). Anticoagulant patient information material is written at high readability levels. *Stroke, 31*(12), 2966–2970. PubMed PMID: 11108757.

Evans, J. A., & Foster, J. G. (2011). Metaknowledge. *Science, 331*(6018), 721–725. PubMed PMID: 21311014. Epub 2011/02/12. eng.

Eysenbach, G., Powell, J., Kuss, O., & Sa, E. R. (2002). Empirical studies assessing the quality of health information for consumers on the world wide web: A systematic review. *JAMA: The Journal of the American Medical Association, 287*(20), 2691–2700. PubMed PMID: 12020305.

Falagas, M. E., Ntziora, F., Makris, G. C., Malietzis, G. A., & Rafailidis, P. I. (2009). Do PubMed and Google searches help medical students and young doctors reach the correct diagnosis? A pilot study. *European Journal of Internal Medicine, 20*(8), 788–790. PubMed PMID: 19892310. Epub 2009/11/07. eng.

Fox, S., & Jones, S. (2009). *The social life of health information* (pp. 1–72). Washington, DC: Pew Internet & American Life Project.

Friedman, D. B., Hoffman-Goetz, L., & Arocha, J. F. (2004). Readability of cancer information on the internet. *Journal of Cancer Education, 19*(2), 117–122. PubMed PMID: 15456669.

Gawande, A., & Bates, D. (2000). The use of information technology in improving medical performance. Part III. Patient support tools. *Med Gen Med., 2*, E12.

GMC. (2009). *Tomorrow's doctors: Outcomes and standards for undergraduate medical education*. London: The General Medical Council.

Guada, J., & Venable, V. (2011). A comprehensive analysis of the quality of online health-related information regarding schizophrenia. *Health & Social Work, 36*(1), 45–53. PubMed PMID: 21446608.

Hanauer, D. A., Fortin, J., Dibble, E., & Col, N. F. (2003). Use of the internet for seeking health care information among young adults. *AMIA Annual Symposium Proceedings, 2003*, 857.

Hanif, F., Read, J. C., Goodacre, J. A., Chaudhry, A., & Gibbs, P. (2009). The role of quality tools in assessing reliability of the internet for health information. *Informatics for Health & Social Care, 34*(4), 231–243. PubMed PMID: 19919300.

Harland, J., & Bath, P. (2007). Assessing the quality of websites providing information on multiple sclerosis: Evaluating tools and comparing sites. *Health Informatics Journal, 13*(3), 207–221. PubMed PMID: 17711882.

Health on the net foundation. *A decade devoted to improving online health information quality* [updated June 29th 2009; cited 2013 February 10th]. Available from: http://www.hon.ch/Global/event_art_pulsations.html

Health on the net foundation. *The HON code of conduct for medical and health web sites (HONcode)* [updated Feb 4th 2013; cited 2013 February 10th]. Available from: http://www.hon.ch/HONcode/Conduct.html

Hider, P. N., Griffin, G., Walker, M., & Coughlan, E. (2009). The information-seeking behavior of clinical staff in a large health care organization. *Journal of the Medical Library Association, 97*(1), 47–50. PubMed PMID: 19159006. Epub 2009/01/23. eng.

Hubbard, B. P., Gomes, A. P., Dai, H., Li, J., Case, A. W., Considine, T., et al. (2013). Evidence for a common mechanism of SIRT1 regulation by allosteric activators. *Science, 339*(6124), 1216–1219. PubMed PMID: 23471411.

Ioannidis, J. P. (2006). Concentration of the most-cited papers in the scientific literature: Analysis of journal ecosystems. *PLoS One, 1*, e5. PubMed PMID: 17183679. Epub 2006/12/22. eng.

Jadad, A. R., & Gagliardi, A. (1998). Rating health information on the Internet: Navigating to knowledge or to Babel? *JAMA: The Journal of the American Medical Association, 279*(8), 611–614. PubMed PMID: 9486757.

Kunst, H., & Khan, K. S. (2002). Quality of web-based medical information on stable COPD: Comparison of non-commercial and commercial websites. *Health Information and Libraries Journal, 19*(1), 42–48. PubMed PMID: 12075849.

Lee, C., Gray, S., & Lewis, N. (2010). Internet use leads cancer patients to be active health care consumers. *Patient Education and Counseling, 81*(Suppl), S63–S69.

Lindberg, D. A., & Humphreys, B. L. (1998). Medicine and health on the Internet: The good, the bad, and the ugly. *JAMA: The Journal of the American Medical Association, 280*(15), 1303–1304. PubMed PMID: 9794299.

Lopez-Jornet, P., & Camacho-Alonso, F. (2009). The quality of internet sites providing information relating to oral cancer. *Oral Oncology, 45*(9), e95–e98. PubMed PMID: 19457707.

Lorig, K. R., Ritter, P. L., Laurent, D. D., & Plant, K. (2008). The internet-based arthritis self-management program: A one-year randomized trial for patients with arthritis or fibromyalgia. *Arthritis and Rheumatism, 59*(7), 1009–1017. PubMed PMID: 18576310. Epub 2008/06/26. eng.

Lorig, K., Ritter, P. L., Laurent, D. D., Plant, K., Green, M., Jernigan, V. B., et al. (2010). Online diabetes self-management program: A randomized study. *Diabetes Care, 33*(6), 1275–1281. PubMed PMID: 20299481. Epub 2010/03/20. eng.

Lustria, M. L. A. (2007). Can interactivity make a difference? Effects of interactivity on the comprehension of and attitudes toward online health content. *Journal of the American Society for Information Science and Technology, 58*(6), 766–776.

Maloney, S., Ilic, D., & Green, S. (2005). Accessibility, nature and quality of health information on the Internet: A survey on osteoarthritis. *Rheumatology (Oxford), 44*(3), 382–385. PubMed PMID: 15572390.

Martins, E. N., & Morse, L. S. (2005). Evaluation of internet websites about retinopathy of prematurity patient education. *The British Journal of Ophthalmology, 89*(5), 565–568. PubMed PMID: 15834086. Pubmed Central PMCID: 1772623.

McKay, H., Feil, E., Glasgow, R., & Brown, J. (1998). Feasibility and use of an internet support service for diabetes. *The Diabetes Educator, 24*, 174–179.

McKay, H. G., King, D., Eakin, E. G., Seeley, J. R., & Glasgow, R. E. (2001). The diabetes network internet-based physical activity intervention: A randomized pilot study. *Diabetes Care, 24*(8), 1328–1334. PubMed PMID: 11473065.

Meric, F., Bernstam, E. V., Mirza, N. Q., Hunt, K. K., Ames, F. C., Ross, M. I., et al. (2002). Breast cancer on the world wide web: Cross sectional survey of quality of information and popularity of websites. *BMJ, 324*(7337), 577–581. PubMed PMID: 11884322. Pubmed Central PMCID: 78995.

Mukherjee, S., Lekli, I., Gurusamy, N., Bertelli, A. A., & Das, D. K. (2009). Expression of the longevity proteins by both red and white wines and their cardioprotective components, resveratrol, tyrosol, and hydroxytyrosol. *Free Radical Biology & Medicine, 46*(5), 573–578. PubMed PMID: 19071213.

Nalliah, S., Chan, S. L., Ong, C. L., Suthan, T. H., Tan, K. C., She, V. N., et al. (2010). Effectiveness of the use of internet search by third year medical students to establish a clinical diagnosis. *Singapore Medical Journal, 51*(4), 332–338. PubMed PMID: 20505913. Epub 2010/05/28. eng.

Nasser, S., Mullan, J., & Bajorek, B. (2012). Assessing the quality, suitability and readability of internet-based health information about warfarin for patients. *Australasian Medical Journal, 5*(3), 194–203. PubMed PMID: 22952566. Pubmed Central PMCID: 3433734.

National Center for Biotechnology Information USNLoM (2005). *PubMed Help*. Bethesda. Available from: http://www.ncbi.nlm.nih.gov/books/NBK3830/

Ni Riordain, R., & McCreary, C. (2009). Head and neck cancer information on the internet: Type, accuracy and content. *Oral Oncology, 45*(8), 675–677. PubMed PMID: 19095486.

Peterlin, B. L., Gambini-Suarez, E., Lidicker, J., & Levin, M. (2008). An analysis of cluster headache information provided on internet websites. *Headache, 48*(3), 378–384. PubMed PMID: 18005143.

Price, S. L., & Hersh, W. R. (1999). Filtering Web pages for quality indicators: an empirical approach to finding high quality consumer health information on the World Wide Web. *Proceedings of the AMIA Symposium,* 911–915. PubMed PMID: 10566493. Pubmed Central PMCID: 2232852.

Rees, C. E., Ford, J. E., & Sheard, C. E. (2002). Evaluating the reliability of DISCERN: A tool for assessing the quality of written patient information on treatment choices. *Patient Education and Counseling, 47*(3), 273–275. PubMed PMID: 12088606.

Renear, A. H., & Palmer, C. L. (2009). Strategic reading, ontologies, and the future of scientific publishing. *Science, 325*(5942), 828–832. PubMed PMID: 19679805. Epub 2009/08/15. eng.

Sajid, M. S., Iftikhar, M., Monteiro, R. S., Miles, A. F., Woods, W. G., & Baig, M. K. (2008). Internet information on colorectal cancer: Commercialization and lack of quality control. *Colorectal Disease, 10*(4), 352–356. PubMed PMID: 17645570.

Silberg, W. M., Lundberg, G. D., & Musacchio, R. A. (1997). Assessing, controlling, and assuring the quality of medical information on the Internet: Caveant lector et viewor – Let the reader and viewer beware. *JAMA: The Journal of the American Medical Association, 277*(15), 1244–1245. PubMed PMID: 9103351.

Siliquini, R., Ceruti, M., Lovato, E., Bert, F., Bruno, S., De Vito, E., et al. (2011). Surfing the internet for health information: An Italian survey on use and population choices. *BMC Medical Informatics and Decision Making, 11*, 21. PubMed PMID: 21470435. Epub 2011/04/08. eng.

Sim, M. G., Khong, E., & Jivva, M. (2008). Does general practice Google? *Australian Family Physician, 37*(6), 471–474.

van der Marel, S., Duijvestein, M., Hardwick, J. C., van den Brink, G. R., Veenendaal, R., Hommes, D. W., et al. (2009). Quality of web-based information on inflammatory bowel diseases. *Inflammatory Bowel Diseases, 15*(12), 1891–1896. PubMed PMID: 19462423.

Wagner, T., Baker, L., Bundorf, M., & Singer, S. (2004). Use of the internet for health information by the chronically ill. *Preventing Chronic Disease, 1*(4), A13.

Weed, L. L. (1997). New connections between medical knowledge and patient care. *BMJ, 315*(7102), 231–235. PubMed PMID: 9253272. Pubmed Central PMCID: 2127165.

Weiss, E., & Moore, K. (2003). An assessment of the quality of information available on the internet about the IUD and the potential impact on contraceptive choices. *Contraception, 68*(5), 359–364. PubMed PMID: 14636940.

Wilson, P. (2002). How to find the good and avoid the bad or ugly: A short guide to tools for rating quality of health information on the internet. *BMJ, 324*(7337), 598–602. PubMed PMID: 11884329. Pubmed Central PMCID: 1122517.

Winker, M. A., Flanagin, A., Chi-Lum, B., White, J., Andrews, K., Kennett, R. L., et al. (2000). Guidelines for medical and health information sites on the internet: Principles governing AMA web sites. American Medical Association. *JAMA: The Journal of the American Medical Association, 283*(12), 1600–1606. PubMed PMID: 10735398.

Chapter 11
Enhancing the Quality of Open Data

Kieron O'Hara

Abstract This paper looks at some of the quality issues relating to open data. This is problematic because of an open-data specific paradox: most metrics of quality are user-relative, but open data are aimed at no specific user and are simply available online under an open licence, so there is no user to be relevant to. Nevertheless, it is argued that opening data to scrutiny can improve quality by building feedback into the data production process, although much depends on the context of publication. The paper discusses various heuristics for addressing quality, and also looks at institutional approaches. Furthermore, if the open data can be published in linkable or bookmarkable form using Semantic Web technologies, that will provide further mechanisms to improve quality.

11.1 Introduction: Open Data

In this paper, I examine the issue of data quality from the point of view of open data. Data quality is generally defined in terms of utility for the data users' purposes (Khan et al. 2002) – in other words, it is a concept relative to the user. Crunching large quantities of data in order to find the weak signals in the noise has become a major industry in the twenty-first century, with claims that it will enable improvements in science (Ayres 2007), drive economic growth (Manyika et al. 2011) and lead to better public service outcomes (Wind-Cowie and Lekhi 2012). The idea of open data is that, as big data and data sharing are so jointly promising, following the logic it makes sense to release datasets to as many people as possible. In theory, this will enable greater innovation in knowledge products and service provision. Current

K. O'Hara (✉)
Electronics and Computer Science, University of Southampton,
Highfield, Southampton SO17 1BJ, UK
e-mail: kmo@ecs.soton.ac.uk

L. Floridi and P. Illari (eds.), *The Philosophy of Information Quality*,
Synthese Library 358, DOI 10.1007/978-3-319-07121-3_11,

practice of keeping data in silos means that products and services cannot easily be developed to place such data in other useful contexts. Yet many application areas require data of many types for a full description, from scientific areas (e.g. climate change or drug design) to the social and political. The main sources of open data are governments and scientific research.

The scientific benefits of sharing data seem clear (Murray-Rust 2008). In non-scientific contexts, it is unlikely that citizens/consumers will consume open data directly. Open data will feed into services, enabling entrepreneurs to create innovative applications (*apps*) which use the data, which are in turn consumed by citizens, organisations, community groups, media analysts and so on. The more heterogeneous the mix, the more creative the app is likely to be. An example might be an app that mashes up data about geography, green spaces, real-time traffic flow, anti-social behaviour and accidents, and outputs a healthy and safe bicycle route between two named points.

The obvious way to make data open is to remove as many legal and technical restrictions as possible. Open data have three principal characteristics: they are (i) available online for download; (ii) machine-readable; and (iii) held under an unrestricted licence, waiving the data owners' rights to monopoly use. Ideally, open data are in open knowledge representation formats; pdf is very restrictive, and requires documents to be scraped for data, while Excel or Word are proprietary so that users need particular software. Better are open formats like CSV, while the ideal, as I shall argue, is an open, linkable format such as RDF (Berners-Lee 2010).

Access or query control, and the use of terms and conditions, are ruled out, because barriers to entry are to be kept as low as possible and reuse is encouraged. However, the apps need not be open – services could be monetised, or restricted to subscribers. Open data removes the possibility of rent-seeking via data monopolies or exclusive access agreements, but if an app is so creative in its use of data that it can support a charge, so be it. In the open data economy income comes from creativity, not rents, so everyone has access to the same data. The system provides a financial incentive for innovation and creativity.

This sketches a hopeful narrative for data quality: open data => extensive critical analysis by a wider user base => crowdsourced data improvement. Even if datasets released are not of the best quality as they are put online, data users and data subjects will soon provide corrections as they scrutinise them. Open data introduce homeostasis into data production, providing a cheap source of feedback on quality.

By releasing open data regularly and getting it out into the user community, quality will – theoretically – increase as comments are received from data subjects, app developers and different departments and agencies who can benchmark against each other. The argument is reminiscent of Hayek's epistemology (Hayek 1945): no-one knows everything, everyone knows something, and everything relevant is known by someone. Open data turns data production into a homeostatic system, with a feedback loop crucially missing in the closed data world.

As an example, consider the National Public Transport Access Node database (NaPTAN), which is the UK Department for Transport record of all points of access to public transport (railway stations, bus stops, ferry ports etc.). The locations of many of the more minor access points (particularly bus stops) were incorrectly recorded on the database. However, the release of NaPTAN as open data enabled apps to be developed that visualised the data and presented them on maps which could be inspected by citizens. Given that everyone knows the real location of a handful of bus stops, and that each bus stop is such that someone knows its real location, the accuracy of NaPTAN has been improved by crowdsourcing corrections via various services (cf. e.g. http://travelinedata.org.uk/naptanr.htm).

There is, however, a paradox with the evaluation of the quality of open data. Most notions of data quality are highly context-dependent (Khan et al. 2002) – to take one obvious example, timeliness is important for quality, but depends on who wants the data when. Yet the point of open data is that they are available for use by anyone for any purpose. Therefore, when publishing open data, officials will need to assess quality in a context-independent way, because the potential contexts of use range over all possibilities.

11.2 Examples

Open data have already proved valuable, and are a major policy initiative of the UK, the US and the EU. In this section, I will look at a couple of examples of open data projects. The first exhibits an instance of one of the utility of open data, while the second shows how quality depends to a large extent on the context in which data are published.

11.2.1 Openness and Utility: Prescribing Analytics

In 2013, the Prescribing Analytics exercise, http://prescribinganalytics.com/, mashed up a large amount of fine-grained open data (about doctors' prescriptions from the NHS Information Centre, Primary Care Trust boundary data from the Office for National Statistics and Clinical Commissioning Group boundary data from the NHS Commissioning Board), and compared different regions as to their usage of generic and more expensive proprietary statins, finding interestingly wide variation (Fig. 11.1). Their analysis on this open but not linked data showed there was real opportunity to save money, by pushing the more generous regions to prescribe more generics (hundreds of millions of pounds just for the statins group). But from the point of view of quality, the study is interesting for its feedback to the system as to what comparisons are useful, and what data are desirable.

In particular, the British National Formulary (the standard reference work in this area, regularly consulted by doctors) does not release data about drug prices under an open licence, so it is not reusable. Hence a lack of openness of one crucial dataset meant that the calculation of the potential saving had to be an estimate. This feedback has been important – the BNF data are available for free within the UK from

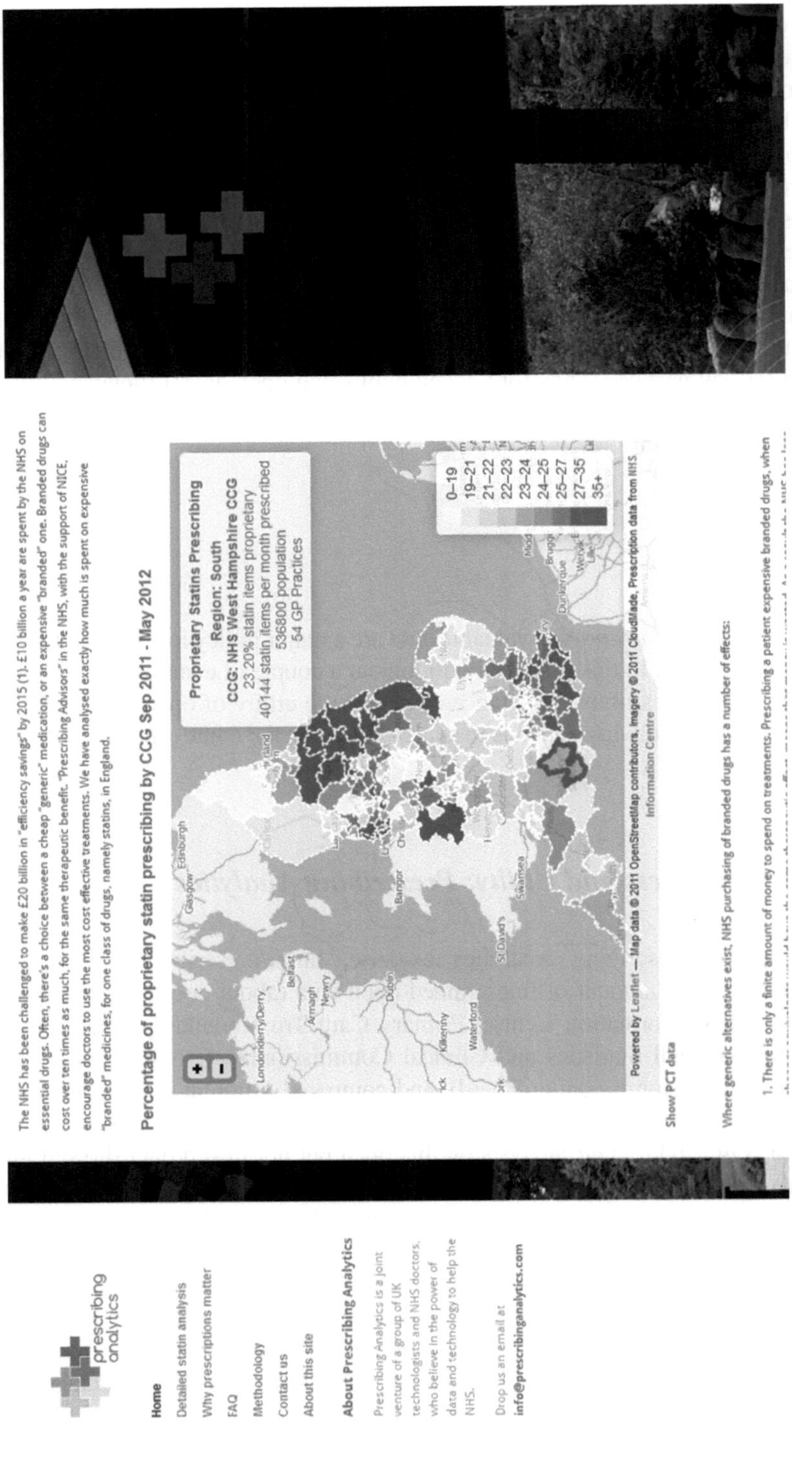

Fig. 11.1 Prescribing analytics

the http://www.bnf.org/bnf/index.htm website, but copyright remains with the holders. However, the value of that dataset has led to an open data app giving open access to BNF data, http://www.openbnf.org/about. Feedback through the system has identified an important dataset, and the Open BNF app represents a first prototype of how that might work. The paradox of the evaluation of open data is resolved by using it in a real-world application.

11.2.2 Representational Intent, Openness and the Data Market: police.uk

As a second example of the way the open data paradox can affect data, let us consider police.uk, a website dispensing open data about crime and criminal justice for the UK's Home Office and Ministry of Justice, which is interesting from the point of view of how data get adjusted to fit particular intentions or requirements (O'Hara 2012a). Police.uk gives out both data and information, setting the data in the context of a map, and providing an increasing set of tools for users to query the data (Fig. 11.2). The figure shows a map of the area around the Southampton University campus, while the numbers refer to the number of crimes on each street in the month covered by the map (December 2012). Clicking on the numbers allows more details about the crimes, including the anonymised outcomes of any investigation. The open data underlying the crime maps (in Excel format under an Open Government Licence) can also be downloaded from police.uk and used to power apps.

Assessment of the quality of the police.uk datasets is affected by the institutional drive to make the data intuitively understandable by locating them on maps. It is also affected by the purpose of police.uk and the purpose of supplying the open data – but different purposes have been adduced for these. Various Home Office ministers have claimed at various times that police.uk is there to (i) inform the public about crime, (ii) hold the police or Police and Crime Commissioners to account for their performance, (iii) enable people or groups to understand and negotiate their community or environment and manage the threat of crime, and (iv) support the development of innovative services which could help drive economic growth. Of course, the value of the data will vary depending on which of these purposes is foremost, and we see the open data paradox in action once more.

police.uk has been an undoubted success, and a leading advertisement for the advantages of open data. The logic of open data suggests that the scope of police.uk should be reduced – in effect to let it wither as the informational app market thrives with the information-consuming public which police.uk has been central in fostering. The logic of success, on the other hand, is to expand it. The result is that the information market struggles against a state-backed information supplier. Furthermore, that supplier, by virtue of the close connections between the site developers and the data publishers, tends to get the data first, and so its output is

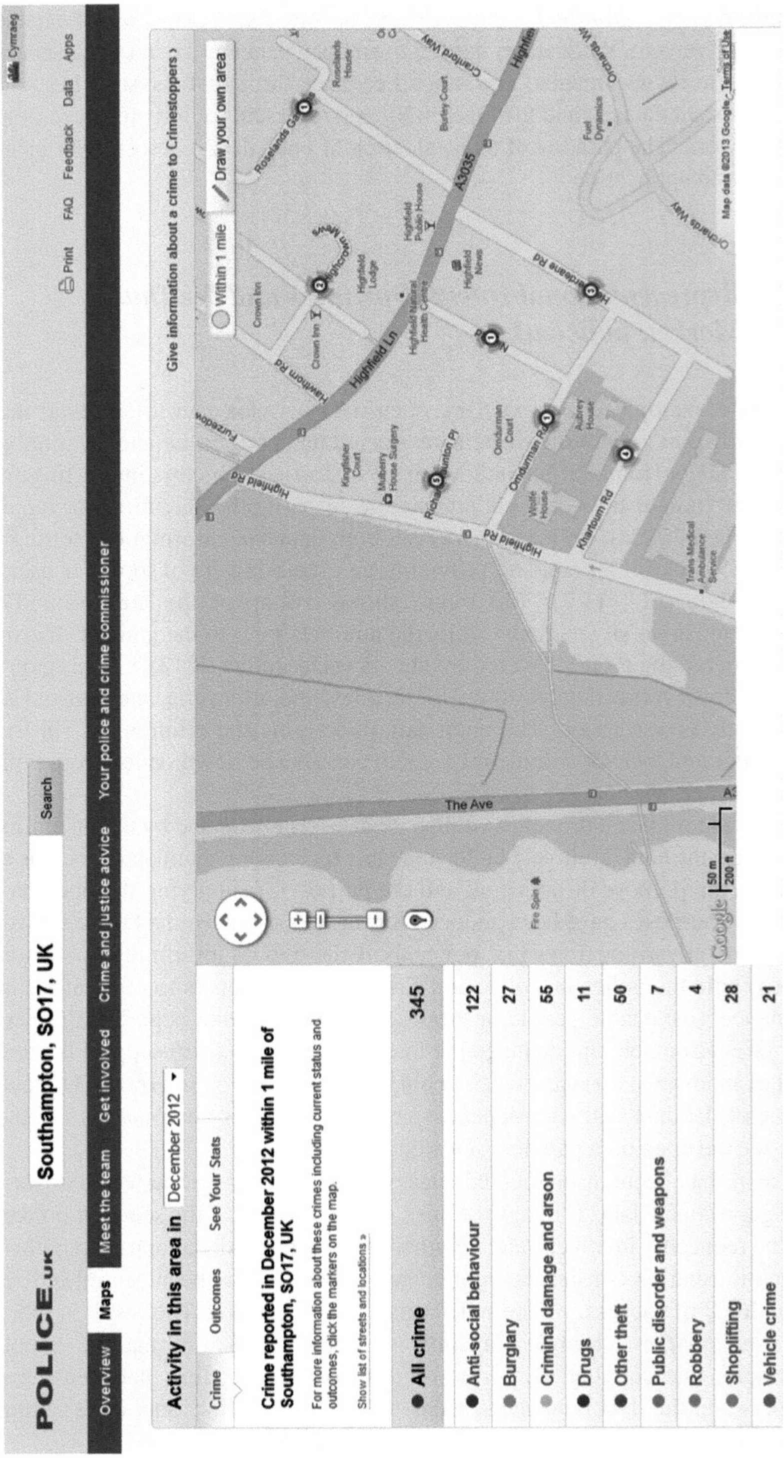

Fig. 11.2 police.uk

more timely. The 'eyeballs' which help improve data quality are often those of the app developers who are competing against police.uk – and hence, although they work to improve quality in their own self-interest, they are simultaneously helping to improve the position of their dominant competitor. This may not always work to their advantage. On the other hand, a difficulty in relying on an information market created by what is in effect the cottage industry of app development is that the continuity of information supply to the public (as opposed to continuity of *data* supply to app developers) may be variable. Yet as noted, one of the main objectives of police.uk is to keep the public informed about crime.

The experience of police.uk is instructive in a number of ways. It was always intended as a mapping site, replacing a previous crime mapper. Maps are powerful platforms for making sense of data, and joining disparate datasets (Shadbolt et al. 2012). There is a strong connection between the data and their location on a map, and the intuitive presentation means that police.uk has received hundreds of millions of hits in its first 2 years.

However, the geography of crime is complex. Consider five types of crime. First of all, some crimes are addressable; a burglary at a house can be placed at a specific address. Second, there are crimes which have approximate addresses – a mugging in a street, for example. Third, there are crimes which have locations but no addresses – a mugging in a public park, for instance. Fourth, there are crimes with uncertain locations, and where the location is actually rather irrelevant – a theft which took place on a train journey. Fifth, there are crimes where location is unknown or irrelevant – identify theft, for instance.

In all but the first of these cases, there are great complexities in locating a crime. Although some can be located at specific postcodes and given precise eastings and northings, many cannot. Many locations will be given as free text on reporting forms, which need interpretation – something unlikely to happen consistently across 43 different police forces in the UK (who provide the data). Yet these interpretations matter – even for crime statistics. Administrative boundaries are often drawn down the centre of streets, so a crime located on a particular street may randomly appear in either one of two administrative units. This problem is exacerbated, as the aim of police.uk is to reveal patterns of crime at street level – and so each crime is associated with a street (as can be seen in Fig. 11.2). All this is exacerbated by the lack of national geodata coding standards.

This geographical indeterminacy also relates to a further factor in the quality of the data, which is that of privacy and data protection. There are a number of issues to do with open data derived from personal data (personal data is not open data, for obvious reasons), which I have explored elsewhere (O'Hara 2011). The relevant point here is that privacy can impinge on data quality. As police.uk was being developed, the UK Information Commissioner's Office (ICO) looked at privacy issues surrounding the display of crimes on a map (Information Commissioner's Office 2010). The potential problem is that, if the location of a crime is X's house, then X is identifiable as the victim of that crime, even if not identified directly in the data. After discussions with the ICO, it was decided to take two privacy-preserving measures.

First, addresses are 'vagued up' – because police.uk was focused on crime at street level, the precise address was not necessary. Hence the snap points on the police.uk map are not exact – they originally covered a minimum of 12 (now 8) postal addresses. It is not known what the average vagueness is (substantially more than 8). This of course impinges on quality by definition, but also there are no metadata to tell the data user how vague the particular location is. Another impact of this on quality is that quite often the exact location of a crime or anti-social behaviour is an important piece of information – telling the user which street corners to avoid, or allowing a campaigner to argue to the authorities that, say, the loss of a street light has led to an increase of crimes in a very small area. And not every type of crime has a privacy implication.

Secondly, the data are aggregated over a month, and released in arrears, to make it harder to associate a crime on the map with an experienced crime in the locality. Hence releases are not very timely, and do not allow the user to make important discriminations (whether crimes are committed at night or during the day, what happens at closing time). It is also likely that the lack of timeliness means that it is harder to help the police; if a citizen sees that a crime has been committed in her neighbourhood yesterday, she would be more likely to be able to report suspicious pedestrians or cars in the area, whereas after a lag of up to 7 weeks, her recall will obviously be less immediate and accurate.

Privacy considerations, where relevant, will have an effect on data quality, and sensitive treatments of privacy that preserve quality as much as possible may require an expensive administrative overhead, compared to the relatively lightweight methods used in police.uk. Privacy will always be an issue with open data, because of the inability to use access controls, consent or use limitation.

11.3 Open Data and the Data Quality Paradox

The meaning of quality for open data is not clear. However, there are certain ways in which to address this issue pragmatically. I begin by looking at some general heuristics, then a specific technological approach (illustrated by an example), and finally consider institutional approaches.

11.3.1 Heuristics to Address the Paradox

No heuristic can produce a magic answer to the quality question, but – especially within the inevitable bounds of resource constraints – there are obvious steps which can be taken, for data users and producers alike, to leverage the possibilities of scrutiny and crowdsourcing.

Data users need to be (a) aware of the limitations of the data they are using, and (b) aware of the potential for limitation as well. Reputable datasets should come with relevant metadata. Research is ongoing into understanding the provenance of data in open systems (Gil et al. 2012), with a W3C working group (http://www.w3.org/2011/prov/wiki/Main_Page) close to producing recommendations for machine-readable data models, ontologies and human-readable notations for provenance metadata. Data should always be used with caution, and when heterogeneous datasets are being linked or mashed up together, the caution needs to be even greater.

For data providers, there are also a number of rules of thumb which can help circumvent the quality paradox. Most obviously, most datasets are developed for specific audiences for specific purposes, but this is likely to make their semantics impenetrable to the outsider; (Shadbolt et al. 2012) gives the example of a valuable dataset of regional health statistics which contained the codes 'SHA code' and 'Org code', meaningless to anyone not *au fait* with the labyrinthine bureaucracy of the NHS. Open data providers should make codings and metadata understandable for general users, not just the primary data users (Shadbolt et al. 2012). There is great value in representing data in both human- and machine-readable forms (Bennett and Harvey 2009).

11.3.2 Linked Data

The mention of machine-readable data models brings us to the Semantic Web, and the possibility of adding technological mechanisms to the problem of open data quality. The protocols and standards of the World Wide Web, initially designed to connect documents, are being reshaped by Semantic Web research to link data directly using knowledge representation languages that represent data using Uniform Resource Identifiers (URIs) (Shadbolt et al. 2006). URIs facilitate linking by giving consistent identifiers to objects, concepts and also relations. By directly connecting data, much more routine information processing can be automated (and therefore can happen at scale). In this world, linking data, or mashing up data from several sources, is widely perceived to increase their value by allowing serendipitous reuse in unanticipated contexts and juxtapositions (Berners-Lee 2010).

Linking also enables some disambiguation – hence, for example, a pair of statements about the *population-of* the UK and the *population-of* Italy can be rendered comparable by a link being made between the two that says that, in the two statements, 'population-of' *means the same thing*.

Linked data raise many practical issues with respect to provenance, ontological alignment, privacy, data protection and intellectual property. One of the most pressing is quality; in the linked data vision, the Web is treated in effect as a large database, as data are brought together from heterogeneous sources and processed together. This leads to an obvious issue of trust in quality; if data processors cannot trust that a dataset is of sufficient quality, they will be reluctant to mash it up with datasets already in use. Quality issues here include such matters as accuracy, correct underlying methodology, timeliness, reliability, consistency of semantics and representation (particularly with time-based series) and format – and it is worth pointing out that these issues apply not only to the datasets themselves, but to their metadata as well.

URIs are valuable instruments – governments in particular have been urged to use URIs to identify each object and concept likely to be used across applications – places, offices, job descriptions, institutions, buildings, post/zip codes, vehicles, pieces of legislation or whatever (Bennett and Harvey 2009; Shadbolt and O'Hara 2013). If data are to be shared across applications, a robust set of common identifiers is necessary; different identifiers developed by different organisations will produce silos (cf. the SHA code example above). The principle of using URIs in data is the same as using them on the Web of documents – a URI refers unambiguously and universally to a particular resource which can be linked to from any document. This simple yet extremely powerful principle can also apply to a piece of data. If we have a piece of data, 'aRb', and 'a', 'R' and 'b' are all assigned URIs, then they can be linked to and looked up using HTTP (Bennett and Harvey 2009).

This will of course not affect the quality of the data in its original setting, where the identifiers would probably be readily understood. However, it *will* affect the quality of a constructed mashed-up or linked dataset created from the original dataset, both by allowing stronger links to be made between data from different datasets, and by removing the possibility of erroneous alignment of different identifiers. Certain datasets can be used as standards. For instance, DBpedia is a linked data rendering of the structured data in Wikipedia (http://dbpedia.org/About) which can act as a reference point for commonly-used objects or concepts (it currently has an ontology with 2.35 m instances, including 764,000 people and 573,000 places).

As an example, consider the portal legislation.gov.uk. The UK's legislation has long been available online. However, this has not always been an unalloyed boon, because it is unclear from the text of legislation whether it is currently in force; legislation is repealed by other laws. Other restrictions are often hidden in the preamble (for instance, in the UK England, Wales, Scotland and Northern Ireland often have different laws), as is the date at which a law comes into force. Acts of Parliaments are often amended, and even worse, the meanings of certain concepts are altered, sometimes subtly, by later Acts. Hence a clause in an Act of Parliament needs context to understand it fully – does it apply to all or part of the UK, is it in force, do the terms in it still mean what they did when written? All this further complicated by UK legislation stretching back some 800 years.

Legislation.gov.uk represents all UK legislation as linked data (Tullo 2011) which means that links can be made between the various pieces of legislation and the concepts they use. Each concept, each law and each particular rendering of each law is identified by a URI. In Fig. 11.3, we see a clause from the Theft Act 1968, defining the maximum sentence for theft. The lozenge by the clause title shows the geographical extent of the law (England and Wales), while a timeline shows how it has been amended – the first point in the line is the passing of the 1991 Criminal Justice Act, and the second point shows when, in 1992, the new law came into force.

Here the data quality has not changed as a result of any change in the data itself. This is the same data that was presented in previous sites with UK legislation.

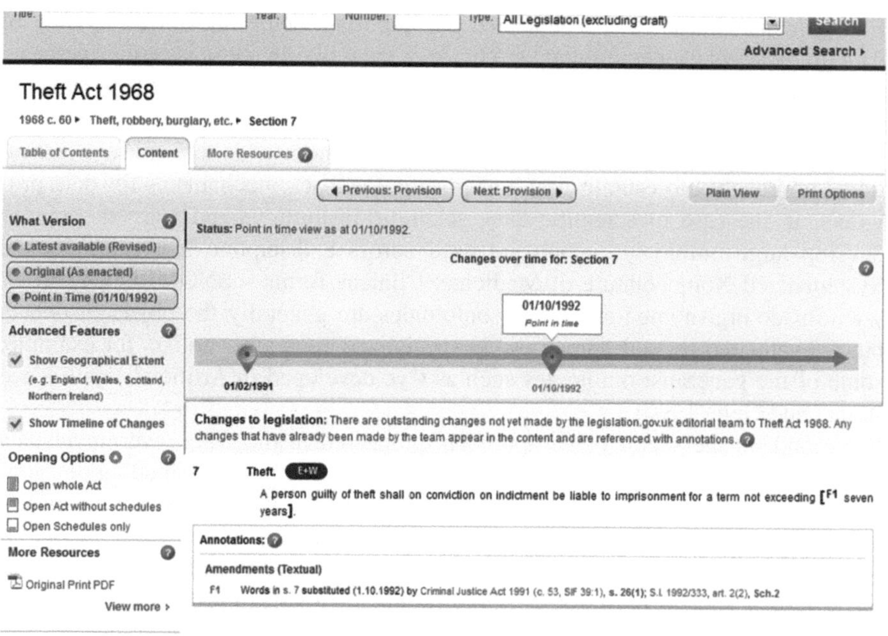

Fig. 11.3 legislation.gov.uk

However, the various user-relative attributes of data that affect quality are clearly affected by the ability to link to other pieces of legislation. A piece of legislation served up in isolation is very useful for a particular type of user (specifically, someone well-versed in understanding the law, who knows how to determine not just the words on the page, but also the complex relations with other laws and institutions that give them their meaning), but the linked version is adaptable to a greater range of users with unanticipated needs. Any legally-defined concept – for example, there are 27 different types of school defined in UK law – can be rendered as a piece of linked data, and then linked to.

Linked data can also help to provide standards. Data about schools from a different source can be linked to the definition of exactly the type of school the data apply to (Tullo 2011). Such standardization programmes are important for improving the quality of data created by integrating datasets.

11.3.3 *Institutional Approaches*

It is important for data providers to engage with likely data users and the development community. Of course, this will not resolve the quality paradox, but will give providers a sense of what data are likely to be valuable to which communities, and why some datasets released under open licence have not been reused very much.

Engagement with users is a neglected but important determiner of the quality of data for different user communities. This may seem like an administrative overhead, but actually can be valuable for quality generally as it adds a source of feedback into the data provision system.

Apart from direct engagement with developers and users, there are four institutional approaches to establishing homeostatic feedback mechanisms for data providers. In the case of scientific data, scientific institutions and learned societies develop large ontologies which are used across a discipline (for example, the Systematized Nomenclature of Medicine: Clinical Terms – SNOMED CT, http://www.ihtsdo.org/snomed-ct/). These ontologies are generally the object of debate but are validated by widespread use in specialized contexts – unlike, for example, some of the generalist ontologies such as Cyc developed in Artificial Intelligence (Lenat and Guha 1990).

Second, in the sector panel approach, pioneered in the UK, government data providers are advised by experts and interested parties, crucially including development and user communities (as well as data subjects, domain experts, technical experts and data protection experts) (O'Hara 2011, 2012b).

Third, a portal can be used to develop communities of practice around particular types of data, such as in Rensselaer Polytechnic Institute's linked open government data portal, which converts government open data to linked data, provides tools and search and fosters communities of data users around types of problem both general (e.g. business) and specific (e.g. clearing up the Gulf of Mexico after an oil spill) (Hendler et al. 2012).

Fourth, once more in the UK, there is the Open Data Institute (ODI – http://www.theodi.org/), an organization that is part-privately, part-publicly funded, which works with government and industry to encourage the publication of open data and the dissemination of best practice, and to work with developers to ensure that data are used and their value fully realized. The mission of the ODI is to bring together providers and developers in a structured environment where there will be support, technical know-how and understanding of best practice.

Such approaches can help side-step the quality paradox, not by changing the data themselves necessarily, but by widening understanding among data providers of what data users might require. This will go beyond the primary user base – i.e. the organization which commissioned the data originally – and will foster a greater understanding of the demand for data. Furthermore, both providers and users will be brought together by the technology of linked data by greater reliance on standards, whether particular reusable ontologies, URI schemes, open, non-proprietorial languages or standard licences.

11.3.4 Cost v Value

The production of linked open data by and with institutions should (done properly) enhance the value of data both in its own right, and by connecting it, perhaps serendiptously, with other relevant datasets (relevant in the context of the user, whose interest in the data may have been unanticipated). However, that does not mean that the costs of producing linked data are low – indeed, many institutions are concerned that there is a steep learning curve and a new set of skills to be learned.

RDF publication is demanding (http://pedantic-web.org). The full stack of Semantic Web technologies and protocols imposes a burden on units with complex data requirements and limited resources. On the other hand, governments' priorities are to increase the amount of open data of whatever form, rather than to maximise linked open data. Excel can be published by exporting data in use, and non-proprietary non-linked formats such as CSV generally require little more than one-off investments in plugins or converters.

On the other hand, linked data has resource and training implications: work must be done on representing data in the ideal format and granularity, assigning URIs, and developing or reusing ontologies. Links need actually to be made to other datasets, and broken links repaired. Small wonder that the low-cost unlinked options using well-known formats like Excel and CSV can look more attractive to cash-strapped managers than RDF.

Yet there are reasons why the cost/benefit analysis should embolden the strategic data manager to ascend the linked data learning curve. First, many government statisticians privately admit that they have too much data to process. Publishing data increases the benefits it will provide and lets users critique and improve it. Crowdsourcing data improvement is a selling point for all open data, and third-party auditing of linked data depends on access (Shadbolt and O'Hara 2013).

Second, linked data's success depends on it being consumed and consumable. This is becoming increasingly important within governments; those agencies which have moved into linked data have found it increasingly useful to consume their own linked open data than to interrogate their own datasets anew each time. Once the initial push occurs, linked data's benefits become clear not only to outsiders but also within the publishing organization. As an example, those using data in the Department for Communities and Local Government find it easier to consume their own linked data via the Local Authority Dashboard (http://opendatacommunities.org/dashboard), an application which presents its own statistics on local issues including finance, housing, and deprivation, and reusing Ordnance Survey geodata and statistics from the Office of National Statistics, than integrate the data manually as they used to. The DCLG is a crucial beneficiary of its own linked data (Shadbolt and O'Hara 2013).

At the moment, institutions remain important; pushes from the top are still needed. Until more linked data is produced, the network benefits will remain nascent. Elsewhere, I have written about bottlenecks in the production and use of linked open government data – dataset catalogues are vital for discovery, while generic metadata standards enable communities other than primary users to benefit. Usable interfaces are needed (Shadbolt et al. 2012). Lightweight, pragmatic methods reduce administrative overhead, while better tools and interfaces should increase uptake (Alani et al. 2008). We also need effective means to track provenance and protect citizens' privacy.

11.4 Conclusion

In the open data world, quality is supposed to be upheld or improved by a series of overlapping communities. Providers benchmark themselves against their fellow providers. App developers need high quality data in order to provide useful and innovative information services to their customers or clients. Data subjects are well-informed at least about the data that concern them. And finally, information consumers are well-informed about their own environment and problems.

The example of crime data shows that the situation is rarely that simple. The market structure is less easily theorised or understood; for historical or political reasons particular representations (maps in this case) are privileged, developers have to compete with a branded government product, and the sustainability of apps' business models has not yet been demonstrated. This lack of sustainability threatens to undermine the feedback loops that are expected to improve quality.

Nevertheless, the situation is not hopeless. As well as a number of heuristics for both data users and data providers, a more rigorous approach to the technology in the use of linked data, and the development of institutions to promote engagement, can all address the open data quality paradox.

Acknowledgements This work is supported by SOCIAM: The Theory and Practice of Social Machines, funded by the UK Engineering and Physical Sciences Research Council (EPSRC) under grant number EP/J017728/1. Thanks are owing to audiences at a number of talks and conferences, including the Information Quality Symposium at the AISB/IACAP World Congress 2012, Birmingham, July 2012.

References

Alani, H., Hall, W., O'Hara, K., Shadbolt, N., Szomszor, M., & Chandler, P. (2008). Building a pragmatic semantic web. *IEEE Intelligent Systems, 23*(3), 61–68.

Ayres, I. (2007). *Super crunchers: How anything can be predicted*. London: John Murray.

Bennett D., & Harvey, A. (2009). *Publishing open government data*. World Wide Web Consortium. http://www.w3.org/TR/gov-data/

Berners-Lee, T. (2010). *Linked data*. World Wide Web Consortium. http://www.w3.org/DesignIssues/LinkedData.html

Gil, Y., et al. (2012). *PROV model primer*. World Wide Web Consortium. http://www.w3.org/TR/prov-primer/

Hayek, F. A. (1945). The use of knowledge in society. *American Economic Review, 35*(4), 519–530.

Hendler, J., Holm, J., Musialek, C., & Thomas, G. (2012). US government linked open data: semantic.data.gov. *IEEE Intelligent Systems, 27*(3), 25–31.

Information Commissioner's Office. (2010). *Crime mapping and geo-spatial crime data: Privacy and transparency principles*. Information Commissioner's Office. http://www.ico.gov.uk/for_organisations/guidance_index/~/media/documents/library/Data_Protection/Detailed_specialist_guides/crime_mapping.ashx

Khan, B. K., Strong, D. M., & Yang, R. Y. (2002). Information quality benchmarks: Product and service performance. *Communications of the ACM, 45*(4), 184–192.

Lenat, D., & Guha, R. V. (1990). *Building large knowledge based systems: Representation and inference in the CYC project*. Reading: Addison-Wesley.
Manyika, J., et al. (2011). *Big data: The next frontier for innovation, competition and productivity*. Washington, DC: McKinsey Global Institute.
Murray-Rust, P. (2008). Open data in science. *Serials Review, 34*(1), 52–64.
O'Hara, K. (2011). *Transparent government, not transparent citizens: A report on privacy and transparency for the Cabinet Office*. Cabinet Office. http://www.cabinetoffice.gov.uk/resource-library/independent-transparency-and-privacy-review
O'Hara, K. (2012a). Data quality, government data and the open data infosphere. In *AISB/IACAP world congress 2012: Information quality symposium*. Birmingham: The Society for the Study of Artificial Intelligence and Simulation of Behaviour. http://eprints.soton.ac.uk/340045/
O'Hara, K. (2012b). Transparency, open data and trust in government: Shaping the infosphere. In *ACM conference on Web Science (WebSci2012)* (pp. 223–232). Evanston: ACM, New York.
Shadbolt, N., & O'Hara, K. (2013). Linked data in government. *IEEE Internet Computing, 17*(4), 72–77.
Shadbolt, N., Berners-Lee, T., & Hall, W. (2006). The semantic web revisited. *IEEE Intelligent Systems, 21*(3), 96–101.
Shadbolt, N., O'Hara, K., Berners-Lee, T., Gibbins, N., Glaser, H., Hall, W., & schraefel, m. c. (2012). Linked open government data: Lessons from data.gov.uk. *IEEE Intelligent Systems, 27*(3), 16–24.
Tullo, C. (2011). Online access to UK legislation: strategy and structure. In M. A. Biasiotti & S. Faro (Eds.), *From information to knowledge* (Frontiers in artificial intelligence and applications, Vol. 236, pp. 21–32). Amsterdam: Ios Press.
Wind-Cowie, M., & Lekhi, R. (2012). *The data dividend*. London: Demos.

Chapter 12
Information Quality and Evidence Law: A New Role for Social Media, Digital Publishing and Copyright Law?

Burkhard Schafer

Abstract Increasingly, judges are asked to act as gatekeepers between law and science, using the rules of admissibility to perform what could be understood as a form of "secondary forensic information quality assurance". To exercise their gate keeping function and to ensure that the jury is only exposed to the "best evidence (possible)", judges rely on other primary gatekeepers, amongst them forensic regulators, scientific communities and academic publishers. This paper addresses how digital media and new forms of publishing are changing the nature of these gatekeepers, focusing in particular on how they change the role of peer review as a major quality assurance mechanism used by the courts at present. Data mining social media also provides us with both quantitatively and qualitatively new information about scientists, scientific communities and the practice of science. This paper argues that the discourse on information quality can be one avenue to make more systematic use of these data, helping to address long-known shortcomings in the justice system.

12.1 Legal Regulation of Information Quality

To the extent that it is studied at all, the interaction between law and information quality is typically seen as a relatively recent development prompted by the desire to regulate the collection, dissemination and use of ever larger amounts of data by public authorities. A paradigmatic example is the highly controversial section 515 of the United States Consolidated Appropriations Act 2001 (better known as the

Research for this paper was supported by the RCUK funded CREATe network, www.create.ac.uk. I'm particularly grateful for the comments and help received from Laurence Diver.

B. Schafer (✉)
SCRIPT Centre for IT and IP Law, The University of Edinburgh, Edinburgh, UK
e-mail: b.schafer@ed.ac.uk

L. Floridi and P. Illari (eds.), *The Philosophy of Information Quality*, Synthese Library 358, DOI 10.1007/978-3-319-07121-3_12,

Data Quality Act). This legislation established a duty for federal agencies to maximize "the quality, objectivity, utility, and integrity of information (including statistical information) disseminated by Federal agencies". Ostensibly, it was a response to the ever increasing amount of data and studies provided by federal agencies. In practice, it had a "chilling effect" on the publication and dissemination of data by federal agencies, which faced increased costs and cumbersome bureaucratic procedures when information that they provided was challenged by industry or political lobby groups. While having been roundly criticized as legislation that does not so much improve information quality as allow parties with commercial interests to suppress it (see e.g. Lacko 2004; Shapiro 2003), it is nonetheless at a conceptual level an interesting example of a legislative attempt to regulate information quality across scientific, administrative and application domains. While there are important examples of earlier regulation of information quality, particularly in the financial sector, these tend to be highly specific regarding the type of data that is regulated, the uses to which it is put and the actors who collect it. The legal regulation of businesses, again in the financial sector in particular, traditionally included duties concerning accounting standards and duties to collect, make accessible and disclose certain information (see e.g. Bushman and Piotroski 2006). The homogeneous nature of the data in question and their very specific use meant, however, that this field of law did not develop an abstract, theoretical notion of information or information quality, instead relying where necessary on the specific understanding developed in business studies and accounting. This paper therefore takes a different perspective, looking at a field of law that for quite some time has grappled with issues of information quality assurance, even if it was not using that precise term – the legal regulation of evidence and proof. This way, we can explore if the emerging discourse on information quality can shed new light on old and intricate legal problems, while at the same time analyse if the concepts and ideas that the law has developed can contribute to the analytical and conceptual vocabulary of information quality research.

Over the last two centuries, scientific progress has changed the nature of both criminal and civil litigation immeasurably. A modern day reader who accesses, for example, the digitised transcript of decisions of the Central Criminal Court of England and Wales, the Old Bailey,[1] may be surprised to see how comparatively recent life-and-death decisions were based on little more than statements by eyewitnesses of often dubious proximity to the crime, statements made by victims and the accused, and "character witnesses" testifying to the character of the accused or the reliability of other witnesses. This changed with the advent of modern forensic disciplines, beginning with forensic medicine and pathology in the early nineteenth century and leading eventually to the foundation in 1909 of the world's first school of forensic science, the "Institut de police scientifique" in Lausanne. Fingerprinting was particularly responsible for introducing forensic science to the public consciousness, although such awareness has led to the "CSI effect", which hypothesises that the depiction of forensic science in popular culture makes it increasingly

[1] These are available from 1674–1913 at http://www.oldbaileyonline.org/

difficult to secure a conviction in a jury trial when no scientific evidence is led (Schweitzer and Saks 2007; sceptical Shelton 2008).

A further relevant development began in the 1980s with the advent of DNA profiling as the new gold standard in forensic science. DNA brought the issue of large forensic databases to the fore. DNA databases play a crucial role not only in the detection of crime, but also in quantifying the match probability between DNA found at the crime scene and that of a suspect. The quality control of data is therefore a problem that evidence law must deal with (see e.g. Gill et al. 1991). The ability to quantify, at least to some degree, the weight of the evidence that is presented in court has resulted in considerable legal debate: is it appropriate for the expert to express their assessment of the weight of the evidence this way, or should this be the sole territory of the judge and jury? What demands should we make of a database to enable that type of assessment to be made? For instance, should we ask for data about specific reference classes? And what problems do we face when the data sets exhibit systematic biases, for example the over-representation of ethnic minorities? (See the discussion in Kaye and Smith 2003). Finally, if it is possible and desirable to quantify evidential weight this way, should the availability of sufficiently good data sets become a precondition to admit a certain type of evidence? This standard was applied in the English case of *R v. T*,[2] which discussed the evidential weight of shoe print evidence and connected the admissibility of statements regarding evidential weight to the existence of sufficiently comprehensive and reliable databases. The outcome of *R v. T* represented a direct linking between more traditional issues of data quality assurance and the law of evidence (see e.g. Redmayne et al. 2011 which, while critical of the verdict, explicitly discusses the issue of data quality for criminal databases).

Apart from using databases to quantify match probabilities, "big data" also plays an increasingly important role in the prosecution of cybercrime and terrorist offences, with "open" social media such as Facebook becoming an important, if controversial, source of policing intelligence (see e.g. Omand et al. 2012).

The growing number of ever-more specialized forensic disciplines and subfields, and their increasing reliance on large data sets, pose further challenges for the administration of justice. From the very beginning, the justice system has been aware of the high degree of persuasiveness that scientific experts command, and the concomitant need to shield jurors from unreliable scientific evidence. This became more pressing once it became clear both that the newly developed forensic methods were not all equally reliable, and that even a reliable theory could result in low quality information in the hands of an incompetent expert. Examples of forensic techniques which were for a time led as evidence in the court setting but were subsequently discredited include polygraphy (Saxe and Ben-Shakar 1999), ear print matching (Broeders 2006) and comparative bullet-lead analysis (Tobin and Duerfeldt 2002). Perhaps more worrying is that some approaches are still in use despite considerable concern about the quality of the data that underpins them, such as forensic bite mark matching (Gundelach 1989; Bowers 2006).

[2] [2010] EWCA 2439.

Legal rules were subsequently developed to assist the courts in the task of screening reliable from unreliable forensic methods, incrementally increasing the gatekeeper function of the judiciary (for an overview see Faigman et al. 2006). Initially, judge–made rules tended to leave the task to safeguard information quality to the internal quality control mechanism of the scientific field in question. One typical early example for such a standard was created in the US decision of *Frye v. United States*[3] from 1923, which had to decide on the admissibility of polygraph tests. Finding that this method did not have the support of the relevant scientific community, the evidence was excluded. Under the *Frye* standard therefore, judges would pay deference to the relevant academic communities and admit scientific evidence provided that the underpinning theories and methods were "generally accepted" by the scientific community. Dissatisfaction with this approach, which was seen as both too forgiving towards popular "junk science", and too rigid towards emerging but legitimate fields of discovery, resulted in legislative change in the Federal Rules of Evidence in 1975. These increased the role of the judiciary as gatekeepers to scientific evidence, requesting from them a more substantive engagement with the quality of the evidence submitted to court. This development reached its temporary zenith with the *Daubert* decision in the US that for the first time gave precise meaning to these new rules. In *Daubert* the question was whether novel methods of testing possible harmful consequences of Bendectin during pregnancy were reliable enough to be admitted in evidence. Even though the methods used by the experts for the plaintiff were at that time not yet "generally" accepted by the scientific community, they were also far from a mere outsider position and had been subject to (some) testing and validation. As we will discuss in more detail below, the court established that the merely formal criterion of "general acceptance" was not sufficient to refuse this evidence if there were other compelling reasons that indicated its validity, forcing the judges to engage with the substance of scientific opinion more than they had done before (Faigman 2013). A watermark in the US, the proposed new rules on expert evidence in England would follow its lead a mere 20 years later. However, these rules often resulted from rather specific problems raised by the issue before the court, and as a result a general conceptual framework did not emerge. One of the arguments developed in this paper is that it can be fruitful to study the heterogeneous and disparate rules of evidence that evolved in common law jurisdictions from the eighteenth century onwards through the prism of information quality research. Considerable controversy still surrounds the regulation of the interface between law and science, a debate to which research in information and data quality should be able to contribute a degree of rigour often absent in legal-doctrinal analysis – especially since the ideas and concepts of (in particular) data quality, which are already part of the everyday practice of forensic experts, are not generally applied in courtrooms. Conversely, legal concepts, shaped over centuries of social practice, may be able to provide useful insights into the emerging discipline of information quality assurance.

[3] Frye v. United States, 293 F. 1013 (D.C. Cir. 1923).

Admissibility decisions in particular are seen in this paper as a form of information quality assurance. The focus will be on two related concepts that courts use to make decisions about the admissibility of evidence, as they are particularly pertinent to issues in information quality research. As we have indicated above, two conceptually different, yet interrelated, questions a judge will ask an expert witness are whether the methods and theories he or she uses have been subject to *peer review* and, if so, whether they are "*generally accepted in the academic community*".

The role of peer review in forensic information quality assurance brings us to the second argument that this paper makes. Peer review is often treated by the law as an abstract notion, disembodied and separate from the actual process of academic knowledge production. By contrast, one of the lessons that lawyers can learn from a philosophically grounded theory of information quality is that the content of information and the process of its construction aren't always easily disentangled. Peer review is intimately connected to a specific business model of academic publishing and to the commercialization of academic results. By challenging these business models, social media and digital publishing also undermine the traditional gatekeepers which judges have relied on in the past. They do, however, also provide the justice system with opportunities to develop much more precise and relevant tools for assessing the quality of expert opinions. The method and frequency with which the press release for a scientific paper is mentioned in science blogs or retweeted might, for example, tell us more about the general acceptance of the ideas expressed in the eventual publication than either the fact that it was peer reviewed by two anonymous academics or that it was cited in another scientific publication. Online citation tools such as Citebase or CiteSeerX have become ubiquitous, and allow for much easier analysis of the impact of individual papers than was previously possible. Whether an idea results in an editing war on Wikipedia, or is quickly accepted and incorporated by experts on their blogs can tell us, in real time, at least *something* about how an idea changes from a contested novelty to a generally accepted idea. Reactions in social media to scientific ideas score highly on "contextual" information quality factors such as timeliness and completeness, but of course raise concerns about familiar "intrinsic" factors such as authoritativeness and objectivity.

Data generated online about science and scientific communities are themselves not produced in a legal vacuum, however. The data that one day can play a role in a trial setting will have been subject to various legal constraints in the process of their compilation. We have already mentioned the Data Quality Act, which enables certain parties to petition an agency to stop disseminating information that they claim does not meet data quality standards. Although perhaps intended to increase the objectivity of information, this nonetheless inevitably reduces its completeness. We will however focus on a more unlikely candidate for the regulation of information quality: copyright law. The role of copyright law in information quality assurance is complex and potentially ambiguous, as we shall see. Copyright, we will argue, has a Janus face. On the one hand it provides the author (who can of course be a scientific researcher) with a degree of control over her work, and with that some protection against distortion, misuse or misrepresentation. Furthermore, the protection of the moral rights of an author is one of the reasons why academics cite their sources (and, as we will see, it also facilitated the historical emergence of peer review), which in turn enables us to ascertain, through citation analysis, the degree to which a paper has been accepted by the relevant academic community. On the other hand, some of the citation tools mentioned above only capture open access information for which the owner has granted a certain type of copyright license. To base our assessment on only this metric could therefore potentially distort the result. Furthermore, keeping information proprietary can also prevent open scrutiny, limit accessibility for users and cause incompatibility of standards – all key information quality concerns. In forensic contexts, this has been discussed particularly in relation to computer forensic tools and methods (Kenneally 2001), where some argue for mandatory open source licensing of novel forensic tools to enable their quality to be ensured through communal programming oversight. For example, the combination of a citation analysis with a web analysis of how often a specific item of information has been accessed could provide a detailed picture of how far a new scientific idea has penetrated its relevant community. If, however, the algorithm for the web analysis was proprietary and its creator's IP rights were enforced, analysing its reliability and objectivity might be difficult or impossible.

To further illustrate the link between copyright and information quality, we will introduce examples of specific forensic disciplines that play a role in copyright litigation. These will be introduced in the next section.

12.1.1 Evidence Law as Forensic Information Quality Assurance

This section will introduce a case study to illustrate some of the ideas that were roughly sketched out above. The example is from copyright litigation, and uses stylometric authorship identification. It should be noted that this analysis is not intended as a criticism of this specific branch of forensics; the sole aim is to highlight those issues of information quality assurance that relate to the assessment of its suitability in a trial setting.

Imagine that an author suspects that a best-selling novel by a competitor is plagiarised from his own work. He sues for copyright infringement. The opposing side acknowledges that there are some similarities, but claims that these are only to be expected, given that they are from the same genre, fictionalise the same historical events and draw on the same traditions, myths and fables. To counter this defence, the original author calls as an expert witness a forensic linguist who carries out a statistical authorship ("stylometric") analysis. One element of this analysis involves a comparison of the two texts to discover if there are any key phrases, expressions or terms appearing in both that are sufficiently unusual in the context of the works' genre, language and time period to suggest that they originate from the same source (based on Olsson 2008; for a criticism of the method see e.g. Rudman 2012).

To decide if this evidence meets the minimum threshold of admissibility (which necessarily comes before any consideration of its weight), the judge may want to ask a number of questions, each of which relates to a problem of information quality:

1. Is the expert truly an expert in the relevant field, and do we have reason to believe that she, personally, is good at what she does?
2. Are the data she used for this specific case reliable, accurate and relevant? In the case of forensic authorship analysis, for example, to evaluate the match probability of the expressions under consideration it is necessary to identify an appropriate reference group, e.g. "novels written on historical fiction in English between 1900 and 2000".
3. Finally, is the abstract theory the expert uses to interpret linguistic patterns as indicative of authorship sound? Is it possible to determine authorship by the statistical distribution of key terms and phrases?

The focus of the analysis in this paper will be on question 3, the abstract evaluation of how sound a theory is. Firstly, however, we will discuss briefly the various issues of information quality that are raised by the other two questions, and also the role of regulation that attempts to ensure or improve appropriate information quality.

For question 1, the court needs information about the expert that is accurate ("does she have the degrees she claims to have?"), objective ("was the accreditation process and the tests she had to undergo objective and rigorous?"), relevant ("does she hold the right type of degree and have pertinent experience?") and complete ("was she debarred from her professional organization, or were there disciplinary proceedings against her?"). For forensic laboratories, ISO norms and similar standards are of increasing importance (see e.g. Penders, and Verstraete 2006; Giannelli 2011; Malkoc and Neuteboom 2007). In this area the similarity between information quality research and forensic regulation is arguably the most obvious, and the results directly applicable.

Courts rely for their evaluation on a mixture of external gatekeepers and regulatory environments, for example legally-mandated registration with a professional body. A second source is of course university degrees, which are in turn attained in environments heavily regulated by laws. Evaluating the answers to the three questions above requires access to reliable information, for example a database of accredited Higher Education institutions – something that in the past has often proved to be surprisingly difficult, at least in the UK. In one famous case, a self-styled expert in "forensic handwriting, drugs, lie detection, toxicology, facial mapping and forensic dentistry" succeeded in giving evidence in more than 700 cases before his lack of qualifications was noticed. Gene Morrison had purported to be a forensic expert with a BSc in forensic psychology, an MSc in forensic investigation, and a PhD in criminology – all of which he bought from "Rochville University", a website that offers "life experience degrees" for money. Many of the reports that his company produced over the years were cut and pasted from the internet.

One final aspect of this case which is relevant for our purposes was the extent to which Morrison was able to "self certify" from the perspective of the legal system. As soon as one court had permitted his evidence, he was able to describe himself as a court-recognized expert, and in every subsequent case where he gave evidence more data were produced to confirm his 'expertise'. Breaking this vicious circle of self-reflexive confirmation that created more and more misleading information about his credentials proved to be the most difficult aspect of identifying him as a fraud. It should have been trivial task to identify concerns about his activities using nothing but information publicly available on the internet. The use of plagiarism detection software such as Turnitin, or even some simple queries of a search engine, would have identified those parts of his reports which were plagiarized from internet sources. In data quality terms, authenticity and provenance should have been easily identified as problematic. Equally straightforward would have been a check of the accreditation status of Rochville University, a check that would have unearthed reports about this organization that clearly expose its illegitimacy. The Morrison case demonstrates that the justice system, with its established hierarchies and reliance on official gatekeepers, needs to find better ways to retrieve, evaluate and utilise the vast amount of information created by digital media, even if that information lacks the *prima facie* reliability of official documentation.

Moving on, question 2 forms a bridge between questions 1 and 3. The issue here is not whether the individual scientist who testifies in a trial is well-qualified, nor is it whether the theory in question is generally sound. Rather, the issue is whether sufficient data are available to apply the theory correctly in the instant case. For example, it is in principle possible to determine the sex of an unknown murder victim from the skeleton by taking certain measurements of the femur and tibia (İşcan and Miller-Shaivitz 1984). The relevant ratios differ between populations, however. To make an accurate determination, therefore, the forensic anthropologist needs sufficiently large data sets from the relevant population of the victim to make an assessment with any degree of reliability. These data also need to be up to date, as these ratios can change over time, for example through changes in diet or lifestyle (discussion of a good example can be found in Steyn and İşcan 1997). Furthermore, to get an appropriate standard the data set needs to comprise a representative sample of the population.

We find similar issues across a variety of heterogeneous forensic disciplines. In forensic linguistics for instance, the frequency of spelling mistakes can be relevant to a determination of whether a ransom note came from the suspect – but again, data are needed to establish appropriate standards and baselines: how often do people from the same population as the suspect, at around the time the note was written, generally misspell certain words?

From an information quality perspective, the data need to be accurate, objective, relevant and timely. These are key concepts in any attempt to define information quality. We have seen in recent years an increasing recognition by the courts that the availability, relevance and accuracy of information about the reference class used by expert witnesses matter crucially in the assessment of the evidential weight of their statements (Tillers 2005; Franklin 2011). While information quality research can therefore provide lawyers with a highly flexible, domain-independent vocabulary to probe expert testimony, it has yet to find its way into legal curricula and training manuals and any discussion of it in a legal context tends to focus only on specific databases and forms of evidence.

Fingerprint and DNA databases raise obvious issues of information quality, and have shaped the legal mindset on these issues to a high degree. They operate often in tightly regulated environments, where ISO standards and accreditation perform a gatekeeper function for the justice system. These create a veil that trial lawyers will not normally be able – or permitted – to pierce. Because of this regulatory environment, the ability of new communication technologies to effect change seems limited. Yet, as the copyright example above shows, the relevant data need not come from "official" state-owned or -regulated databases or from tightly-controlled scientific studies (as in the femur example mentioned above). To determine, for example, how widespread the use of a certain term in a specific community is, mining the data contained in posts on an online community frequented by members of the target population can provide pertinent information. In the absence of official gatekeepers, these informal, unregulated, often user-generated data sets pose more intricate legal issues. Lawyers need to acquire the skills and knowledge to query them more effectively, including the right kind of analytic vocabulary – something which information quality research can provide.

The compilation of relevant data sets by hobbyists and "amateur scientists" which the digital revolution has enabled will become much more widespread in the future (Grand et al. 2012 for the quality assurance implications of this; see Neis and Zipf 2012 for an application of these ideas). Equally, forensic re-use of data from non-forensic sources will become more relevant. Returning to the copyright example above, in order to generate the reference class of expressions in published English books of the relevant genre, Google Books or Project Gutenberg could provide the necessary database, even though of course it was neither project's aim to establish a database of changing English word usage for forensic purposes. The digital revolution will therefore play a more prominent role in the future, especially by providing non-regulated, informal datasets that nevertheless allow forensic practitioners to establish base rates and standards.

The example of Google Books as a database that allows the standardisation of a baseline for similarity between texts also demonstrates the interaction of information quality with (copyright) law. Firstly, copyright law influences to some degree which books are made available in the database, as copyright holders can prevent the inclusion of their work. This could, therefore, theoretically result in a biased sample. Secondly, forensic scientists carrying out data mining on the database for this new, secondary application could be, in the absence of a license, violating the copyright of both Google and the original authors/publishers (where applicable). At present Google encourages the development of secondary usage through custom third-party applications, but there is of course no guarantee that this liberal access policy will continue, or for how long. In an effort to create a more permissive copyright environment, the UK government intends to amend the Copyright, Designs and Patents Act 1988 to permit people who already have the right to access a copyright work to copy it for the purposes of data mining, provided the analysis and synthesis of the content of the work is solely for non-commercial research.[4] With the recent privatisation of the forensic services market however, such a defence would not be available any longer for many forensic scientists based in the UK. In terms of information quality, copyright restrictions can therefore impact on the completeness, accuracy and availability of the relevant information. There is no broad cross-jurisdictional exception for "furthering the interests of the justice system" in copyright law, although in the UK section 45 of the Copyright Designs and Patents Act 1988 specifies a defence for "anything done for the purposes of parliamentary or judicial proceedings". This provision would, however, in all likelihood not extend to the investigative stage of a criminal investigation, let alone to the work of a commercial provider of forensic services as part of the preparation of a civil lawsuit. Similarly, despite providing broader and more flexible exceptions than European copyright law, the United States' "fair use" doctrine would also most likely not permit such a defence. Copyright law therefore places at least some limitations on the ways in which the increasing amount of digital information can be utilised in forensic contexts.

It is, however, the third question which has attracted the greatest amount of concern and academic scrutiny. It is here that difficult conceptual questions are raised about the interface between the law and science: how can lawyers, untrained in the sciences, nonetheless determine that a specific method or scientific theory is "sound enough" to be at least admissible at trial, and how can they assess the reliability of the evidence should this threshold have been met (see e.g. Risinger 2000; Haack 2003)? Before we address this question in more detail, we note that the three questions posed above introduce two very different types of concern about information quality in forensic settings. The first concern deals with information quality on the substantive level of the scientific discipline itself and was captured in the second question. Typical queries are of the form: "are the data in the DNA database of sufficient quality to allow the experts to assign evidential weight to a match?"; "are the data harvested through Google Books biased towards a certain genre?" or "do we have sufficiently large fingerprint databases to say with confidence that no two prints are identical?" By contrast, the second concern deals with meta-information about the scientific process itself. Judges and lawyers rely on gatekeepers that provide data *about* science to do their job. This in turn raises the question of what demands should reasonably be made of the quality of those data. If, for example, "acceptance in the scientific community" is one of the proxies judges use to evaluate the reliability of a forensic theory, what sort of data do judges require to determine if indeed that theory has been accepted? It is at this meta-level that changes in publication and science communication practices, brought about by the digital revolution, will have their greatest impact.

We now turn to the ways in which theories about the production of scientific knowledge found their way into evidence law, the obstacles they currently face, and the contribution that research in information quality can make to alleviating these problems.

---

[4] http://www.ipo.gov.uk/techreview-data-analysis.pdf

12.2 Who Reviews the Reviewers? Gatekeeping After *Daubert*

As noted in the introduction, courts have been profoundly affected by scientific progress. While eyewitness statements had been the main source of evidence for thousands of years, within a few decades disciplines such as forensic chemistry, DNA analysis, and forensic dactyloscopy emerged as main competitors, imbued with the aura of scientific objectivity and rigour. The process accelerated in the twentieth century, with fingerprints and in particular DNA becoming the new gold standards for evidential rigour. Initially, the limited number of both forensic practitioners, who were generally affiliated with universities, and forensic disciplines themselves meant that judges were required to master only a fairly manageable amount of information in order to assess those expert witness' credibility. Self regulation by the forensic profession(s) along with the formal training and accreditation of practitioners established another layer of gatekeeping on which the judiciary began to rely – we mentioned above the foundation of the "Institut de police scientifique" at the University of Lausanne in 1909 as the first school of forensic science in the world. 1948 saw the foundation of the American Academy of Forensic Sciences. Soon dedicated, peer-reviewed journals began to emerge that catered for these forensic research communities, such as the *Journal of Forensic Sciences*, which has been published since 1948 by the AAFS. The importance for evidence law of these gatekeepers became apparent in the 1923 US decision of *Frye*[5] that laid down the standards for admissibility of scientific evidence for the next few decades. The court in *Frye* opined that

> Just when a scientific principle or discovery crosses the line between the experimental and demonstrable stages is difficult to define. Somewhere in this twilight zone the evidential force of the principle must be recognized, and while the courts will go a long way in admitting experimental testimony deduced from a well-recognized scientific principle or discovery, the thing from which the deduction is made must be sufficiently established to have gained general acceptance in the particular field in which it belongs.[6]

The trust placed in the self-regulatory potential of the scientific enterprise implied by the phrase "general acceptance in the relevant scientific field" leaves much of the information quality assurance or "gatekeeping" function with bodies outside the judicial system. No attempt was made in *Frye* to define "general acceptance in the scientific community", and while subsequent courts offered various diagnostic criteria, the term remained largely ill-defined. Having the relevant science published in peer-reviewed journals has, however, been one strong indicator, as the case of *Berry v. CSX Transportation, Inc.* 709 So.2d 552, 569 (1998) showed:

> while the existence of numerous peer-reviewed published studies does not guarantee that the studies are without flaws, such publication alleviates the necessity of thorough judicial scrutiny at the admissibility stage

[5] *Frye v. United States* 293 F. 1013.

[6] *Ibid.* at 1014.

Problems with this definition persisted however. Conceived as a conservative test intended to keep untested ideas away from the jury, it was soon criticized for being on the one hand overly exclusive, preventing novel but promising and ultimately sound ideas from being heard, and on the other hand subject to "avoidance by self-certification" though more and more specialized groups of experts. We can, for example, assume that all polygraph examiners believe in the validity of polygraph testing. If "particular field" is narrowly defined, then we would have to conclude that it is therefore generally accepted – by other polygraphists. Only if we widen our scope and include, for example, cognitive scientists or behavioural psychologists do we find "external" validation – or lack thereof. In the 1993 case of *Daubert v. Merrell Dow Pharmaceuticals* 509 U.S. 579 (1993) the United States Supreme Court finally reacted to increased concern about low quality ("junk") science (Huber 1993) by amending and refining the criteria for admissibility. The Justices held that in order for expert evidence to be admissible, the following five criteria have to be met:

1. Is the theory falsifiable, refutable, and/or testable?
2. Has it been subjected to peer review and publication?
3. Is the actual or potential error rate known?
4. Are there standards and controls concerning the operation of the scientific technique?
5. Is the theory generally accepted by a relevant scientific community?

It is criteria 2 and 5 that concern us here. While *Daubert* is a US decision, none of these principles is linked intrinsically to concepts from US law and they can be seen as an attempt to provide common sense guidance to solve a problem that all jurisdictions face: How can judges control the quality of the information presented in a courtroom when it originates in disciplines of which they have little or no knowledge? This is part of a deeper paradox of evidence law identified by Mnookin: We only admit expert evidence if it is helpful, that is, if it covers ground the jury (or judge) cannot cover on their own. But if the trier of fact by definition lacks the knowledge that the expert provides, how can they rationally evaluate the expertise on offer (Mnookin 2007, p. 1012)? This epistemic conundrum is structurally related to the paradox of explanation (Boden 1961) and also Moore's paradox of analysis (see e.g. Balcerak 2012). In the classical philosophical form of these paradoxes, the question is how an analysis or an explanation can be both correct and informative. On the one hand, a good analysis or explanation should not be circular. X can't be explained by X. On the other hand, a good analysis should also be correct, and that means in particular that it should be meaning preserving. If it is meaning preserving however, then we should be able to substitute the analysis/explanation in any context where the original concept is used, without change in meaning of the discussion. This however would mean that the explanation/analysis can't also be truly informative.

While there have been several proposals in the philosophical literature on how to resolve this paradox, we face here a more modest task. Shifting the analysis to a meta-level and interrogating not the science directly, but the process by which it was produced and evaluated, is in principle a way to overcome our problem. Judges and juries do not need to know the specific forensic discipline under discussion; they do not even need to know much about generic but substantive tests of the quality of expert statements, such as analysing the statistical methods employed by the expert. Rather, they only need to know enough about the ways in which scientific knowledge claims are produced, communicated and their quality controlled – each of which can be judged by criteria that apply across scientific disciplines. The problem with the *Frye* test is therefore not so much that it asked a meta-question to determine the information quality of scientific expert statements, but that it asked the wrong question, or at least gave insufficient guidance for an answer. The general acceptance test in *Frye*, and its reformulation in *Daubert's* criteria 2 and 5, were insufficient to address the root cause of the increase in low quality science in courts. By putting their faith in peer review and the quality control of dissemination strategies that lead to "general acceptance", the courts created rules that may have worked in the nineteenth century when the number of journals was limited, set up costs for new publications were high, and the number of expert communities was limited. By the end of the twentieth century, however, this landscape had been fundamentally altered (Judson 1994). By focusing on forensic theories in the abstract, rather than the process of scientific knowledge production itself, judges missed an opportunity to take into account just how much what we understand as science depends on specific structures for the publication and dissemination of knowledge – how even in science the medium is part of the message. An information quality approach that focuses on the meta-attributes of scientific knowledge claims can partly address this failing in forensic contexts. It avoids on the one hand the explanatory paradox, and on the other requires the trier of fact to apply only criteria which are relevant across the range of scientific disciplines, without necessarily having a detailed understanding of their substance.

The relation between peer review and *Daubert* has been masterly dissected by Susan Haack (Haack 2006), who added to the increasing number of voices expressing concern about peer review as the gold standard of quality control in science (see e.g. Triggle and Triggle 2007; Enserink 2001). It is only possible to summarize her analysis briefly, focusing on those aspects relevant for our argument. Haack distinguishes between peer review in the narrow sense – pre-publication peer review – and peer review in the broader sense, that is to say the continuing process of analysing, scrutinising, testing and criticising of scientific ideas after publication. Post-publication peer review by contrast becomes relevant mainly as an element of assessing acceptance within the scientific community. Despite the well understood limitations of this method, judges tend to focus on pre-publication peer review and have also exhibited confusion about peer review methodologies (Chan 1995). Studies on the limitations of (pre- and post-publication) peer review are by now well established; its failings to identify intentional fraud are in particular well documented, as is its susceptibility to manipulation by commercial interests (Haack 2006, pp. 21–33). *Daubert* itself, a case concerning birth defects caused by the drug Bendectin, illustrates one particular shortcoming: None of the existing *published* studies indicated a problem. But this does not mean that there were no studies indicating a problem. Publication bias – the difficulties involved in publishing negative results – is a well-known problem in traditional academic publishing (see e.g. Fanelli 2012). For forensic applications that operate in an environment where burdens of proof are allocated asymmetrically, this is particularly problematic – the ability to show that a certain method fails, at least under some conditions, may be the only thing needed to create the reasonable doubt necessary for an acquittal. In this respect, however, the digital revolution and the changes in publication practice that it has enabled provide a remedy. Reduced set up costs have enabled the launch of several journals that are dedicated solely to the publication of negative results, including the *Journal of Negative Results in Biomedicine*, the *Journal of Negative Results – Ecology & Evolutionary Biology* and the *Journal of Articles in Support of the Null Hypothesis*. Citation of these journals will, however, follow very different patterns to the citation of journals containing positive results, and further research would be needed on how to combine citation scores for both types of result in order to determine how they can in combination provide a better, quantifiable proxy for the term "acceptance in the scientific community".

Generally, we claim that most problems with the *Frye* criterion exploit limitations in the way in which academic knowledge was traditionally produced. If there are only a few "choke points" in a quality assurance system, manipulation is considerably easier than in a distributed, networked environment. New forms of science communication in the distributed and networked world of the internet change the nature of post-publication peer review, and with that have the potential to shift the burden from peer review in the narrow sense back to the older *Frye* criterion of general acceptance, except now with the possibility of using data mining techniques to provide it with a more rigorous quantitative underpinning.

One important enabler for the emergence of peer review in the seventeenth century were copyright concerns. In response to scientists' interest in receiving due credit for their work, the Royal Society of London began to record the date on which it received a letter announcing an experiment (Zuckerman and Merton 1971, p. 70; Haack 2006, p. 792). This enabled them to adjudicate claims of "priority", in effect giving publications a time stamp as a first piece of important meta-data. Soon, these time stamps were accompanied by further information showing that other members of the society had read and recommended the submission. Thus the foundations for peer review were laid. Scientists' desire to receive acknowledgement for their work, which is in the present day enshrined in the copyright law concept of the "moral rights of the author", also provides a basis for tracking the spread of an idea within the relevant scientific community. Bibliometrics can provide an evidential basis for the claim that an idea has gained "acceptance in the relevant scientific community" by following how it has been cited. Digital publication combined with citation tools such as Citeseer can then provide judges with "second level" information to evaluate the reliability of scientific expertise.

Following the lead of the Royal Society many other prominent scientific publications implemented early forms of peer review. Nevertheless it remained common for individual editors to establish journals primarily as outlets for their own research (Burnham 1990), and "review by editor only" remains a procedure that can be found in well established and highly influential journals. It was only in the twentieth century that peer review became the gold standard for quality control, albeit that that function occasionally hid the fact that part of its rationale was the rationing of a scarce recourse – space in print journals (Burnham 1990, p. 1236). However, as the cost of producing journals fell and the profits earned from publishing them rose, both the scarcity rationale and peer review more generally became less and less justified. Today, one of the criticisms of peer review points out that, eventually, most manuscripts submitted for publication find their home "somewhere", even if only in a low-quality journal. The proliferation of journals also means that it becomes more and more difficult for anyone not deeply involved in the relevant scientific area to check if results have been replicated, criticised, falsified or otherwise scrutinised already. The digital revolution prima facie increases this pressure. The minimal cost involved in establishing a new journal, as well as policy-driven initiatives like "author pays" (Harley et al. 2007) which is promoted as the "gold standard" of open access publishing by inter alia the UK government, have resulted in an exponential growth of online open access journals. Some of these border on the fraudulent while some are essentially vanity publishers with limited, if any, quality control.

On the other hand, digital publishing provides us with novel tools to trace the reception of scientific ideas by their respective communities. Citation analysis and bibliometrics are greatly facilitated by online tools. Google Scholar, with its snipped preview function, allows even an untutored user to quickly gain an idea of how often a paper has been cited, and if the citation was in a positive or negative light, information that previously was only accessible to the experts working in the area. Download statistics in open archives such as SSNR provide additional information about the impact of a publication, as can website analytics (number of visits, their duration, number of downloads etcetera). Information retrieval models which are based on the "popularity" of an item – expressed by, for example, the number of links to it (this is roughly akin to Google's PageRank technology) or even the number of "likes" it receives on a science blog entry – are other approximations of the concept of "acceptance in the scientific community". Finally, social media-based publishing, including the often maligned Wikipedia, shows an approach to publication that blurs the borders between pre- and post-publication peer review. To assess the quality of information in a Wikipedia entry, the user can also refer to the "talk pages" to ascertain if any major controversies took place in the process of the article reaching its present form. Some innovative science publishers have by now spotted the potential of this approach; PLOS One in particular decreased the parameters for peer review in exchange for adding a "comment" function to their articles which enables readers to submit criticism and commentary more efficiently and openly. Further, in contrast with the traditional publication of a refutation in separate journal, such responses are directly accessible by readers who may not have access to that other journal. The "Frontiers in..." series of journals, now part owned by the publishers of Nature, has taken a different route. There, the referees of accepted papers are fully acknowledged in the publication, the aim of which is not just to reward the referees (and protect their IP), but also to increase transparency. Now it can be ascertained if the referees had a personal stake in the theory, or if there are other reasons to doubt their independence or qualifications. This enhances, in information quality terms, the believability of the work they have refereed, and also the objectivity and reputation of the work's conclusions themselves.

Another problem identified by Haack is the difficulty to determine if an article has been subsequently withdrawn or corrections to it have been published due either to misconduct or to mistakes. It is even more problematic to trace all those studies that have relied on such papers (Haack 2006, p. 803; see also Pfeifer and Snodgrass 1990). Here too, digital publication models and the ability to trace digital objects through their meta-data can be of assistance. PubMed, for example, allows searching for retracted publications in medical journals. Reduced start up costs for journals, identified as a potential problem for forensic information quality assurance above, can also play a positive role.

So far, the legal system makes only minimal use of these new tools. The fact that an article has been cited is sometimes offered as evidence for it having been accepted in the academic community, but that conclusion is usually reached with little systematic evaluation (Haack 2006, p. 45). Questions that should be asked, but rarely are, include: "How many citations are typical for that discipline?"; "Are the citations positive or negative?"; "Do they cite the article because it is new, interesting and dealing with an unsettled issue, or do they cite it because it is by now part of the canon?"; "Are the authors who cite the paper colleagues, collaborators or students of the author, or does the paper penetrate the wider academic community?" and "Is there evidence for a small community of researchers citing each other, or is a diverse and sizeable community convinced by the results?".

Thus it can be seen that data mining, link analysis and computational bibliometrics provide an as-yet-untapped resource for the justice system that has the potential to address many of the limitations Haack and others have identified. These new approaches in turn need to be quality controlled, and it is this quality control that information quality research should facilitate.

12.2.1 Peer Review 0, Social Media 1

Haack's final conclusion on the value of peer review is sceptical. Pre-publication peer review fails even for the modest goal of ensuring minimal quality standards while, despite the great potential of post-publication peer review, there is insufficient guidance for judges on how to access and evaluate it. We have argued above that many of Haack's concerns are tightly bound to the limitations of traditional, physical publishing, and that digital publication, whilst also facilitating an increase in the amount of low quality information that must be filtered, also provides tools that address many of the issues currently encountered. In particular, it increases the amount of meta-information which can be used for assessing credibility and communal acceptance. Social media, science blogging and online discussion groups run by the research community (such as the old UseNet groups) are all places where academics leave digital traces that can be data mined; the information and patterns so discovered can help form a more robust answer to the question of whether an idea has found general acceptance.

The controversy around NASA's recent "alien life" débâcle illustrates this point nicely[7]: NASA had with great fanfare announced a press conference that coincided with the publication by some of its researchers in the highly reputable journal Science of a paper which gave rise to speculation about the discovery of alien life. While the content of the paper was in reality much less exciting than some had hoped, what was more worrying were the perceived shortcomings in its methodology. Within hours, leading researchers in the field weighed in by posting commentary on blogs which included their real names and in many cases their university affiliation. Others took the initiative of bringing these highly critical comments together on one site, with contributors discussing and further elaborating on each other's criticism. The authors of the Science paper initially refused to answer these criticisms, insisting instead that any detractor should subject their opposition first to peer review and publication – a process that would have been time consuming and would have reduced the traceability of the criticism for readers considerably. Not only did they misjudge the mood of the scientific community, they also failed to realize that the critics were the very same people who, under the traditional system, would otherwise be acting as referees for papers like the one under discussion, albeit anonymously and with much less accountability. The degree of public discontent with the findings forced Science to alter its own approach, whereby it permitted the publication of the "collated" criticism in the journal, and made subsequent articles criticising the initial findings freely available on their website. Eventually, further studies, now published through traditional channels, confirmed that it was not possible to replicate the results, and although calls for a retraction have increased the article as of today remains unchanged on their website.

What does this show for Daubert? The original findings met that case's requirement that they be published in a high-quality, peer-reviewed journal. However, those following the discussion in the blogosphere immediately realised that the paper was not accepted in the relevant academic community, and that the quality of the peer review had explicitly been called into doubt. In this case, the pattern that can be identified in the unregulated, non-peer-reviewed "grey" literature provides a much clearer idea both of how to assess the reliability of the paper and of the attitude of the wider scientific community towards it. Tools to mine information from blogs, and in particular to identify arguments and how they are taken up or rejected by communities, are being developed, thus placing the process of extracting this type of information on an even firmer footing. Making sense of these information sources and making sound conclusions about their quality requires new skills, but the changing nature of academic communication which is being driven by digital media also promises new solutions to the shortcomings of current courtroom practice identified by Haack and others.

[7] For a blog based account, see http://scienceblogs.com/worldsfair/2010/12/21/parallel-universes-arsenic-and/; or see Zimmer. "The Discovery of Arsenic-Based Twitter" (Slate.com, 27 May 2011, http://www.slate.com/articles/health_and_science/science/2011/05/the_discovery_of_arsenicbased_twitter.html).

12.3 Conclusion: Reviving Frye Through Information Quality Research

Let us recap. One of the earliest attempts to find a single unifying principle for the law of evidence was the "best evidence" rule, formulated by the eighteenth century jurist Gilbert (Twining 1994, p. 188). While this approach later fell out of favour to be replaced by a multiplicity of rules that seem impervious to reduction to a single principle, in 1806 a North Carolina court nonetheless pronounced that there is but one decided rule in relation to evidence, and that is, that the law requires the best evidence (cited from Langbein 1996, p. 1174). The rule remains valid today, if reduced to one amongst many of equal rank. It was evoked most recently in computer forensics to determine the status of electronic copies of documents (see e.g. Grossman 2004). The role of the judge then is to ensure that the jury hears the best possible evidence, which we can reformulate for our purposes as the evidence with the highest information quality. For Gilbert, this meant the use of original documents rather than copies, or of direct witness testimony rather than that of someone he confided in (hearsay). But with the advent of modern science in the courtroom, and the proliferation of scientific disciplines, this task has become all but impossible. No lawyer can possibly master and keep up to date with all the scientific theories he or she might encounter in court over a professional lifetime.

A first attempt to assist the judge in this task was the *Frye* test, which relied on the scientific community as gatekeepers. Acceptance in the scientific community, for which peer-reviewed publication was one important indicator, became a proxy for information quality. The interest of the judge in such an approach shifted to what was referred to above as "second-level information quality": how good is the information *about* scientific theories and the scientists that use them, such that people trained in law can make a rational decision on whether or not to believe them. This approach turned out to be insufficient, and was subsequently amended by the *Daubert* criteria. These added substantive tests of scientific theories, which nonetheless ought to apply across all forensic disciplines. They require from the judge an understanding of statistics, experiment design and epistemology. However, 20 years after the *Daubert* decision, it has become apparent that this change has had limited success at best. Judges rarely evoke the additional criteria; there is strong evidence both that they feel insufficiently prepared for substantive scrutiny of often complex scientific ideas, and also that the *Frye* test (as incorporated into *Daubert*) still exerts considerable influence. But if *Frye* was problematic in the 1980s, then its original formulation is even less tenable today. Dramatic changes in the way in which academic knowledge is produced and disseminated have undermined the old gate keeping functions and quality control mechanisms that the court in *Frye* relied upon. These were brought about by a mix of political, economic and technological change. Economically, the number of universities, academics and researchers increased with the opening of higher education in the 1960s. This also created a greater market for academic publications, a proliferation of journals and increased pressure on peer review. Politically, the "evaluation culture" in higher education further incentivised academics to publish more and to show "relevance" through application of their theories in, for example, forensic settings. Finally, from a technological perspective, the Internet and digital publishing reduced the start up costs for new journals, which were thus able to react to the greater demand for publication outlets. This paper has tried to argue tentatively that while the implementation of *Frye* was fraught with problems, the thinking behind it was sound. Second-order information quality proxies are indeed what the judiciary should be focusing on. The procedural aspects of scientific knowledge production are much closer conceptually to lawyers' procedural thinking than substantive scientific investigation. Most lawyers will have experienced during their studies the need to evaluate the credibility of published articles or court decisions, or may even have gained first-hand editorial experience. Such relevant skills apply across a wide range of forensics disciplines, and also enable lawyers to avoid the "paradox of explanation", since the evaluation of the information quality of a source is in this approach independent from the source's substantive content.

This does not mean that the task is trivial, however. The problem becomes instead identifying the right type of meta-information about the scientific process, ensuring its quality and equipping judges with the necessary skills to interpret the data. In this approach, new forms of academic publishing models become a potential ally and a rich source of relevant information. While "acceptance in the academic community" is currently the least well-defined, and least exact, of the concepts used in evidence law to determine admissibility, the use of modern bibliometrics and data mining has the potential to give this concept a highly precise underpinning. These techniques are not limited to digital versions of traditional publications but can also be used to analyse the reception of scientific ideas in the wider community through blogs, wikis and other user-generated content. In the long run this also opens up exciting opportunities for new forms of scrutiny and democratic participation in domains that were previously the preserve of a small cadre of 'experts', something particularly suited for the jury in its role as democratic guarantor of the legitimacy of the trial. The exponential increase in the availability of meta-data that can facilitate quality assessments, and new tools to make use of it, opens up avenues for outsiders to gauge how a scientific theory is perceived by those in the relevant field, something that previously only those intimately engaged in the scientific process themselves could assess – with all the potential of partisanship that this entails. While the demise of traditional forms of gatekeeping in science creates a host of new challenges for the interface between it and the law, the need to sort through the plethora of voices and to establish which should be treated as authoritative, the increased volume of meta-information about science and scientists, together with new tools to mine, combine and interpret these data, can also provide an alternative approach to forensic information quality assurance.

The "acceptance in the scientific community" criterion is potentially put on a new and more solid footing through easier traceability of post-publication peer (and non-peer) review, more open and transparent processes of deliberation about science, and the possibility of more easily quantifying and visualising how ideas are received by, and spread through, scientific communities. Information quality research can help to develop the necessary analytical vocabulary for this task. By recasting the legal-evidential question of admissibility to one of information quality assurance, tools developed in the computer science community can further enhance the way in which the legal system deals with the increased role of science in legal fact finding. For the justice system, this requires that new skills be imparted to judges and party advocates to enable them to utilise these new sources of information. However, the communication between evidence scholarship and information quality research is not a one way street. Information is rarely, if ever, created in a legal vacuum, and the relevant legal rules can in turn have an impact on the quality of the information that is generated. Intellectual property rights, and copyright in particular, can both hinder and assist these new models of forensic information quality. They hinder the evolution towards such models when digital meta-data about scientific ideas and theories are treated as proprietary and shielded from being openly accessible (and subject to scrutiny of the way in which they are collected), or when copyright is used to limit new forms of digital comment and critical annotation of research results, for example prohibiting the "fisking" of a problematic article in the blogosphere on copyright grounds. This was attempted by Scientific America who tried to take down a very detailed criticism of one of their publications, claiming copyright violation.[8]

Intellectual property rights can further these models of information quality by providing authors with new methods to disseminate their ideas via creative commons licenses, open access publishing and self-archiving, thus restoring to them some of the control over their intellectual efforts that had previously been lost to commercial publishers and their walled gardens. Copyright also enables quality control by the author, preventing distortions and certain types of misrepresentation. The study of how different ownership regimes of data contribute to information quality should become a field for research that crosses the boundaries between law and information science. Finally, it opens up the question of how appropriate copyright regimes can be created to further the social goal of reliable judicial fact finding. Extending the UK's limited copyright exceptions to include material produced during the forensic evaluation and gathering of data in the preparation of a case is a radical but nonetheless necessary step if this new information is to have maximum impact on the administration of justice. Creating an exception for the use of certain types of data use along the lines envisioned by the data mining exceptions (i.e. it serves non-commercial research purposes) could act as a blueprint for more ambitious public interest exceptions to copyright. Conversely, professional societies of forensic scientists and their codes of conduct can play a regulatory role in prescribing under which conditions forensic practitioners should make their datasets available through open source models, and where proprietary approaches are legitimate. In this way, our analysis comes full circle, to legal regulation of information quality similar to that attempted by the Data Quality Act, but hopefully with a view to it being informed by the more sound theoretical underpinnings that might be provided by a new discourse in forensic information assurance.

[8] http://www.greenspirit.com/lomborg/pages.cfm?num=3

References

Boden, M. (1961). *The paradox of explanation* (Vol. 62). Proceedings of the Aristotelian Society, The Aristotelian Society (Blackwell).

Bowers, C. M. (2006). Problem-based analysis of bitemark misidentifications: The role of DNA. *Forensic Science International, 159*, S104–S109.

Broeders, A. P. A. (2006). Of earprints, fingerprints, scent dogs, cot deaths and cognitive contamination – A brief look at the present state of play in the forensic arena. *Forensic Science International, 159*(2), 148–157.

Burnham, J. C. (1990). The evolution of editorial peer review. *JAMA: The Journal of the American Medical Association, 263*(10), 1323–1329.

Bushman, R. M., & Piotroski, J. D. (2006). Financial reporting incentives for conservative accounting: The influence of legal and political institutions. *Journal of Accounting and Economics, 42*(1), 107–148.

Chan, E. J. (1995). The brave new world of *Daubert*: True peer review, editorial peer review, and scientific validity. *New York University Law Review, 70*, 100.

Enserink, M. (2001). Peer review and quality: A dubious connection? *Science, 293*(5538), 2187–2188.

Faigman, D. L. (2013). The *Daubert* revolution and the birth of modernity: Managing scientific evidence in the age of science. *University of California Davis Law Review, 46*(3), 893–931.

Faigman, D. L., et al. (2006). *Modern scientific evidence*. St. Paul: West Group.

Fanelli, D. (2012). Negative results are disappearing from most disciplines and countries. *Scientometrics, 90*(3), 891–904.

Franklin, J. (2011). The objective Bayesian conceptualisation of proof and reference class problems. *Sydney Law Review, 33*, 545.

Giannelli, P. (2011). Wrongful convictions and forensic science: The need to regulate crime labs. *North Carolina Law Review, 86*, 163.

Gill, P., Evett, I. W., Woodroffe, S., Lygo, J. E., Millican, E., & Webster, M. (1991). Databases, quality control and interpretation of DNA profiling in the Home Office Forensic Science Service. *Electrophoresis, 12*(2–3), 204–209.

Grand, A., et al. (2012). Open science a new "trust technology"? *Science Communication, 34*(5), 679–689.

Grossman, A. M. (2004). No, don't IM me-instant messaging, authentication, and the best evidence rule. *George Mason Law Review, 13*, 1309.

Gundelach, A. (1989). Lawyers' reasoning and scientific proof: A cautionary tale in forensic odontology. *The Journal of Forensic Odonto-stomatology, 7*(2), 11.

Haack, S. (2003). Inquiry and advocacy, fallibilism and finality: culture and inference in science and the law. *Law, Probability and Risk, 2*(3), 205–214.

Haack, S. (2006). Peer review and publication: Lessons for lawyers. *Stetson Law Review, 36*, 789.

Harley, D., et al. (2007). The influence of academic values on scholarly publication and communication practices. *Journal of Electronic Publishing, 10*(2). doi: http://dx.doi.org/10.3998/3336451.0010.204.

Huber, P. W. (1993). *Galileo's revenge: Junk science in the courtroom*. New York: Basic Books.

İşcan, M. Y., & Miller-Shaivitz, P. (1984). Determination of sex from the tibia. *American Journal of Physical Anthropology, 64*(1), 53–57.

Judson, H. F. (1994). Structural transformations of the sciences and the end of peer review. *JAMA: The Journal of the American Medical Association-US Edition, 272*(2), 92–95.

Kaye, D. H., & Smith, M. E. (2003). DNA identification databases: Legality, legitimacy, and the case for population-wide coverage. *Wisconsin Law Review, 3*, 413.

Kenneally, E. (2001). Gatekeeping out of the box: Open source software as a mechanism to assess reliability for digital evidence. *Virginia Journal of Law and Technology, 6*(13). http://www.vjolt.net/vol6/issue3/v6i3-a13-Kenneally.html. Accessed 4 Oct 2013.

Lacko, M. V. (2004). The data quality act: Prologue to a farce or a tragedy? *Emory Law Journal, 53*, 305.
Langbein, J. H. (1996). Historical foundations of the law of evidence: A view from the Ryder sources. *Columbia Law Review, 96*, 1168–1202.
Malkoc, E., & Neuteboom, W. (2007). The current status of forensic science laboratory accreditation in Europe. *Forensic Science International, 167*(2), 121–126.
Mnookin, J. L. (2007). Expert evidence, partisanship, and epistemic competence. *Brooklyn Law Review, 73*, 1009.
Neis, P., & Zipf, A. (2012). Analyzing the contributor activity of a volunteered geographic information project—The case of OpenStreetMap. *ISPRS International Journal of Geo-Information, 1*(2), 146–165.
Olsson, J. (2008). *Forensic linguistics*. London: Continuum.
Omand, D., Bartlett, J., & Miller, C. (2012). Introducing Social Media Intelligence (SOCMINT). *Intelligence and National Security, 27*(6), 801–823.
Penders, J., & Verstraete, A. (2006). Laboratory guidelines and standards in clinical and forensic toxicology. *Accreditation and Quality Assurance, 11*(6), 284–290.
Pfeifer, M. P., & Snodgrass, G. L. (1990). The continued use of retracted, invalid scientific literature. *Journal of the American Medical Association, 263*(10), 1420–1423.
Redmayne, M., Roberts, P., Aitken, C., & Jackson, G. (2011). Forensic science evidence in question. *Criminal Law Review, 5*, 347–356.
Risinger, D. M. (2000). Navigating expert reliability: Are criminal standards of certainty being left on the dock. *Albany Law Review, 64*, 99.
Rudman, J. (2012). The state of non-Traditional authorship attribution studies—2012: Some problems and solutions. *English Studies, 93*(3), 259–274.
Saxe, L., & Ben-Shakhar, G. (1999). Admissibility of polygraph tests: The application of scientific standards post-Daubert. *Psychology, Public Policy, and Law, 5*(1), 203.
Schweitzer, N. J., & Saks, M. J. (2007). The CSI effect: Popular fiction about forensic science affects the public's expectations about real forensic science. *Jurimetrics, 47*, 357–364.
Shapiro, S. A. (2003). Information quality act and environmental protection: The perils of reform by appropriations rider. *William & Mary Environmental Law & Policy Review, 28*, 339.
Shelton, D. E. (2008). The 'CSI-effect': Does it really exist. *National Institute of Justice Journal, 25*, 1–7.
Steyn, M., & İşcan, M. Y. (1997). Sex determination from the femur and tibia in south African whites. *Forensic Science International, 90*(1), 111–119.
Tillers, P. (2005). If wishes were horses: Discursive comments on attempts to prevent individuals from being unfairly burdened by their reference classes. *Law, Probability and Risk, 4*(1–2), 33–49.
Twining, W. (1994). *Rethinking evidence: Exploratory essays*. Evanston: Northwestern University Press.
Tobin, W. A., & Duerfeldt, W. (2002). How probative is comparative bullet lead analysis. *Criminal Justice, 17*, 26.
Zuckerman, H., & Merton, R. K. (1971). Patterns of evaluation in science: Institutionalisation, structure and functions of the referee system. *Minerva, 9*(1), 66–100.

Chapter 13
Personal Informatics and Evolution in the Digital Universe

On the Selection of Information, Archival Design, and Retrospective Analysis

Jeremy Leighton John

> *It would be quite possible for the machine to try out variations of behaviour and accept or reject them in the manner you describe and I have been hoping to make the machine do this. This is possible because, without altering the design of the machine itself, it can, in theory at any rate, be used as a model of any other machine, by making it remember a suitable set of instructions. The ACE is in fact, analogous to the 'universal machine' described in my paper on computable numbers.*
>
> *A. M. Turing, in a letter written from the National Physical Laboratory, Teddington, to the cyberneticist W. R. Ashby, circa November 1946, discussing the Automatic Computing Engine. The original letter is in the William Ross Ashby Archive among the manuscript collections of the British Library.*

Abstract The quantities of personal information being created, retained and analysed through digital technologies have received prominent attention. The present contribution considers information quality in this context. What is it, and how should it be assessed? What factors influence it? The chapter outlines a field of personal informatics concerned with all aspects of personal information over its entire lifecycle, directing attention primarily at its digital manifestations. Although the value of personal digital information to contemporary research is broadly understood, its future value for historical and scientific research warrants more consideration. Personal information is becoming increasingly essential to the digital lives of individuals, as a means of not only conducting day-to-day activities but also for ensuring identity, memory and reflection, as well as a possible emancipation of individual creativity and personal legacy. It is anticipated that there will be

J.L. John (✉)
Department of Digital Scholarship, British Library,
96 Euston Road, London NW1 2DB, UK
e-mail: jeremy.john@bl.uk

L. Floridi and P. Illari (eds.), *The Philosophy of Information Quality*,
Synthese Library 358, DOI 10.1007/978-3-319-07121-3_13,

a move towards individuals retaining and curating (increasingly with supervised automation) their own personal digital archives. On the other hand both technology and information are subject to ever more rapid change, increasing unpredictability and complexity. The handling of significant volumes of information while ensuring crucial aspects of quality are sustained could benefit from evolutionary insight. At the same time personal digital archives in the wild offer an important opportunity to further our ability to discern the limits and possibilities of retrospective analysis and our understanding of evolutionary processes themselves.

13.1 Introduction

13.1.1 Digital Life Information and (Personal) Big Data

People are creating, obtaining and holding vastly more information than ever before. It is possible to envisage people capturing almost every action of their lives through ubiquitous computing, life tracking and personal telemetry, and through the proliferation of countless computing devices and sensors connected to diverse networks. With the United Nations anticipating a world population of 7.3 billion people by 2016, Cisco forecasts that there will be more than ten billion devices, generating 130 exabytes of mobile data per year (Reuters 2012).

Increasingly information is being captured by individuals for their own benefit, monitoring their own health, digitally securing and enhancing their own home and private space, recording their daily activities and use of time, attending to their portfolio and creative productions, and generally accumulating numerous informational belongings as digital objects (Doherty and Smeaton 2008; Rodden 2008; Bell and Gemmell 2009; Laursen 2009; Olguin Olguin et al. 2009; O'Reilly 2009; Wolf 2009). Detailed and ongoing personal genomic, metabolic, neurological, physiological and immunological information may become commonplace for individuals.

Through their daily digital lives people will passively record information about their environment both social and natural as well as actively engaging with information collection for social research, community history and citizen science (Hopper and Rice 2008; Kanjo et al. 2008; Steed and Milton 2008; Kwok 2009).

Of special interest is the emerging use of technologies of information beyond sound and vision: remote haptic exchanges conducted with distant relatives. Three-dimensional printing of ornaments, clothing, tools and devices already raises the possibility of being able to recreate the artefacts of an individual's home using the digital information that prescribes these objects.

At the same time this personal information is expected to help personalise technology so that it serves the individual more effectively and efficiently (Anonymous 2011).

There is also the possibility of increasing emancipation of individuals and of personal legacy. The personal digital archive may be the ultimate resource for personally enhanced usability, which in turn motivates the curation of the personal archive (John et al. 2010).

New theories, principles and practices are essential in order to foster and sustain the utility of personal information along with individuals' trust (Jones and Teevan 2007; Rodden 2008; O'Hara et al. 2008; Mayer-Schönberger and Cukier 2013).[1]

13.1.2 Personal Curation

Memory institutions are already caring for archives of personal digital objects derived from writers, scientists, artists, and sociopolitical reformers. At the British Library these personal digital objects are referred to as eMANUSCRIPTS (eMSS), and found on an individual's hard drives, optical discs, floppy disks, memory sticks, online storage locations and webmail and social networking sites.

Personal informatics is the study of all aspects of personal information including such topics as privacy protection, the reuse of personal digital archives, personalized usability and personal information management (John 2012). Although personal information exists in public, charitable and philanthropic archival repositories as well as governmental and commercial databases, the present paper is focusing on what happens outside the repository, on 'archives in the wild' (personal digital information held locally at home or remotely in the cloud by individuals themselves and their families).

Childhood and adulthood reminiscence, ancestral origins and artefacts, and personal memory and family history manifested in archival objects, are of profound importance to individual, familial and cultural well being and self-esteem, to a sense of belonging (Hobbs 2001, 2012; Etherton 2006).

The necessity of suitable management and curation of personal digital belongings has been repeatedly emphasised (Marshall et al. 2006; Marshall 2007, 2008a, b). Similarly in a book entitled "The future of looking back", Richard Banks (2011) has reflected on how we should keep things when things are digital. The Memories for Life project deliberated technological implications (O'Hara et al. 2006). Thoughts of archival academics and practitioners have been documented by Cal Lee (2011) and coauthors in a book entitled "I, Digital".

The Digital Lives Research Project anticipated *archival* personal information management (*archival* PIM) (Williams et al. 2009; John et al. 2010) assimilating the importance placed by archivists on the entire lifecycle of the archive and its sustained reuse during an individual's life and beyond.

The value and quality of personal information for the individual and for future heritage, history and science will depend on the effectiveness of personal curation.

---

[1] The project InterPARES Trust is currently researching the issue internationally: Trust in Digital Records in an Increasingly Networked Society, www.interparestrust.org, based at the Centre for the International Study of Contemporary Records and Archives at the University of British Columbia.

13.2 Information and Evolution

13.2.1 Information as an Asset and as a Concept

Along with being an asset, information is also a concept. The abstraction of information provides a powerfully unifying perspective. Accordingly, theories of information and computation have been advanced in the physical sciences (e.g. Szilard 1929; Brillouin 1956; Zurek 1990; Bennett 1999; Landauer 1999; Nielsen and Chuang 2000; Bekenstein 2003; Lloyd 2006).

Most influential of all has been Turing's 'universal machine', the conceptual seed from which the era of digital information originated (Turing 1936). Befittingly, but dauntingly, this success in informational and computational theory has led to prodigious quantities of information of varying quality being produced. So much so that a philosophy of information (PI) has been summoned: "a philosophy of information design and conceptual engineering" to explain and foster the informational foundations of contemporary society and its intellectual environment (Floridi 2011).

This dual role of information as a means of elucidating and condensing universal and general processes and as a prodigious resource matches the two aims of the paper.

The first aim of the chapter is to reflect on the design of systems for digital curation and utilisation of information, specifically in the context of an individual's digital belongings. The second aim of the paper is to contemplate a natural history and theory of personal information that could help yield a better understanding of the digital universe.

Correspondingly, this chapter suggests that evolutionary approaches can be helpful. The two aims may be mutually beneficial since a better understanding of evolutionary processes in the digital universe may inform the design of systems for curating and using personal information.

13.2.2 Objective 1: Designing Adaptable Information Systems?

> The design of adaptive systems will be among the key research problems of the 21st century (Frank 1996; but see also Frank 1997).

13.2.2.1 Evolution as Technology

Evolution as a technology may take three forms. The two most common are: (i) adopt or modify a biotic technology (build a house using logs); and (ii) mimic or model a biotic technology (build a cooling system following the structural principles of a termite nest). A third way (iii), is to design a technology using evolutionary and adaptive principles (let the design of houses evolve through selection of variants of an already successful design) (for hints of building evolution see Li 2012 and references therein). Bioinspired engineering (Thakoor et al. 2002) may combine more than one approach along with human design creativity.

An important proof of concept for the scaling up of long term storage of information entailing the synthesis of DNA (deoxyribonucleic acid) has been conducted by storing Shakespeare's sonnets, Martin Luther King's most celebrated speech, and other major cultural information (Goldman et al. 2012). But just as significant as this direct use of DNA is the evolutionary process that has sustained this molecular system for millions of years (John 2008).

The possible place of evolutionary theory and methodology in digital preservation has been discussed in earlier publications (e.g. see John 2008; Doorn and Roorda 2010). These papers invoke the concepts of digital objects being selected and of archives and repositories functioning like digital organisms. The present paper expands on these ideas.

13.2.2.2 Evolutionary Algorithms, Opportunistic Networks

In nature, the role of information quality (IQ) and its protection and management can be examined at various scales from molecular and cellular through physiological and organismal to populational and ecological: from DNA repair to organismal reproduction and the metapopulation.

Can we contrive a system of evolutionary design for a personal information management system that allows individuals to better curate their information to suit their requirements?

One approach is to apply the evolutionary techniques being used for creating multifarious adaptive technologies. Systems and evolutionary biology is already providing a rich source of practical inspiration (Frank 1997; Bentley 1999; Nolfi and Floreano 2000; Forbes 2004; Kim and Ellis 2008; Crowcroft 2008; Allen et al. 2008). In computing and communication technology, selection-and-variation algorithms have been applied in both software and hardware evolution (Goldberg 1989; Mitchell et al. 1992; Thompson 1996, 1998; Sipper 1997; Stoica 2002; Goues et al. 2012), strikingly in the design of antennae for spacecraft (Hornby et al. 2011); see also Bentley (2007) and Sakellariou and Bentley (2012) for simulation of biological computation in digital design.

A strong insight into the future impact of ubiquitous computing and its social involvement of the individual is the use made of personal information ("the history of social relations among users") by dynamic network routing protocols (Allen et al. 2008; Boldrini et al. 2008; Hui et al. 2008; Lu et al. 2009; Xu et al. 2009).

13.2.2.3 Systems for Space, Systems for the Future

One of the requirements of long-term space missions is for independent capability of technologies with persistent automation. This requirement for a fail proof system that is sustainable far away from its origins, is not unlike the requirement of repositories to prepare for a time when there is no going back to obtain directly digital assets and equipment that may no longer exist.

Autonomicity is commonly invoked (e.g. Sterritt et al. 2006); but an early serious examination of the possibility of self-replicating and self-dependent entities (artificial life or evolving technology) was published in a report of NASA (National Aeronautics and Space Administration) in the 1980s (Bradley 1982; Freitas Jr and Gilbreath 1982). The whole volume is an intriguing mix of the theoretical and practical. One of its concerns was the design of a self-replicating lunar factory.

It also considers the transmission of information from one generation to the next: "Normally, the parent should transmit all information necessary for offspring to do their jobs and to construct further offspring in turn, but beyond this simple requirement there are many alternatives. For example, a parent machine might augment its program during its lifetime with some valuable information, and this augmented part of the program could then be transmitted to its offspring" (Freitas Jr and Gilbreath 1982, p. 199), this notion being reminiscent of epigenetic and Lamarckian processes (e.g. Jablonka and Lamb 1995, 2005).

The NASA report espouses reliable data management, database systems, information retrieval and management services, and notes the role of the user (pp 362–363). More recent research makes explicit use of evolutionary computation for decision-making including planning of trajectories (Sultan et al. 2007; Abdelkhalik 2013), scheduling (Johnston 2008), and command and data handling architectures (Terrile et al. 2005), and for the design of tools for simulating systems under varying circumstances (Kordon and Wood 2003).

The present paper considers the functioning of the archive in its own right, highlighting the application of evolutionary principles to the design of the archive itself.

13.2.2.4 Design for Curation as Retrospection

Personal information management has been geared towards use of current information in the present or from the immediate past. As time goes by in the digital era there will be a requirement for sustained memory of a life and of earlier lives in a family. This in turn calls for design for retrospective analysis, for systems that would facilitate the identification and reconstruction of past events and states. What kinds of personal digital objects will maximise or optimise future retrospection, and for what kinds of retrospection?

Both of these improvements in the design of personal information systems (everyday curation and sustained retrospection) would affect the quality of personal information and its use.

13.2.3 Objective 2: Observing Personal Information and Advancing Theory

13.2.3.1 Naturalistic Observation

How is personal information actually used by individuals? When and why? How does its use unfold in the wild, how does changing culture and society impact the creation, acquisition and reuse of digital information by people?

The *modus operandi* of ethology and behavioural ecology combines naturalistic observation with modest experimental manipulation in the wild. The use of sensors and the like is one approach for the study of personal information, in virtual and real worlds. Just as an understanding of the ecology and the vicissitudes of life for migrating birds benefits the design of sound conservation strategies (Wood 1992; see also Pullin 2002) so an awareness of the ecology of personal information can fortify digital preservation and curation strategies.

One of the most powerful toolsets is the progressively sophisticated suite of phylogenetic methods, and it is this avenue of research that is briefly elaborated at the end of this paper. For another naturalistic perspective see Naumer and Fisher (2007).

13.2.3.2 Evolution as an Informational Theory

The conjunction of information and biological systems is widely evident. It is common for introductory texts to portray genetics as an informational science (for example Mushegian 2007; and Hartwell et al. 2008, pp. 440–444). Similarly the immune system may be perceived from an informational perspective (Orosz 2001).

But the other side of the coin is that among the most potent of informational processes is natural selection. Most notably, of course, the concept of natural selection has been successful in explaining natural design, the design of material organisms that function through information and bring about its flow.

A further sign of the *puissance* of natural selection theory is its adoption in the physical sciences. Just as physics has become interested in informational perspectives, it has also begun to explore the selection process (Zurek 2004; Blume-Kohout and Zurek 2006; Smolin 1992, 1997, 2006; see also Gardner and Conlon 2013).

These theories remain subject to debate but it is entirely conceivable that selection theory can be extended into new realms and in new ways, and beyond the purely biotic.

13.2.3.3 Cultural Evolution: eMSS as Memes

Culture has been broadly described as "all that individuals learn from others that endures to generate customs and traditions" shaping human lives (Whiten et al. 2011). Access to numbers of personal digital archives pertaining to individuals' lives would advance future studies of culture and human behaviour as well as histories of individuals per se.

A special scientific benefit is that an understanding of personal information in the wild may engender valuable insights into evolutionary processes themselves, into a complex and dynamic system of digital objects and archives.

Among the most influential expressions of cultural information replication has been the concept of memes, loosely replicators of cultural information (Blackmore 1999, 2009; Dennett 1995, especially pp. 335–369; Shennan 2002).

A distinct advantage of treating eMSS as (putative) units of selection is that they are essentially discrete, and less nebulous than, say, an 'idea'. Analysis of eMSS may be conducted by investigating the information held in computer storage media. The brain is not available for analysis in the same way.

Another productive approach has been to focus on words or languages as discrete units (e.g. Pagel 2008, 2009a) and in due course it would be useful to combine the two perspectives.

13.3 Natural History of Personal Information

> These cards were, he told me, his 'extra-bodily grey matter'. He couldn't remember all the papers he had read, so this was his only record of many of them. The note cards were extremely valuable to me as a way into the literature
>
> Laurence D. Hurst (2005) on W. D. Hamilton

13.3.1 Personal Informatics and the Past: Portable Media and the Hand

The importance of looking to the past is that it helps to distinguish longstanding and consistent patterns from more ephemeral and fortuitous events. It is illuminating to reflect on the ancient relationship between humans and information, and to heed its materiality.

A key aspect of personal informatics is the use of information by individuals in the past, how it impacted their lives, the way informational culture shapes human lives and nature. While individuality and identity become less clear the further back in time one goes, there are glimpses of the individual even from prehistory. Blombos Cave in South Africa has yielded artefacts of symbolic meaning from more than 70,000 years ago: red ochre pieces with distinctly engraved markings. There were also ornamental shells that would have been held together, and worn by an individual "very possibly for more than a year" (Henshilwood 2007) (see d'Errico 1998; Jacobs and Roberts 2009; d'Errico and Stringer 2011).

One of the most profound junctures in humanity's past occurred when our ancestors began to store symbolic information beyond the confines of the brain (Donald 1991, 1998; Renfrew and Scarre 1999) (for overviews, see Dunbar 2004; Jablonka and Lamb 2005; Stringer 2012; Pagel 2013). The reprised making of tools and refining of ornaments – long considered the originating acts of human creativity – leads to devices and structures that bear information, hints of how each was made, used and admired: traces of material design.

Fundamental to the design of tools and the use of information is the human hand. To this day, it plays a role in thinking and inventing, the enduring legacy being that sketching, diagramming, and the manipulation of artefacts are integral to learning, problem-solving and hands-on experimentation. Some experimental scientists feel that they are better able to think while working with their hands in the laboratory.

Beyond tool making and ornament manipulation, hand gestures may have anticipated spoken language, and through lingering markings and impressions in sand and clay, on wood and stone, may also have led the way to drawings, signs, tallies, and ultimately scripts (Corballis 2002, 2009; Barbieri et al. 2009; Gentilucci and Dalla Volta 2008; Ingold 2007), while grooming may have been important in social interaction (Dunbar 1996). Many mathematicians develop and map their thinking with a pen in hand; and recent research has suggested that gesturing by teachers during mathematics lessons can aid learning (Cook et al. 2013). From the interplay of hand and mind came the means to create, to discover, and to make sense of nature through metaphor, model and experiment (McNeill 1992; Kellogg 1994; McNeill 2000; Goldin-Meadow 2003; Murray 1987; Kendon 2004; Barbieri et al. 2009; see also Ma and Nickerson 2006; Takken and William Wong 2013).

Likewise the formation of wide ranging networks of conceptual, informational and personal exchange and communication is also evident in the Palaeolithic. Bone and ivory flutes from more than 35,000 years ago, found in southwestern Germany, point to the existence of music and instrument playing which would, along with early figurative art and personal ornaments, contribute – it is suggested – to group cohesion and large social networks through shared experiences of music and storytelling (Adler 2009; Conard et al. 2009; Cook 2013).

Archaeologists, anthropologists, human scientists and palaeontologists are still deliberating the precise pathways of prehistory, but the significant entwining of technology, materiality, information and selection is palpable.

13.3.2 The Materiality of the Digital Revolution

Earlier informational revolutions such as the origin of language itself, the creation of scripts and alphabets, the use of parchment and paper, and the arrival of the printing press, and the progressive empowerment of individuals that these advances brought, have been defining moments intimately linked to agricultural, industrial and sociopolitical revolutions (e.g. Grayling 2008; John et al. 2010; Bentley and O'Brien 2012). With the emergence of the digital era, another phase in the evolution and development of human society has begun; and it is a material one too.

The electronic processor is rightly acclaimed but the unsung heroes of the digital revolution are the hard drive and the magnetic tape. As Andrew Blum (2013) has forcefully reminded us, the Internet has definite underlying materiality in wires, optical fibres, wireless transmitters, and switches, but most of all in the storage devices in the servers. Matthew Kirschenbaum's (2008) textual study of electronic technology strongly accentuated the materiality of digital media (see also Ross and Gow 1999).

The power of digital technology is expressed in its informational global reach, instantaneous communication and capacity for replication, even proliferation. It has become possible for many people to create useful devices and beautiful artefacts, and it is only the beginning. The emergence of even more influential and pervasive technologies is imminent (Floridi 2010).

Decidedly material digital technologies are emerging on a scale not yet seen. People will create personalised 3D products (Gershenfeld 2007; Anonymous 2013) including clothes, shoes, ornaments, furniture and guitars as well as specialist architectural models and robotic hands; future prospects extend to medicines and foods as well the building of structures (White 2013). Advancing sensory and motor capabilities and remote informational communications may become haptic and olfactory. Individuals themselves may become bionically augmented, meaning that their natural capabilities arising from their DNA and developmental environment may be enhanced by digital technologies that are tuned to the individual through personal information.

A reminder of the impact of the combination of information, materiality and design is the rapid predominance of handheld devices with touchscreen, involving subtle and dexterous finger control (for further haptic and gestural integration, see Chen et al. 2009; Heikkinen et al. 2009).

13.3.3 Creator, Curator and Consumer

There are three basic components to sustainable information systems: (i) information content created and packaged as a digital object (the seat of innovation, originality, variety, knowledge), (ii) information storage and delivery entities and their management or curation (the basis of material existence); and (iii) the environment of the information including users, consumers (the motivating source of replication).

Information quality (IQ) depends on the creation, curation and consumption of information: the information itself, the means of storage and care, and the environment.

This relationship cannot be addressed satisfactorily without considering the nature of the storage and delivery entities, their structure and behaviour, their design. It is the material, though transitory, being that actually bears and shares the information.

13.4 Information Quality of a Digital Archive

13.4.1 Lifecycles for Life Histories

Before discussing personal information quality (pIQ), it is helpful to outline the archival lifecycle in a general way. Archival theory and practice stress the *entirety* of the lifecycle, an exhortation that warrants the attention of anyone who wants to advance personal information management (Williams et al. 2009).

The lifecycle model of the Digital Curation Centre (Higgins 2008) may be briefly summarized as follows:

- conceptualise (conceive data, capture and storage);
- create or receive (both data and metadata – administrative, descriptive, structural, technical);
- appraise and select (evaluate and select data for long term curation);
- ingest (transfer information to repository);
- preservation actions (including retention of authenticity, validation and provision of suitable data structures and file formats);
- store (store data securely);
- access, use and reuse (ensure day-to-day access with robust controls);
- transform (create newly modified or subsets of data for specific purposes, e.g. publication).

A number of aspects of quality can be identified and arranged in seven classes: factors that influence the composition of an archive and the nature of the personal digital objects themselves (Table 13.1). The outline vaguely follows the archival process; but since the entire lifecycle needs to be contemplated throughout in order to balance costs and benefits, it is an imperfect mapping.

For a comprehensive discussion of the classification of information quality, see Illari and Floridi (2012).

13.4.2 Seven Components of Personal Information Quality (pIQ)

1. Digital Integrity and Sustainability

The first question in contemplating an original digital object or its exact replicate concerns its integrity. Can the individual bits be identified correctly? If the file system is understood – in other words, if the way files are organised on the disk or other media is recognised – copying the file may be straightforward, but the binary information of the file itself still needs to be interpreted. Can the digital object be rendered, even partially?

A related question concerns the resilience and robustness of the digital object itself. How vulnerable is it in the face of corruption as 1, 2, 3 or more bits are altered? Some experiments where one or more 'bits' of a file are corrupted at random have been conducted to assess the robustness of file formats: so-called 'shotgun experiments' (designed by Manfred Thaller, Universität zu Köln).

There is the further question of the sustainability of the means of rendering. Important research has been done in creating sustainable systems for emulating (usually virtually) the original hardware and software so that exact replicates of the original digital objects can be rendered and perceived dynamically and interactively (Kol et al. 2006; Lorie and van Diessen 2005; van der Hoeven et al. 2005; Konstantelos 2010).

Table 13.1 Some of the components of personal information quality (pIQ)

(1) Digital integrity and sustainability
Originality of digital object
Quality of digital replicate
(2) Exactness, completeness and ornamentation
Completeness of the digital object originally obtained
Quality of migration and of digital facsimile
Quality of emulation and emulating system
(3) Authenticity: digital object
Provenance of object
Provenance of metadata including embedded date and time
(4) Privacy
Identification and elucidation of privacy requirement
Quality of privacy protection, eg anonymisation, redaction
(5) Property and individual creativity and origin of digital object
Identification of intellectual property (IP) requirement
Quality of IP protection
Extent to which identity of creator is attributed
Extent to which use of personal information is recorded
(6) Appraisal, value and cost
Cost of reception, care and use
Personal and family value including story telling
Awareness of scholarly, scientific and cultural value including aesthetics
Scale
(7) Usability, reliability and availability
Quality of sustainable usability, manageability
Searchability and finding potential
Potential for analysis, interpretability of archive
Currency of technology, modernity
Interoperability, integration

2. Exactness, Completeness and Ornamentation

How complete is the digital object? If it is an exact replicate it is in itself complete (although the fidelity of the rendering system still has to be considered). If it is not an exact replicate, what is the nature and extent of its incompleteness? It may be a digital facsimile that is limited in some coherent way. Text may be readable even if some information reflected in the layout has been lost. Images may be viewable even if colours are not precisely accurate. On the other hand style, layout and behavior – ornamental aspects – of the digital object may be critical for proper and full interpretation.

In order to address issues of media degradation and technological obsolescence digital preservation practice strives for the interoperability of files – so that future – as well as present systems can function with digital objects from diverse sources. To counter technological obsolescence, a digital object may be 'migrated', a term for the conversion of a digital object from one file format to another one that is, hopefully, more interoperable: for example, an open file format that is freely understood.

The file format is a model that gives meaning to this binary information. It determines the way elements of the content information are arranged, provides locations for information about the specific file, and allows the file to be processed by software in its technical environment so that the content is expressed appropriately (Heydegger 2009; Abrams 2006, 2007; Barve 2007; Todd 2009). An illuminating study of the long term preservation risks of a file format is that for Adobe Portable Document Format (PDF) (van der Knijff 2009).

There is therefore the quality of the process of migration and of the resulting digital object to consider. How faithfully is the original look and feel retained? Migration frequently will result in a loss of information, in style or functionality, and there is the concern that repeated migration may be necessary over the years and decades (as software and hardware continue to evolve), and consequently there may be a tendency to erode the original information. This is one justification for keeping the digital replicates.

Similarly, how faithfully does the emulating system render a replicate of the original digital object? The quality of the perceptual experience will depend on the quality of the emulation. One of the aims of the KEEP project: Keeping Emulation Environments Portable was to create an Emulation Access Platform that will allow static and dynamic digital objects to be rendered accurately.

Many of the technical, policy and outreach issues of digital preservation are being addressed by the Alliance for Permanent Access (APARSEN), the Digital Preservation Coalition (DPC), and the Open Planets Foundation (OPF).

3. Authenticity

One of the most significant challenges in enabling the future use of personal information is in ensuring the authenticity of the digital information, it being demonstrably free of inadvertent or deliberate change. It is perhaps the quality that is held paramount in the archival and repository profession. Clearly, authenticity is absolutely essential for research of many kinds, from historical and critical scholarship to the social and natural sciences.

Provenance is a longstanding criterion with analogue objects such as paper documents, and it is being applied in the digital context. Less attention has been given to the provenance of embedded metadata and the typically latent content of digital objects such as dates and times. One key approach that has become increasingly accepted is the application of digital forensics procedures and tools (John 2008, 2012; Duranti 2009; Kirschenbaum et al. 2010; Woods et al. 2011).[2]

One of the benefits of *bona fide* forensic software is that there is an expectation that standards are met. Forensic tools and procedures are tested on behalf of organisations such as the National Institute of Justice in the USA as well as digital forensic professionals, and reliability is routinely considered in law courts. The use of writeblocker hardware prevents the examination computer from making alterations to the collection hard drive. Cryptographic hash values may serve as 'digital

[2] Open source forensic tools specifically for the archival community are being adopted and developed by the BitCurator project, www.bitcurator.net.

fingerprints' for each and every digital object including the entire disk itself. Digital capture may be undertaken twice with distinct systems to check that the same hash values are obtained on both occasions.

4. Privacy

The most sensitive quality of personal digital objects is that which concerns their social and individual origins: specifically privacy, and the ability of people to control their personal information. Traditionally, in order to preserve necessary confidentiality, archivists have put aside papers for agreed durations either at the request of originators or in the interest of third parties as judged by the archivist.

Clearly digital information raises new issues that require careful consideration both inside a repository and beyond. Even outside a public repository, individuals may be faced with responsibility to take steps to protect the privacy of friends, family and other people. How private is the content of a digital object, and what kind of privacy is entailed? Whose privacy is to be protected and for how long? There are notable and much discussed challenges concerning privacy and data protection (O'Hara and Shadbolt 2008; Floridi 2010; Mayer-Schönberger 2009; O'Hara 2010; O'Hara et al. 2008).

The phrase 'personal information' may refer to information that a person owns, created, or collected. Private information is in some sense restricted in access or limited in lawful or rightful availability. Some private and confidential information might not be legally owned by the individual, as with medical information held by health practitioners in the UK.

Some identifying information needs to be public in order for identity to be established and secured. Yet the identity of an individual, the persona and personality have critically private elements. There is a distinction between intimacy and private domesticity. Luciano Floridi (2006) outlines the possibility that a person's nature is constituted by that person's information, and suggests that this "allows one to understand the right to informational privacy as a right to personal immunity from unknown, undesired or unintentional changes in one's own identity as an informational entity". He notes that one way of looking at it is that "you are your information", from which it follows that "the right to informational privacy… shields one's personal identity. This is why informational privacy is extremely valuable and ought to be respected". As Kieron O'Hara and colleagues (2008) recall: one of John Locke's (1632–1704) insights was that personal identity is reflected in the continuity of a person's memory and mind over the decades.

5. Property and Individual Creativity

Who owns the object, and who is entitled to copy the object and its contents? Who has the right to avail oneself of the information? What are the conditions of rightful use and manipulation of personal information?

An essential consideration in the curation of personal information and cultural heritage is intellectual property (IP) (Charlesworth 2012). Much discussion about IP concerns corporate ownership of it, notably by publishing companies; but personal intellectual property and the recognition of personal creativity is important to individuals too. It may be useful to distinguish between due recognition and a tangible or financial reward. One or the other may suffice, or both may be desired by an individual.

It has been proposed that individual creativity in combination with corresponding property rights has played a vital role in creating sustainable economic prosperity in the past and present (e.g. Acemoglu and Robinson 2013). The impact of the digital revolution itself might be interpreted in a similar vein: the unleashing of personal and cooperative innovation.

Copyright is often justified as a means of rewarding and encouraging creativity, and stimulating production, dissemination and distribution of creative objects. In the digitally networked era, replication of useful digital objects can be driven by many individuals, which might promote the object's long term existence as well as its reach.

Is it possible to find a way to reward or recognize creators of useful objects without artificially constraining replication? So that individuals are inclined to participate, to collaborate with others. The origins of ideas and any created entity gain much greater credence and attention when convincingly documented.

Besides the creative works of individuals, information *about* individuals is being harnessed more and more. The recording of the exploitation of digital information objects could be combined with systems of reward. A Personal Information Trust (PIT) would reward individuals when marketers use their personal information (Schull 2007): although the tracking and measurement of such use can invoke privacy issues. Who should have access to an individual's genome and other novel forms of personal information?

Organisations such as Facebook and Google can argue that they reward individuals for making use of their personal information through the online services that they offer. This relationship could be made more transparent. What is the actual value of a specific individual's personal information to the company, in terms of utility or financial return, and what is the true cost and benefit to the individual?

6. Appraisal, Value and Cost

For an archive to be looked after, individuals need to be aware of its value and the value of sundry digital objects. To whom is it valuable? Who can benefit from it in the future? Which digital objects should be sequestered for later in the individual's life? Individuals might be aware of the future or even current value to science and scholarship and want to cater for this eventuality. Yet people sometimes dispose of historically valuable items that on first impression seem trivial such as informal appointment notes, airline tickets and shopping receipts (all of which may help to place an individual in time and space), while retaining less valuable collections of published papers that already exist in libraries. Individuals might want to leave their personal archive to the benefit of particular individuals such as family, friends and professional colleagues. What would those benefits be?

Concerning objects created by individuals and their families, which ones should be kept? Should interim versions of created objects be retained? If so, which ones and for how long? Many people may not be inclined to hang on to variants (e.g. John et al. 2010): settling for the final edited version of a family photo instead of the original.

The other side of the equation is cost. What is the cost of capture, care and use of digital objects? The cost implications of an archive can be challenging to predict even for institutional repositories although significant efforts are being made to improve the costing of long term digital preservation (Kaur et al. 2013).

Individuals differ in the quantities and qualities of digital objects hoarded and in the care taken in sustaining the quality and curating the quantity. Priorities can be expected to vary among individuals, partly due to differing consequences of cost. While authenticity, completeness and context may be important to an individual as well as professional researchers and curators at a repository, these qualities may sometimes seem less important to people during their daily lives.

In the context of digitization (digital capture of analogue source such as digital photography of paper documents) there has been a debate about the merits of file formats that entail lossy or lossless compression compared with those that are uncompressed such as TIFF files which impose significant storage costs (Martin and Macleod 2013; Palmer et al. 2013).

The composition of a personal archive may be changed not only through the injection (acquisition or creation) of new digital objects but also the simple abandonment of existing digital objects. Faced with complex personal archives, files may simply be lost (followed later by a deletion that might remain unrecognised); for example, a survey of personal information management by individuals found that a major cause of data loss is an inability to find files (John et al. 2010, pp. 23 and 42; see also Williams et al. 2009). A kind of 'error-prone copying' of an archive where people inadvertently delete files may manifest itself.

Repositories apply a mixture of appraisal strategies including a collection development policy and curatorial expertise. There is always a risk of overlooking objects that would have been deemed valuable in the decades ahead. Although much effort is made in appraisal, there is an element of chance, of diversity of opinion among curators, of serendipity.

Key life stages of an individual are critical to a personal archive such as when the individual is born, leaves home, obtains a partner, moves the family home, purchases new technology, changes employment, retires and dies. The timing of events is often fortuitous due to the unpredictable nature of life and death, and such phases may or may not chime with an individual's resources or appreciation of the value of digital objects.

While access to numerous and diverse personal archives would benefit research of all kinds, there is the question of how many generations would families keep objects derived from their ancestors. The extent of the interest in family history suggests that people might do so for many generations but it will depend on future digital preservation and archival technologies and practices.

7. Usability, Reliability and Availability

A quality that has become profoundly important in the digital era is the manageability of personal digital objects by individuals themselves. The ease with which the information itself may be cared for and reused by individuals is critical. Usability research is commonly directed towards the design of devices, even personal digital technologies; but the management, curation, and use and reuse of personal information to the benefit of an individual in the longer term warrants much more study (for a recent slant see Lindley et al. 2013).

Of course, in the context of archives in the wild, the originator and the user will typically be one and the same person during that individual's life. Nonetheless access to such personal data for, say, social and medical research could take place through a trusted mediation that protects the privacy of third parties as well as the originator, while concomitantly assuring researchers concerned about authenticity.

A digital object and the emulating environment may yield an authentic experience but consequently be slow and awkward to use by current standards. It is necessary to consider the quality of modernity, of being able to interact with a digital object and take advantage of the most modern and up-to-date technologies and techniques.

The utility of a single object to anyone who might conceivably use it, will be affected by the availability of other digital objects? The value of any unit of information will be influenced by its degree of interoperability and its possible integration within and across diverse participating archives.

The incidental and organic nature of many personal archives – commonly unsystematic and varyingly disorganized – has in the past been what makes them so valuable to curators and, in due course, historical scholars. This often means that important contextual information is preserved.

The personal digital objects have to be stored somewhere. Up to now this has mostly been locally, at home. Many people prefer to use commercial services in the cloud. Wherever the information is located, improvements in curation in the wild would undoubtedly be beneficial.

13.4.2.1 A Catalogue of Criteria for Sustained Storage and Access

The Network of Expertise in long-term STORage (nestor) has published a catalogue of criteria for trusted digital repositories (nestor 2009). For the present paper it may serve as a model with which to consider briefly the possible issues for personal digital archives in the wild. Its focus is the entire digital repository.

The document considers three broad topics – namely (i) organizational framework; (ii) object management; and (iii) infrastructure and security – within which it poses a series of questions about the digital repository:

- Defined goals?
- Identified its designated community and granted adequate access?
- Legal and contractual rules observed?
- Nature of organization is appropriate (adequate financing, sufficient numbers of staff, long-term planning, reacting to change)?
- Continuation of preservation tasks is ensured even beyond existence of the digital repository?
- All processes and responsibilities have been defined and documented?

- Ensures integrity and authenticity during ingest, storage and access?
- Strategic plan in place?
- Accepts, stores and uses digital objects according to defined criteria?
- Permanently identifies its objects, records adequate metadata for content, technical aspects, context, changes, and usage?
- Preserves package structure of digital objects (e.g. complex compound entities such as a web site or email file containing attachments or files with accompanying metadata and manifests)?
- Adequate information technology for object management and security?
- Infrastructure protects the digital repository and its digital objects?

The list is directed at official digital repositories that might seek certification; but it is revealing to reflect on the same criteria in the context of a personal digital archive. Despite the formality of the criteria the issues raised by them are pertinent to archives in the wild and will have to be addressed by designers of personal curation systems.

13.4.3 Significant Properties of Digital Objects (and Archives)

In digital preservation there is the concept of 'significant properties' (Giaretta et al. 2009; Hockx-Yu and Knight 2008). It is useful to distinguish between the 'property' of a digital object (e.g. customary colour of the letters in a text document) and the 'value' of this property (black), which when taken together represent a 'characteristic' of the digital object (Dappert and Farquhar 2009). When file format migration takes place for preservation purposes, care should be taken, it is argued, to ensure that 'significant properties' of the digital object are respected with specific 'values' retained; other characteristics may be abandoned. The concept is driven by the identification of those characteristics of a digital object and its environment that influence the object's usefulness and the way it is used (Dappert and Farquhar 2009).

From the perspectives of information quality and evolution it is deeply apposite, for it is proposed that significant quantitative and qualitative characteristics of a digital object and rendering environment (e.g. software) should be selected for preservation according to the wishes of its user community, notably the researchers who would consult it at a repository (or, one might add, the individuals themselves in the case of archives in the wild).

The challenge lies in the unpredictable way that the environment and priorities of the user base can be expected to change in the future; ultimately the stakeholders are not just the users of the present.

The concept is normally associated with digital *files* but a similar notion could be identified in the way entire *archives* are created and curated. With personal archives in the wild, the question will be which digital objects are sustained and which are neglected, abandoned?

13.5 Unifying Personal Information Quality (pIQ)

13.5.1 Fit for Purpose

The discussion of significant properties leads very naturally to the purpose of information. It has been argued that "no IQ dimension is completely independent of purpose" (Illari and Floridi 2012; see also Illari 2012). The notion of 'fit for purpose' is broadly accepted as a general definition for IQ, with an assessment of the relationship between an information system and the purpose of the user being essential (Illari and Floridi 2012).

This concurs with the thinking of evolutionary biologists and behavioural ecologists. In a famous paper the Nobel Prize winner Nikolaas Tinbergen (1963) outlined four categories of questions about behaviour. The point can be applied to other characteristics of living organisms such as morphology and physiology not just behaviour. Thus the nature of a bird's wings may be explained in terms of its mechanical operation, its ontogenetic development, its phylogenetic history or its functional purpose (for continuing discussion see Reeve and Sherman 1993; Barrett et al. 2013).

The use of 'intentional language' might seem unfortunate but it is a longstanding practice in evolutionary biology (Grafen 1999) serving as a kind of shorthand that highlights the way natural selection leads to the creation of 'contrivances' that have a purpose (for a philosophical examination of intentionality see Dennett 1987). Thus it is legitimate to ask what are the functional purposes of a bird's wing – and to answer that one of them is 'to fly'. One might even state that it is designed for flight just as the book titles *The Blind Watchmaker* (Dawkins 1986) and *Plan and Purpose in Nature* (Williams 1996) allude to natural selection as a designer.

The adaptive linking of organismal design to the genetic information lies, of course, at the core of behavioural ecology and evolutionary theory (Davies et al. 2012). It is sometimes apparent that the purpose for which a characteristic originally arose in the organism's phylogenetic past is no longer the purpose for which the characteristic is currently maintained. A basic instance of this phenomenon is *preadaptation* where a characteristic (which may already be adaptive) turns out to be useful for another purpose and is subsequently selected for this new purpose.

The second consideration that engages the concept of personal information quality in an overarching way is the pace of technological change. As Pogue (2012) observes: "nothing changes faster, and more unpredictably, than consumer technology".

This propensity for change represents a fundamental problem for maintaining the utility of personal information. The primary importance of authenticity, provenance and genuine contextual richness in archival valuation heightens this challenge.

To conclude this section, two unifying concepts can be identified in pIQ (if not IQ) – purpose and pace – and both might be said to point to the necessity of an adaptable archival system: since evolution is so effective in designing systems with structures and behaviours that are pragmatically fit for purpose in the face of some unpredictablility – due to a mercurial environment and the unforeseeable effectiveness of a design.

13.5.2 Purposeful Rendering and Codependency

As the practice of digital preservation shows all too perspicuously, it is the rendering of a digital object that is useful. The object may be worse than useless if it cannot be rendered. Commonly, a rendering of a digital object is the result of synergy with other digital objects (core software, additional plug-ins, drivers and so on). The utility of a digital object such as a WAV file may not be known until it is rendered. One can distinguish between eMSS as software (useful in association with digital objects that can be rendered by the software) and other, personally created, eMSS (useful in association with digital objects that can render them). Many people have the same software (commercial or open source and free), identical digital objects, shared among individuals. Many people create their own digital objects, unique or few in replicate, personally created. It is obvious that the software and the personally created digital object influence each other, influence their mutual value; but personally created digital objects influence each other too, their perceived utility: for example, travel photos (represented by TIFF files) may be even more useful if careful notes (represented by DOC files) have been contemporaneously taken during the journeying.

13.5.3 Information Flow from Environment to Organism

An interesting concept in the context of information quality is Fisher Information. Using it, physicist B. R. Frieden has promoted a form of variational principle, Extreme Physical Information (EPI) (Frieden 1998) as “a practical tool for deriving probability laws” (Frieden et al. 2001) – as a metric that characterises the informational outcome of observation, loosely the ‘flow’ of information from phenomenon to observation, whereby a phenomenon is deemed to bear a level of intrinsic or bound information while actual observation of the phenomenon yields a level of extrinsic information.

The evolutionary biologist Frank (2008) has interpreted Fisher Information within population genetics in another way, suggesting that information within the environment is captured by the genome through the process of natural selection. The informational relationship between organismal and environmental complexity has been explicated in several ways (Krakauer 2011; Adami 2012; Frank 2012).

13.5.4 Coevolutionary Cycling as Preservation

In natural populations (of mammals and birds for example) at any given moment there is considerable variation among individuals and some of this variation is of more or less random origin. At some later time, it transpires that some of this chance variation proved to be effective, useful, in the environment that exists at this later time.

It not being possible to predict the future in detail, a modicum of random variation serves as a protection against erratic environmental change, and as a means of capitalising on unanticipated opportunities.

Evolutionary adaptation of a curatorial system may seem worthwhile, and as such there would be a place for unsystematic, haphazard variation. Appraisal of future value is a challenge and the role of serendipity is well recognized by the archival community – in a sense therefore the concept of natural selection is rationalizing this approach. Even so, there is a critical question. Where should the element of chance be incorporated?

A common distinction can be made between collection items (conserved and not willingly subject to change) and the curatorial system (open to improvements, and subject to change); but in scenarios where there are numbers of identical collection objects in the population (e.g. among archives in the wild) it may be possible to construct a frequency-dependent selection system where combinations of collection items may be subject to change with timely selection for objects that are becoming increasingly at risk of extinction, becoming rare.

The Red Queen theory suggests that antagonistic coevolutionary interactions may occur between the populations of two or more species, and help to explain the maintenance of genetic variants (Hamilton et al. 1990; Judson 1995; Salathé et al. 2008; Decaestecker et al. 2007; Ebert 2008).

In the archival scenario, it would not be the digital object itself but the *combination* of digital objects that would be allowed to vary with a degree of randomness. This might entail some form of automated oversight that would catch a digital object before its final extinction.

Thus alongside the propensity for change that is the affliction of digital preservation comes the propensity for proliferation, for there to exist numerous copies, and *therein may lie a solution.*

13.6 Behavioural Ecology of Archives in the Wild

> Houses are amazingly complex repositories. What I found, to my great surprise is that whatever happens in the world – whatever is discovered or created or bitterly fought over – eventually ends up, in one way or another, in your house.
>
> Bill Bryson (2010)

13.6.1 Personal Digital Objects as Replicators

A person's hard drive may well contain operating system and application software belonging to the individual. Thus even in the wild an active 'living' personal archive will have within it, systems necessary for the creation, acquisition and reuse of eMSS (e.g. word processing and photo editing) and (more optimistically) for protection and handling of the archive (e.g. antivirus, backup, digital preservation, and personal photo management software).

For initial purposes of analysis the associated software may be considered as eMSS and as part of the personal archive, because: (i) of the many component files within software suites that bear personal data that personalise the behaviour of the software for the individual; and (ii) of the necessity of the software for rendering the files at all.

A unique analogue object can only be passed on as a single entity. Thus a paper notebook of a mother or diary of a father could be inherited by only one individual such as a daughter. By contrast, exactly identical copies of a digital picture or digital video of a family gathering can be received by all of the sibling offspring (John 2009).

Many eMSS may well be inherited at or near the end of the lives of parents, one set from each parent. Sons and daughters may need to select from these archives some eMSS but not others. Couples will bring together into the family, archives from different lineages.

With limited resources, individuals and families will face the question of how much and what to keep. Siblings may choose to keep different things and after several generations there may be a series of related archives with shared ancestries (John 2009; John et al. 2010). A conceivable outcome is that after some generations, a population of people will hold manifold personal digital archives that share a number of identical (or very nearly identical) personal digital objects.

The scenario just outlined is a thought experiment, a model to be explored for understanding processes and purposes.

An eMS is, of course, not strictly atomistic, since it has elements that may be modified. It is possible that existing digital objects may be modified, inadvertently or deliberately; for example, a son might prefer to crop aesthetically some sky from his version of a family photo of an outing to the beach, while a daughter might prefer her video recording of a seasonal party at a lower resolution due to digital storage constraints. Moreover, just as others have supposed that computer viruses might be designed to evolve spontaneously, Schull (2007) suggests that digital objects themselves can be evolvable: "it is a matter of time before they are 'genetically engineered' to evolve and adapt through a process very much like natural selection".

The present paper draws attention to the digital object as a whole in order to gain a discrete objectivity and to contemplate a digital curation system that does not alter the digital objects in its care. Thus one might ask what was the effect of this digital object and why did it have this effect? This is not, of course, to presuppose that digital objects are immutable.

Of more immediate interest than seeing if eMSS evolution does in the fullness of time arise spontaneously is to simulate possible scenarios to understand what might occur under different circumstances and policies.

An approach that warrants further activity is the simulation and modelling of adaptive informational designs with digital organisms (Marijuan 1994; Wilke and Adami 2002; Ostrowski et al. 2007; for a sense of what is becoming possible see Yedid et al. 2012).

13.6.2 Distinctions

There are deeply significant differences between organic evolution and the cultural system of eMSS and personal archives. Seven stand out for immediate notice: (i) not only may eMSS be readily shared beyond the family and its descendants; but (ii) this may take place at any time (with people receiving eMSS from diverse sources); and again (iii) eMSS are actively and commonly created by individuals during their lives.

Fourthly, (iv), the size of the collection of eMSS may well increase or vary significantly during an individual's life. That said, there may be common constraints on size. Just as horizontal transmission or gene exchange is found in nature, significant variation in genome size is evident in biology as are the costs of large genomes that limit their ultimate size.

A fifth distinction (v) is due to the fact that an eMS may well harbour not only instructive information (sensu Floridi 2010) of the kind attributed to genes, but other kinds of information, notably semantic.

There may be a sense in which the eMS in the context being contemplated may impel, incite or instigate (semantically or otherwise) but perhaps we shall only be able to say that this eMS represents an instruction if it represents a functional role that serves a *consistent* purpose.

A sixth (vi) distinction is that with a set of inviolable eMSS a "mutation" might involve the replacement of one digital object by another. Unlike with DNA, where one might contemplate four bases, there could be many more contending digital objects.

Finally, (vii) a difference between this scenario and biological systems is that the computer operating system is not commonly (or legally, with the exception of open source software such as Linux) replicated along with the personal data files or digital objects, but tends to be purchased anew and obtained from the software supplier. Yet, the replication can be done in this way and from the point of view of an initial model perhaps it might as well be since the code is in fact frequently identical across personal archives (as is attested by the hash libraries that identify software files) (see John 2012): such code might be deemed to be 'conserved' since so many personal digital collections depend on it.

13.6.3 The Importance of the Phenotype and Adaptation

It is one thing for there to be replicating eMSS subject to selection but adaptation refers to the phenotype: with, at the organismal level, morphological, physiological or behavioural characteristics emerging from the corresponding genotype.

For David Haig (2006) the essence of adaptation is that "a material gene's phenotypic effects such as the protein it encodes influence the probability that the material gene, or its replicas, will be copied", while Andy Gardner (2009) has emphasised the importance of 'design' and 'purpose' at an organismal level in evolutionary adaptation: "Put another way, the problem of adaptation is to account for the empirical fact that living material is packaged into units of common purpose (organisms)".

Typically the phenotype has been conceived as an organismal characteristic such as tail length, wing size, bone circumference, and eye colour. With the arrival of molecular biology this changed. Francis Crick (1958) commented that the sequences of amino acids in proteins are arguably "the most delicate expression possible of the phenotype of an organism and that vast amounts of evolutionary information may be hidden away with them". At the same time, other biologists look further than an individual organism's immediate structure invoking a concept of 'extended phenotype' pointing to a beaver's dam of tree branches, a bowerbird's colourful bower constructions, a bird's nest of twigs, and so on (see Dawkins 1982).

Away from organic biology, the phenotype may manifest itself in ways far beyond proteins or any biologically organic structure as with electronic circuitry (Thompson 1998). With the arrival of 3D printers, it is possible to envisage that individuals will design personalised physical artefacts (from furniture and kitchenware to clothing and ornaments). The eMSS that bear the creative design information will determine the production and nature of the 3D artefacts. Similarly haptic actuators and other physical components of the 'internet of things' (e.g. Gershenfeld et al. 2004) may be informationally tuned to the personal experiences and preferences of individuals.

The question is: could a system of replicating eMSS conceivably yield recognisable adaptation?

13.6.4 Computation as Phenotypic Expression

Phenotypic expression of the eMSS might be seen simply as their rendering. Stuart Kauffman (1990) likened the genotype to the 'program' and the phenotype to the actual 'computation': "For many programs, it is well known that there is no short cut to 'seeing the computation' carried out beyond running the program and observing what it 'does'". Similarly, Worden (1995) comments: "The genetic code specifies the phenotype, just as a computer program specifies a computation".

Recall that in our definition of eMSS we were careful to include the software that exists in the possession of the individual.

A computation is the outcome of some data and a program. A word document, photo or video file is not a program but is a set of data which in combination with appropriate software yields a computation that manifests as a specific rendering – perceived as a set of stylised and arranged text, a still image or a moving image existing on a computer screen or as an artifact emanating from a printer.

In an initial exploratory analysis, each eMS might be treated as a black box, as a fundamental entity that may influence material expression. It is not difficult to envisage the software within a personal archive as a kind of 'memome' with many of these digital objects occurring in many other archives, but what about the remaining digital objects, the family photos, the financial records, and the writer's drafts. Of course, this gets to the heart of the problem of simply seeing a parallel between eMSS and genes: the putative 'memome' though faintly reminiscent of a genome is not evidently a coherent whole.

Still it seems reasonable to suppose that some eMSS (more precisely, their information) would persist over many generations, whereas the physical aspects of the archive are ephemeral in the same way that DNA information persists, whereas the cells, organisms and groups of organisms are transient beings.

13.7 The Digital Archive as an Organism?

13.7.1 A Preliminary Scenario

In digital preservation a distinction is drawn between the dark archive (which holds the core information) and the delivery system that typically operates with copies deemed suitable for access. The dark archive is the 'germline' from which the archive can be regenerated and sustained. The delivery and management systems interact somatically (and sometimes reproductively) with the external environment. (For an informal explanation of a dark archive in the context of digital preservation see Hilton et al. 2013.)

An archive expressing eMSS as computations and renderings that serve useful functions receives resources necessary to continue being useful. Individuals may be crucial agents in an archive's environment allowing it to produce more archives. Just as the archive is a component of the information ecosystem so are people. One has only to contemplate the effect for people of losing the entirety of their digital belongings. Individuals, if they are sensible, might want to oblige their personal archive by providing resources that ensure the continued existence of their essential digital belongings.

When the archive functions actively and dynamically, it is behaving as if it is an 'organism'. By caring for and making available its primary contents it thereby ensures the continuing availability of resources for the archive as a whole. One can see people and their archives as '*symbionts*', each person and archive having closely interweaved mutual interests.

Considering the systems biology analogy, one can observe that organisms have various response systems for protection or resource acquisition, and for managing ontological development, for which parallels have been noted on occasion in computational research. Among the most vivid are information foraging theory (Pirolli 2007) and computer immunity (Kephart 1994; Hofmeyr and Forrest 1999; de Castro and Timmis 2002; Forrest and Beauchemin 2007).

13.7.2 OAIS: Open Archival Information System

An archive is of course more than the simple sum of the objects contained within it. There is the orchestration of processes and policies.

The most widely accepted scheme for the preservation of digital collections including archival material is the OAIS reference model (Open Archival Information System) (CCSDS 2002). It is concerned with three prominent flows of information: ingest, storage and data management, and access. It incorporates components for administration (day-to-day), preservation planning, and overall management (policy within a broader policy domain). Mandatory responsibilities include ensuring that the information is independently understandable to the Designated Community, and that the information is "preserved against all reasonable contingencies". A key aspect is the preparation and retention of preservation metadata (Lavoie and Gartner 2013).

A central role is quality assurance. A repository is a system that maintains a carefully controlled internal environment while at the same time interacting with the external environment (receiving new information and making available information, frequently in response to external events). Likewise, biological cells and organisms need to maintain consistent internal conditions despite the changing external environment.

Many of the purposes which the OAIS model serves are akin to the functions found in an organic organism, and it is reasonable to suppose that an OAIS compliant archive may be deemed to be packaged into a unit of common purpose.

At present the OAIS model is far from universal in the wild but it must surely be a wide aim of the archival profession to encourage and enable individuals to manage their personal information properly. To this extent the notion of an archive functioning in a way that is reminiscent of an organism may not be implausible. Obviously an OAIS compliant personal archive is not in that state due to natural selection of eMSS and adaptation.

Could the functioning archive be set up as a kind of *preadaptation*? Bearing in mind that evolutionary processes can facilitate the improvement of design, could suitable archival policies and digital preservation protocols that exploit evolutionary principles be implanted – in the context of long term missions in space if not of personal archives in the wild?

13.8 Questions of Complexity and Adaptability

13.8.1 Consistency and Coherence

There are a number of considerations, some of which are fundamental.

1. It might seem more "natural" if it could be supposed that the individual's computer hardware is a phenotypic expression of the personal archive. The trouble is that the code, the 'memome', for hardware devices themselves resides elsewhere

(under the auspices of manufacturers AMD, HP, Intel, Nvidia, Seagate, Western Digital, and so on). Still, it raises intriguing questions. How much does this matter? Why? Do concepts of extended genotype ('memotype') or phenotype or symbiosis have any bearing?

2. The absence of a clear ontology, a period of 'purposeful' development is important. The environment influences the development of an organism from birth and youth to maturity and death (West-Eberhard 2003; Bateson and Gluckman 2011; see also Oyama 2000; Hagen and Hammerstein 2005). Of course, the archive does develop during the life of an individual, often increasing in richness and scale. Presumably, the way it proceeds is influenced by the eMSS that exist within it as well as by its environment. But can the ensuing development ever be recognisably purposeful in an adaptive sense? The significance of ontogeny has been recognised in evolutionary computing such as genetic algorithms and in both software and hardware evolution (Bentley and Kumar 1999; Bentley 1999; Stanley and Miikkulainen 2003; Devert 2009) and the approach might be similarly adopted in archival design.
3. A fundamental issue is that the same identical eMSS may have different effects in different 'organismal' archives. The record of an identical email or letter may affect the owners of diverse archives differently. Even in biological organisms identical genes may produce diverse effects. On the other hand, the same beautiful landscape photograph may invoke a universal sense of wonder in everyone who possesses one in his or her archive. Moreover, prevailing software may engender similar effects. But Haig's (2012) conclusion concerning the 'strategic gene' is crucial in this context: "Genes are 'indefinite hereditary replicators'… that accumulate functional information about what works in the environment to a much greater extent than other candidates for the replicator role". It points to both the manner in which diverse genes need each other (to accumulate functional information) and the expectation of some consistency in what individual genes do. As West and Gardner (2013) put it: "Adaptations are underpinned by multiple genes distributed across the genome".
4. Although eMSS do not operate together with the same near unity as the genes that form a living organism, there are hints of inherited cohesiveness, as when a computer boots up and earlier ancestral code manifests itself as the rudimentary text of the booting computer, and in a manner that is reminiscent of the way ontogeny can recapitulate phylogeny in living organisms. The parallel is imperfect but it demonstrates that ancestral code is conserved, and passed from one generation to the next. The ontogenetic recapitulation of phylogeny, evidently due to the hurdles of trait complexity and developmental disruption, has been revealed with digital organisms (Clune et al. 2012).
5. An important topic is the possibility of linkage, which is classically portrayed as neighbouring genes on a linear chromosome that are unlikely to be separated by crossing over. Genetic linkage disequilibrium does not necessarily reflect a direct physical linkage but rather a tendency for some genes to be inherited together and to function together. It is possible to envisage some eMSS being inherited together. Photos may be held together in a folder. Both Microsoft Windows and Apple OS X have default folders for storing documents and pictures. Components of software are often stored together as in special folders such as the Program Files

directory in Microsoft Windows. Personally created files such as individualistic designs for a studio in the garden may be even more useful if pictures of the building and landscaping are transferred together as well as the software to display 2D and 3D models and even to 'print' the structure itself.

Nonetheless these hypothetical situations are far from the natural complexity and spontaneous dynamism of organic life.

It has been argued in the biological context that relatedness caused by common ancestry is special as it "unites the interests of genes across the genome, allowing complex, multigenic adaptations to evolve" (West and Gardner 2013). Could this idea be simulated and tested within the context of digital objects rather than biotic genes? Should it be factored into design for personal curation?

What kinds of policies or strategies are people likely to adopt when at the end of a life, for instance, a personal digital archive is inherited? What ways of conducting sorting, selecting and combining files might be best adopted?

13.8.2 On the Demeanour of Information

Information is actively gathered during the life of an 'organism' (in this scenario, the 'organismal' archive), which at first glance proffers the disconcerting notion of an 'organism' collecting the equivalent of genetic material. Something passably reminiscent does happen in nature, during virus infections for instance. Equally, living organisms receive through the senses information that is not genetic in nature. Some of the garnered information may be retained in the memory of the brain. Information does not only instruct or impel. It may simply inform. The notion of foraging for information points to information as a resource. It is necessary therefore to ascertain when an eMS might be behaving like a gene and when it is serving as a resource that 'informs' but does not 'genetically' (or memetically) instruct or impel.

Humans may hold information outside the brain in external media, which serve as *aides-memoire* during everyday life. Critically an almost defining aspect of humanity's use of information is that it can be passed from one generation to another not only through brain to brain but through media to media or brain. External media serve as an extended memory during and beyond the lifetime of an individual.

A conceptually straightforward distinction can be made between information that beneficially passes from one generation to another, representing information flow, and entirely novel information that is ephemeral, obtained during and confined to a single life. Much information in the brain does not flow to the next generation but some does. Much information in the genome does flow to the next generation but some does not. Regardless of whether the information of an eMS is transmitted through numbers of generations or originates and terminates in a single generation it will likely have a purpose but not necessarily an *adaptive* one.

Biologically, genetic information is directed towards the creation and maintenance of temporary organization (sometimes referred to as a 'vehicle'). To instruct in this way, elements of genetic information work together.

The outstanding question would be whether personal archival information flowing between generations might be deemed to instruct, with elements of it working together and directed towards archival organisation?

To answer questions of this kind it will be necessary to disentangle various types of information: instructive versus informative, sustainable information flow versus ephemeral information, informational coherence and collaboration versus inconsistency and idiosyncrasy. The discernment of environment, phenotypic vehicle, genotypic replicator in the ecology of personal information may not be straightforward without careful modelling.

The subtlety is exemplified by an observation tendered by Haig (2012) in the biological context: "A body can be viewed as the collectively-constructed niche of the genes of which it is an extended phenotype. Among the most important parts of a gene's environment are the other molecules with which it interacts. Other genes, even other alleles at the same locus, are parts of a gene's social environment.... On the other hand, any factor that is experienced by a gene, but not by its alternative, belongs to the gene's phenotype, not its environment".

This is exactly where a philosophy of information could be helpful, and where the intersection between information quality and evolutionary perspectives will be fruitfully maintained. The subject has a bearing on an understanding of the demarcations and nature of the organism, the replicator and their evolution.

Some challenges are reminiscent of those in the biology of natural organisms. Disentangling these factors may help to shed light on the role of epigenetic and genetic as well as memetic transmission.

It may be possible after some generations of personal digital archiving to ask a series of questions. Which eMSS get replicated from one generation to the next more than others? And why? Do any eMSS make it more likely that an 'organismal' archive contributes to descendant personal archives?

In the meantime there are numerous imponderables concerning the future environment: digital rights, privacy, intellectual property. An effective way to assess possible implications is to model the manifold scenarios (e.g. with alternative policies) using computer simulation. Moreover, some information technologists and inventors are already turning to evolutionary technology. Schull (2007) has addressed the specific issue of digital rights management by invoking biological principles: "It will be interesting to see how well rights management languages can be designed for adaptive evolution".

13.8.3 Evolutionary Selection and Replicator Dynamics

> Can we wonder… that nature's productions should be far 'truer' in character than man's productions; that they should be infinitely better adapted to the most complex conditions of life, and should plainly bear the stamp of far higher workmanship?
>
> Charles Darwin (1859)

Although anthropomorphic information and selection seems even more complex than the more familiar biological evolution, it does have to be addressed: to understand selection processes more fully, in informational (Krakauer 2011), cultural (Pagel 2009b) and archival contexts (John 2009), *and to design better information systems*. The triumvirate of variation, selection and replication is found everywhere and calls for comprehensive explanation.

Many people recognise selection processes in cultural and memetic evolution that are redolent of natural selection. But there is in biological nature a tightness of function and efficiency of resource use and direction of purpose that seems more evident than in the cultural context. Darwin made a similar point (see quotation) when comparing domestic selection with natural selection, and yet despite the perceived lower ‘workmanship’, domestic selection was sufficiently pertinent for Darwin to use it to strongly bolster the argument he was advancing for the existence of natural selection.

Domestic selection is different from cultural selection in the way that it is founded directly on biological organisation that first arose through *bona fide* natural selection. Of course when life began the greatest sophistication of complex life would lie in the future. The possibility of prebiotic replication and “the importance of recombination and recycling in the advent of living systems” (Valdya et al. 2013; see also Nowak and Ohtsuki 2008; Derr et al. 2012) accentuate the elemental nature of early systems.

But in the case of archives, highly complex design can exist anyway. It is essential that a personal archive in the future will be efficient, orderly and a closely integrated whole directed towards the purpose of providing the individual with the means to make the most of information.

Just as the organizational complexity of domestic organisms is founded on the adaptive capabilities of wild organisms evolved through natural selection, evolvable archival functionality and complexity might be founded on archival personal information management systems initially constructed through human intelligence.

A compelling place for the notion of a broad concept of ongoing selection, variation and replication to be examined is in the generality of descent with modification.

13.9 Ancestral Information and the Present

13.9.1 Descent with Modification and Digital Phylomemetics

The phenomenon of descent with modification or roughly imperfect replication with error or change over a number of generations is a general one that is not confined to life and the corresponding tree of life. Besides language, phylogenetic approaches are being adopted by archaeologists for the study of artefacts such as stone tools (Shennan 2002; O'Brien and Lee Lyman 2005; Mace and Holden 2005; Forster and Renfrew 2006).

Of greatest resonance in the present context is the phylogenetic study of manuscripts and folk tales (Spencer et al. 2006; Windram et al. 2008; Howe and Windram 2011; Tehrani 2013; see also Kraus 2009 for discussion of textual scholarship).

The phylogenetic approach has been used successfully with computer software and malware (Goldberg et al. 1998; Carrera and Erdelyi 2004).

It seems plausible that future personal archives will show descent with modification. Some archives can be expected to be more similar to each other than others due to shared ancestry, and this relationship could be mapped using phylogenetic analysis. A future researcher may ask why some eMSS have remained in existence for many generations and even spread while others have not done so?

In principle it might be possible therefore to surmise from extant archives the composition of ancestral archives. Thus retrospective analysis allows ancestral states to be recovered to some extent. How many extant personal archives would be needed for an effective analysis, for a significant reconstruction of an ancestral archive? What information can be retrieved in this way, and what information is irrecoverable? What aspects of information quality affect the potential for retrospection?

But the really interesting questions are (i) how can future retrospective analysis be made optimally effective through the careful selection and care of digital objects by the curation system, and (ii) how can the design of the curation system itself be advanced by retrospective analysis of its own functioning?

13.10 Conclusions

Almost every action and aspect of a life could be captured through a plethora of personal digital technologies: from personal genome to daily activity pattern. Personal information may be used to dynamically customise technologies, fine tuning them for the purposes and preferences of the individual. There is the possibility of very many people wanting to pass on their personal digital legacies to subsequent generations, and – through a trusted mediator – to share their personal information for the benefit of medical and social research.

There is, therefore, an ongoing requirement for effective systems for both the curation and utilisation of personal digital archives. An evolutionary perspective, specifically involving adaptive design, could benefit both the everyday sustenance of the personal archive and the application of retrospective analysis to reconstruct past events and states. A frequency-dependent selection for digital objects as they become less common might offer a practical model.

As well as challenging its handling and interpretation, the proliferation of digital information, frequently replicated and dispersed across the digital universe, offers new landscapes within which to explore the processes of variation, selection and replication, and advance the understanding of evolution and complex systems.

The two goals may be mutually beneficial since a better appreciation of evolutionary processes may inform the design of systems for curating and using personal information.

In contemplating a framework for information quality, this paper therefore counsels an evolutionary approach that embraces materiality along with selection, variation and replication.

A diversity of real personal archives in the wild might in time encourage a natural history of archive function, from which to glean empirical observations and theoretical insights; but computer simulation, mathematical exploration and philosophical analysis are likely to play a significant role in modelling and elucidating possible scenarios for the foreseeable future.

Acknowledgments This contribution is a continuation from the AHRC funded research conducted by the Digital Lives Research Project led by the British Library, Grant Number BLRC 8669. I thank Kieron O'Hara for pointing out useful papers and for alerting me to the existence of the rich vein of research being undertaken in the guise of information philosophy. I am profoundly grateful to Phyllis Illari and Luciano Floridi for inviting me to the Information Quality symposium and for making it possible for me to contribute to this volume. Phyllis Illari kindly gave me useful and highly pertinent comments on an earlier draft. Special thanks to Christiane Ohland. The opinions expressed are mine and do not necessarily reflect the policy of my employer.

References

Abdelkhalik, O. (2013). Autonomous planning of multigravity-assist trajectories with deep space maneuvers using a differential evolution approach. *International Journal of Aerospace Engineering, 2013*, 11pp.

Abrams, S. (2006). *Knowing what you've got. Format identification, validation, and characterization.* Paper presented at the DCC/LUCAS joint workshop, University of Liverpool, 30 Nov–1 Dec 2006.

Abrams, S. (2007). File formats. *DCC digital curation manual,* version 1.0, Oct 2007, Digital Curation Centre, http://www.dcc.ac.uk/resources/curation-reference-manual

Acemoglu, D., & Robinson, J. A. (2013). *Why nations fail. The origins of power, prosperity and poverty*. London: Profile Books.

Adami, C. (2012). The use of information theory in evolutionary biology. *Annals of the New York Academy of Sciences, 1256*, 49–65.

Adler, D. S. (2009). The earliest musical tradition. *Nature, 460*, 695–696.

Allen, S. M., Conti, M., Crowcroft, J., Dunbar, R., Liò, P., Mendes, J. F., Molva, R., Passarella, A., Stavrakakis, I., & Whitaker, R. M. (2008). *Social networking for pervasive adaptation.* Paper presented at the Second IEEE international conference on self-adaptive and self-organizing systems workshops (SASO 2008), Venice, 20–21 Oct 2008.

Anonymous. (2011). Beyond the PC. Special report. Personal technology. *The Economist*, 8 Oct 2011

Anonymous. (2013). Print me the head of Alfredo Garcia. 3D printing with paper. *The Economist*, 10 Aug 2013.

Banks, R. (2011). *The future of looking back. Microsoft research.* Redmond/Washington, DC: Microsoft Press.

Barbieri, F., Buonocore, A., Dalla Volta, R., & Gentilucci, M. (2009). How symbolic gestures and words interact with each other. *Brain & Language, 110*, 1–11.

Barrett, L., Blumstein, D. T., Clutton-Brock, T., & Kappeler, P. M. (2013). Taking note of Tinbergen, or: the promise of a biology of behaviour. *Philosophical Transactions of the Royal Society B, 368*(20120352).

Barve, S. (2007). *File formats in digital preservation*. Paper presented at the International conference on semantic web and digital libraries (ICSD 2007), 21–23 Feb 2007, pp. 239–248 (eds. A. R. D. Prasad & Devika P. Madalii).

Bateson, P., & Gluckman, P. (2011). *Plasticity, robustness, development and evolution*. Cambridge: Cambridge University Press.

Bekenstein, J. D. (2003). Information in the holographic universe. *Scientific American, 289*(2), 48–55.

Bell, G., & Gemmell, J. (2009). *Total recall. How the e-memory revolution will change everything*. New York: Dutton.

Bennett, C. H. (1999). Quantum information theory. In A. J. G. Hey (Ed.), *Feynman and computation. Exploring the limits of computation* (pp. 177–190). Cambridge, MA: Perseus Books.

Bentley, P. J. (1999). *Evolutionary design by computers*. San Francisco: Morgan Kaufman Publishers.

Bentley, P. J. (2007). Systemic computation: A model of interacting systems with natural characteristics. *International Journal on Parallel, Emergent and Distributed Systems: Special Issue, 22*(2), 103–121.

Bentley, P., & Kumar, S. (1999). Three ways to grow designs: A comparison of embryogenies for an evolutionary design problem. *Genetic and evolutionary computation conference (GECCO 1999)*, Morgan Kaufmann, pp. 35–43.

Bentley, R. A., & O'Brien, M. J. (2012). Cultural evolutionary tipping points in the storage and transmission of information. *Frontiers in Psychology, 3*, article 569, 14pp.

Blackmore, S. (1999). *The meme machine*. Oxford: Oxford University Press.

Blackmore, S. (2009). The third replicator. *New Scientist*, 1 Aug 2009.

Blum, A. (2013). *Tubes. Behind the scenes at the internet*. London: Penguin.

Blume-Kohout, R., & Zurek, W. H. (2006). Quantum Darwinism: entanglement, branches, and the emergent classicality of redundantly stored quantum information. *Physical Review, A 73*(062310).

Boldrini, C., Conti, M., & Passarella, A. (2008). Exploiting users' social relations to forward data in opportunistic networks: The HiBOp solution. *Pervasive and Mobile Computing, 4*(5), 633–657.

Bradley, W. E. (1982). First attempt to define a self-replicating system (A personal note contributed by W. E. Bradley, June 1980). In R. A. Freitas Jr and W. P. Gilbreath (eds.) *Advanced automation for space missions. Proceedings of the 1980 NASA/ASEE summer study sponsored by the National Aeronautics and Space Administration and the American Society for Engineering Education*, University of Santa Clara, Santa Clara, 23 June–29 Aug 1980. NASA Conference Publication 2255. Washington, DC: National Aeronautics and Space Administration.

Brillouin, L. (1956). *Science and information theory*. New York: Academic.

Bryson, B. (2010). *At home. A short history of private life*. London: Transworld Publishers.

Carrera, E., & Erdelyi, G. (2004, September). Digital genome mapping – Advanced binary malware analysis. *Virus bulletin conference*, Chicago, pp. 187–197.

CCSDS. (2002). *Recommendation for space data system standards. Reference model for an open archival information system (OAIS)*, CCSDS 650.0-B-1, Blue Book, Jan 2002, Consultative Committee for Space Data Systems.

Charlesworth, A. (2012). Intellectual property rights for digital preservation. *DPC technology watch report series, 12–02*, Digital Preservation Coalition, http://www.dpconline.org/advice/technology-watch-reports

Chen, Q., Cordea, M. D., Petriu, E. M., Varkonyi-Koczy, A. R., & Whalen, T. E. (2009). Human-computer interaction for smart environment applications using hand gestures and facial expressions. *International Journal of Advanced Media and Communication, 3*(1/2), 95–109.

Clune, J., Pennock, R. T., Ofria, C., & Lenski, R. E. (2012). Ontogeny tends to recapitulate phylogeny in digital organisms. *The American Naturalist, 180*(3), e54–e63.

Conard, N. J., Malina, M., & Münzel, S. C. (2009). New flutes document the earliest musical tradition in southwestern Germany. *Nature, 460*, 737–740.
Cook, J. (2013). *Ice age art. The arrival of the modern mind.* London: British Museum Press.
Cook, S. W., Duffy, R. G., & Fenn, K. M. (2013). Consolidation and transfer of learning after observing hand gesture. *Child Development, 84*(6), 1863–1871 (Nov and Dec 2013).
Corballis, M. C. (2002). *From hand to mouth. The origins of language*. Princeton: Princeton University Press.
Corballis, M. C. (2009). Language as gesture. *Human Movement Science, 28*, 556–565.
Crick, F. H. C. (1958). The biological replication of macromolecules. *Symposia of the Society for Experimental Biology, 12*, 138–163.
Crowcroft, J. (2008). Engineering global ubiquitous systems. *Philosophical Transactions of the Royal Society of London Series A, 366*, 3833–3834.
d'Errico, F. (1998). Palaeolithic origins of artificial memory systems: An evolutionary perspective. In C. Renfrew & C. Scarre (Eds.), *Cognition and material culture: The archaeology of symbolic storage* (pp. 19–50). Cambridge: McDonald Institute for Archaeological Research/University of Cambridge.
d'Errico, F., & Stringer, C. B. (2011). Evolution, revolution or saltation scenario for the emergence of modern cultures. *Philosophical Transactions of the Royal Society B, 366*, 1060–1069.
Dappert, A., & Farquhar, A. (2009). Significance is in the eye of the stakeholder. In M. Agosti, J. L. Borbinha, S. Kapidakis, C. Papatheodorou, & G. Tsakonas (eds), *13th European conference on digital libraries (ECDL 2009), research and advanced technology for digital libraries*, Corfu, 27 Sept–2 Oct 2009.
Darwin, C. (1859). *On the origin of species by means of natural selection or, the preservation of favoured races in the struggle for life*. London: John Murray.
Davies, N. B., Krebs, J. R., & West, S. A. (2012). *An introduction to behavioural ecology* (4th aufl.). Chichester: Wiley.
Dawkins, R. (1982). *The extended phenotype. The long reach of the gene*. Oxford: Oxford University Press.
Dawkins, R. (1986). *The blind watchmaker*. Harlow: Longman Scientific & Technical.
de Castro, L. N., & Timmis, J. (2002). *Artificial immune systems. A new computational intelligence approach*. London: Springer.
Decaestecker, E., Gaba, S., Joost, A. M., Raeymaekers, R. S., Van Kerckhoven, L., Ebert, D., & De Meester, L. (2007). Host-parasite 'Red Queen' dynamics archived in pond sediment. *Nature, 450*, 870–873.
Dennett, D. C. (1987). *The intentional stance*. Cambridge, MA: MIT Press.
Dennett, D. C. (1995). *Darwin's dangerous idea. Evolution and the meanings of life*. London: Penguin.
Derr, J., Manapat, M. L., Rajamani, S., Leu, K., Xulvi-Brunet, R., Joseph, I., Nowak, M. A., & Chen, I. A. (2012). Prebiotically plausible mechanisms increase compositional diversity of nucleic acid sequences. *Nucleic Acids Research, 40*(10), 4711–4722.
Devert, A. (2009). *Building processes optimization: Towards an artificial ontogeny based approach.* Doctor of philosophy thesis, 21 June 2009, Ecole Doctorale d'Informatique, Université Paris-Sud 11, Institut Nacional de Recherche en Informatique et en Automatique (INRIA).
Doherty, A. R., & Smeaton, A. F. (2008). *Automatically segmenting lifelog data into events*. Paper presented at the ninth international workshop on image analysis for multimedia interactive services (WIAMIS 2008), Klagenfurt: IEEE Computer Society, pp. 20–23.
Donald, M. (1991). *Origins of the modern mind. Three stages in the evolution of culture and cognition*. Cambridge, MA: Harvard University Press.
Donald, M. (1998). Hominid enculturation and cognitive evolution. In C. Renfrew & C. Scarre (Eds.), *Cognition and material culture: The archaeology of symbolic storage*. Cambridge: McDonald Institute for Archaeological Research/University of Cambridge.
Doorn, P., & Roorda, D. (2010). *The ecology of longevity: The relevance of evolutionary theory for digital preservation*. Paper presented at the digital humanities conference 2010 (DH2010), King's College London.

Dunbar, R. I. (1996). *Grooming, gossip and the evolution of language*. London: Faber and Faber.
Dunbar, R. (2004). *The human story. A new history of mankind's evolution*. London: Faber and Faber.
Duranti, L. (2009). From digital diplomatics to digital records forensics. *Archivaria, 68*, 39–66.
Ebert, D. (2008). Host-parasite coevolution: Insights from the Daphnia-parasite model system. *Current Opinion in Microbiology, 11*, 290–301.
Etherton, J. (2006). The role of archives in the perception of self. *Journal of the Society of Archivists, 27*(2), 227–246.
Floridi, L. (2006). Four challenges for a theory of informational privacy. *Ethics and Information Technology, 8*, 109–119.
Floridi, L. (2010). *Information. A very short introduction*. Oxford: Oxford University Press.
Floridi, L. (2011). *The philosophy of information*. Oxford: Oxford University Press.
Forbes, N. (2004). *Imitation of life. How biology is inspiring computing*. Cambridge, MA: MIT Press.
Forrest, S., & Beauchemin, C. (2007). Computer immunology. *Immunological Reviews, 216*, 176–197.
Forster, P., & Renfrew, C. (2006). *Phylogenetic methods and the prehistory of languages* (McDonald Institute monographs). Cambridge: McDonald Institute for Archaeological Research/University of Cambridge.
Frank, S. A. (1996). The design of natural and artificial systems. In M. R. Rose & G. V. Lauder (Eds.), *Adaptation* (pp. 451–505). London: Academic.
Frank, S. A. (1997). The design of adaptive systems: optimal parameters for variation and selection in learning and development. *Journal of Theoretical Biology, 184*, 31–39.
Frank, S. A. (2008). Natural selection maximises Fisher information. *Journal of Evolutionary Biology, 22*, 231–244.
Frank, S. A. (2012). Natural selection. V. How to read the fundamental equations of evolutionary change in terms of information theory. *Journal of Evolutionary Biology, 25*, 2377–2396.
Freitas Jr, R. A., & Gilbreath, W. P. (1982). *Advanced automation for space missions*. Proceedings of the 1980 NASA/ASEE summer study sponsored by the National Aeronautics and Space Administration and the American Society for Engineering Education, University of Santa Clara, Santa Clara, 23 June–29 Aug 1980. NASA Conference Publication 2255. Washington, DC: National Aeronautics and Space Administration.
Frieden, B. R. (1998). *Physics from Fisher information*. Cambridge: Cambridge University Press.
Frieden, B. R., Plastino, A., & Soffer, B. H. (2001). Population genetics from an information perspective. *Journal of Theoretical Biology, 208*, 49–64.
Gardner, A. (2009). Adaptation as organism design. *Biology Letters, 5*, 861–864.
Gardner, A., & Conlon, J. P. (2013). Cosmological natural selection and the purpose of the universe. *Complexity, 18*(5), 48–54.
Gentilucci, M., & Volta, R. D. (2008). When the hands speak. *Journal of Physiology – Paris, 102*, 21–30.
Gershenfeld, N. (2007). *FAB. The coming revolution on your desktop – From personal computers to personal fabrication*. New York: Basic Books.
Gershenfeld, N., Krikorian, R., & Cohen, D. (2004). The internet of things. *Scientific American, 291*(4), 46–51.
Giaretta, D., Matthews, B., Bicarregui, J., Lambert, S., Guercio, M., Michetti, G., & Sawyer, D. (2009). *Significant properties, authenticity, provenance, representation information and OAIS*. Proceedings of the sixth international conference on the preservation of digital objects (iPRES 2009), California Digital Library.
Goldberg, D. E. (1989). *Genetic algorithms in search, optimization, and machine learning*. Reading: Addison-Wesley Publishing Company.
Goldberg, L. A., Goldberg, P. W., Phillips, C. A., & Sorkin, G. B. (1998). Constructing computer virus phylogenies. *Journal of Algorithms, 26*(1), 188–208.
Goldin-Meadow, S. (2003). *Hearing gesture. How our hands help us think*. Cambridge, MA: The Belknap Press of the Harvard University Press.

Goldman, N., Bertone, P., Chen, S., Dessimoz, C., LeProust, E. M., Sipos, B., & Birney, E. (2012). Towards practical, high-capacity, low-maintenance information storage in synthesized DNA. *Nature, 494*, 77–80.

Goues, C. L., Weimer, W., & Forrest, S. (2012). *Representations and operators for improving evolutionary software repair*. Genetic and evolutionary computation conference (GECCO 2012), Philadelphia.

Grafen, A. (1999). Formal Darwinism, the individual-as-maximising-agent analogy and bet-hedging. *Proceedings of the Royal Society of London B, 266*, 799–803.

Grayling, A. C. (2008). *Towards the light. The story of the struggles for liberty and rights that made the modern West*. London: Bloomsbury Publishing.

Hagen, E. H., & Hammerstein, P. (2005). Evolutionary biology and the strategic view of ontogeny: Genetic strategies provide robustness and flexibility in due course. *Research in Human Development, 2*(1&2), 87–101.

Haig, D. (2006). Gene meme. In A. Grafen & M. Ridley (Eds.), *Richard Dawkins. How a scientist changed the way we think* (pp. 50–65). Oxford: Oxford University Press.

Haig, D. (2012). The strategic gene. *Biology & Philosophy, 27*(4), 461–479.

Hamilton, W. D., Axelrod, R., & Tanese, R. (1990). Sexual reproduction as an adaptation to resist parasites (a review). *Proceedings of the National Academy of Sciences USA, 87*, 3566–3577.

Hartwell, L. H., Hood, L., Goldberg, M. L., Reynolds, A. E., Silver, L. M., & Veres, R. C. (2008). *Genetics. From genes to genomes* (3rd edn. Aufl.). New York: McGraw-Hill.

Heikkinen, J., Rantala, J., Olsson, T., Raisamo, R., Lylykangas, J., Raisamo, J., Surakka, V., & Ahmaniemi, T. (2009). Enhancing personal communication with spatial haptics: Two scenario-based experiments on gestural interaction. *Journal of Visual Languages and Computing, 20*, 287–304.

Henshilwood, C. S. (2007). Fully symbolic sapiens behaviour: Innovation in the Middle Stone Age at Blombos Cave, South Africa. In P. Mellars, K. Boyle, O. Bar-Yosef, & C. Stringer (Eds.), *Rethinking the human revolution: New behavioural and biological perspectives on the origin and dispersal of modern humans* (McDonald Institute Monographs, pp. 123–132). Cambridge: McDonald Institute for Archaeological Research.

Heydegger, V. (2009). Just one bit in a million: On the effects of data corruption in files. In M. Agosti, J. L. Borbinha, S. Kapidakis, C. Papatheodorou, & G. Tsakonas (eds.) *13th European conference on digital libraries (ECDL 2009), research and advanced technology for digital libraries*, Corfu, 27 Sept–2 Oct 2009.

Higgins, S. (2008). The DCC curation lifecycle model. *The International Journal of Digital Curation, 3*(1), 134–140.

Hilton, J. L., Cramer, T., Korner, S., & Minor, D. (2013). The case for building a digital preservation network. *EDUCAUSE Review*, July/Aug 2013.

Hobbs, C. (2001). The character of personal archives: reflections on the value of records of individuals. *Archivaria, 52*, 126–135.

Hobbs, C. (2012). *Centring ideas of personal digital context on behaviour and mindset*. Paper presented at the International seminar and symposium: unpacking the digital shoebox: the future of personal archives, The University of British Columbia, Vancouver, 15–17 Feb 2012.

Hockx-Yu, H., & Knight, G. (2008). What to preserve? Significant properties of digital objects. *International Journal of Digital Curation, 3*(1), 141–153.

Hofmeyr, S. A., & Forrest, S. (1999). Immunity by design: an artificial immune system. *Genetic and evolutionary computation conference (GECCO 1999)*, Orlando, pp. 1289–1296.

Hopper, A., & Rice, A. (2008). Computing for the future of the planet. *Philosophical Transactions of the Royal Society A, 366*, 3685–3697.

Hornby, G. S., Lohn, J. D., & Linden, D. S. (2011). Computer-automated evolution of an X-band antenna for NASA's Space Technology 5 mission. *Evolutionary Computation, 19*(1), 1–23.

Howe, C. J., & Windram, H. F. (2011). Phylomemetics – Evolutionary analysis beyond the gene. *PLoS Biology, 9*(5), 5. e1001069.

Hui, P., Crowcroft, J., & Yoneki, E. (2008). *Bubble rap: social-based forwarding in delay tolerant networks*. Paper presented at the 9th ACM international symposium on mobile ad-hoc networking and computing (MobiHoc08), Hong Kong.

Hurst, L. D. (2005). Sex, sexes and selfish elements. In M. Ridley (Ed.), *Narrow roads of gene land* (Last words, Vol. 3, pp. 89–97). Oxford: Oxford University Press.

Illari, P. (2012). *IQ and purpose*. Paper presented at the AISB/IACAP world congress, University of Birmingham.

Illari, P., & Floridi, L. (2012, November 16–17). IQ: purpose and dimensions. In *Proceedings of the 17th International Conference on Information Quality (ICIQ 2012), Conservatoire national des arts et métiers* (pp. 178–192). Paris. http://iciq2012.cnam.fr

Ingold, T. (2007). *Lines. A brief history*. Abingdon: Routledge.

Jablonka, E., & Lamb, M. J. (1995). *Epigenetic inheritance and evolution. The Lamarkian dimension*. Oxford: Oxford University Press.

Jablonka, E., & Lamb, M. J. (2005). *Evolution in four dimensions. Genetic, epigenetic, behavioral, and symbolic variation in the history of life*. Cambridge, MA: MIT Press.

Jacobs, Z., & Roberts, R. G. (2009). Human history written in stone and blood. *American Scientist*, July-Aug 2009.

John, J. L. (2008). *Adapting existing technologies for digitally archiving personal lives. Digital forensics, ancestral computing, and evolutionary perspectives and tools*. Paper presented at the iPRES 2008 conference. The fifth international conference on preservation of digital objects, The British Library, London, 29–30 Sept 2008. http://www.bl.uk/ipres2008/presentations_day1/09_John.pdf

John, J. L. (2009). The future of saving our past. *Nature, 459*, 775–776.

John, J. L. (2012). Digital forensics and preservation. *Technology Watch Report, 12–03*. Digital Preservation Coalition, http://www.dpconline.org/advice/technology-watch-reports

John, J. L., Rowlands, I., Williams, P., & Dean, K. (2010). *Digital lives. Personal digital archives for the 21st century >>an initial synthesis*. Digital Lives Research Paper: Project funded by Arts & Humanities Research Council, britishlibrary.typepad.co.uk/files/digital-lives-synthesis02-1.pdf

Johnston, M. D. (2008). An evolutionary algorithm approach to multi-objective scheduling of space network communications. *International Journal of Intelligent Automation and Soft Computing, 14*, 367–376.

Jones, W., & Teevan, J. (Eds.). (2007). *Personal information management*. Seattle: University of Washington Press.

Judson, O. P. (1995). Preserving genes: A model of the maintenance of genetic variation in a metapopulation under frequency-dependent selection. *Genetical Research, 65*(3), 175–191.

Kanjo, E., Benford, S., Paxton, M., Chamberlain, A., Fraser, D. S., Woodgate, D., Crellin, D., & Woolard, A. (2008). MobGeoSen: Facilitating personal geosensor data collection and visualization using mobile phones. *Personal Ubiquitous Computing, 12*, 599–607.

Kauffman, Stuart, A. (1990). Requirements for evolvability in complex systems: Orderly dynamics and frozen components. In W. H. Zurek (ed.), *Complexity, entropy, and the physics of information* (pp. 151–192). Santa Fe Institute Studies in the Sciences of Complexity. Cambridge, MA: Perseus Publishing.

Kaur, K., Herterich, P., Schmitt, K., Schrimpf, S., & McMeekin, S. (2013). *Report on cost parameters for digital repositories (D32.1): APARSEN: Alliance Permanent Access to the Records of Science in Europe Network*, http://www.alliancepermanentaccess.org

Kellogg, R. T. (1994). *The psychology of writing*. Oxford: Oxford University Press.

Kendon, A. (2004). *Gesture. Visible action as utterance*. Cambridge: Cambridge University Press.

Kephart, J. O. (1994). *A biologically inspired immune system for computers*. Proceedings of artificial life IV, the fourth international workshop on the synthesis and simulation of living systems, MIT Press.

Kim, J. T., & Ellis, R. (2008). Systems biology and artificial life: Towards predictive modeling of biological systems. Editorial introduction. *Artificial Life, 14*(1), 1–2.

Kirschenbaum, M. G. (2008). *Mechanisms. New media and the forensic imagination*. Cambridge, MA: MIT Press.

Kirschenbaum, M. G., Ovenden, R., Redwine, G., & Donahue, R. (2010). *Digital forensics and born-digital content in cultural heritage collections*. Washington, DC: Council on Library and Information Resources, http://www.clir.org/pubs/abstract/reports/pub149

Kol, N. J. C., van Diessen, R. J., & van der Meer, K. (2006). An improved universal virtual computer approach for long-term preservation of digital objects. *Information Services & Use, 26*, 283–291.

Konstantelos, L. (2010). Preservation of dynamic and interactive content by use of binary translation and virtualisation – A methodology for experimentation. In Planets Deliverables, TB/6-D1F (PA/6-D12), 63 pp, Final Report, 6 Apr 2010, Planets – Preservation and Long-term Access through NETworked Services (PLANETS), http://www.planets-project.eu/docs/reports

Kordon, M., & Wood, E. (2003). *Multi-mission space vehicle subsystem analysis tools*. IEEE aerospace conference 2003, Big Sky, Montana, 8 Mar 2003, http://trs-new.jpl.nasa.gov/dspace

Krakauer, D. C. (2011). Darwinian demons, evolutionary complexity, and information maximization. *Chaos, 21*(037110), 11 pp.

Kraus, K. (2009). Conjectural criticism: Computing past and future texts. *DHQ: Digital Humanities Quarterly, 3*(4), 24 pp.

Kwok, R. (2009). Phoning in data. *Nature, 458*, 959–961.

Landauer, R. (1999). Information is inevitably physical. In A. J. G. Hey (Ed.), *Feynman and computation. Exploring the limits of information* (pp. 77–92). Cambridge, MA: Perseus Books.

Laursen, L. (2009). A memorable device. *Science, 323*, 1422–1423.

Lavoie, B., & Gartner, R. (2013). Preservation metadata. *DPC technology watch series 13–03, second edition*. Digital Preservation Coalition, http://www.dpconline.org/advice/technology-watch-reports

Lee, C. A. (2011). *I, Digital. Personal collections in the digital era*. Chicago: Society of American Archivists.

Li, L. (2012). The optimization of architectural shape based on genetic algorithm. *Frontiers of Architectural Research, 1*, 392–399.

Lindley, S., Marshall, C. C., Banks, R., Sellen, A., & Regan, T. (2013). Rethinking the web as a personal archive. *International world wide web conference (WWW 2013)*, Rio de Janeiro.

Lloyd, S. (2006). *Programming the universe*. New York: Knopf.

Lorie, R. A., & van Diessen, R. J. (2005). *Long-term preservation of complex processes*. IS&T archiving conference 2005, 26–29 Apr 2005. Washington, DC: Society for Imaging Science and Technology, http://www.imaging.org/ist/conferences/archiving/index.cfm

Lu, Y.-E., Roberts, S., Liò, P., Dunbar, R., & Crowcroft, J. (2009). *Size matters: Variation in personal network size, personality and effect on information transmission*. Paper presented at the IEEE international conference on social computing (Social COM), Vancouver.

Ma, J., & Nickerson, J. V. (2006). Hands-on, simulated, and remote laboratories: A comparative literature review. *ACM Computing Surveys, 38*(3), Article 7.

Mace, R., & Holden, C. G. (2005). A phylogenetic approach to cultural evolution. *Trends in Ecology and Evolution, 20*(3), 116–121.

Marijuan, P. C. (1994). Enzymes, automata and artificial cells. In R. Paton (Ed.), *Computing with biological metaphors* (pp. 50–68). London: Chapman & Hall.

Marshall, C. C. (2007). How people manage personal information over a lifetime. In W. Jones & J. Teevan (Eds.), *Personal information management* (pp. 57–75). Seattle: University of Washington Press.

Marshall, C. C. (2008a). Rethinking personal digital archiving, part 1. Four challenges from the field. *D-Lib Magazine, 14*(3/4), Mar–Apr 2008.

Marshall, C. C. (2008b). Rethinking personal digital archiving[,] part 2. Implications for services, applications, and institutions. *D-Lib Magazine, 14*(3/4), Mar–Apr 2008.

Marshall, C., Bly, S., & Brun-Cottan, F. (2006). *The long term fate of our digital belongings: Toward a service model for personal archives*. IS&T archiving conference 2006, Society for Imaging Science and Technology, Ottawa, 23–26 May 2006, http://www.imaging.org/ist/conferences/archiving/index.cfm

Martin, S., & Macleod, M. (2013). *Analysis of the variability in digitised images compared to the distortion introduced by compression*. Paper presented at the iPRES 2013 conference. The tenth international conference on preservation of digital objects, Lisbon, 3–5 Sept 2013, http://purl.pt/24107/1/iPres2013_PDF/iPres2013-Proceedings.pdf

Mayer-Schönberger, V. (2009). *Delete. The virtue of forgetting in the digital age*. Princeton: Princeton University Press.
Mayer-Schönberger, V., & Cukier, K. (2013). *Big data. A revolution that will transform how we live, work and think*. London: John Murray.
McNeill, D. (1992). *Hand and mind: What gestures reveal about thought*. Chicago: Chicago University Press.
McNeill, D. (2000). *Language and gesture*. Cambridge: Cambridge University Press.
Mitchell, M., Forrest, S., & Holland, J. H. (1992). *The royal road for genetic algorithms: Fitness landscapes and GA performance*. Toward a practice of autonomous systems: Proceedings of the first European conference on artificial life, MIT Press, Cambridge.
Murray, D. M. (1987). *Write to learn*. New York: Holt, Rinehart and Winston.
Mushegian, A. R. (2007). *Foundations of comparative genomics*. London: Academic.
Naumer, C. M. & Fisher, K. E. (2007). Naturalistic approaches for understanding PIM in Personal Information Management. In W. Jones & J. Teevan (Ed.), (pp. 76–88). Seattle: University of Washington Press.
Nestor. (2009). *Catalogue of criteria for trusted digital repositories*. Frankfurt am Main: Working Group Trusted Repositories – Certification, nestor materials 8, version 2, Network of Expertise in long term STORage and accessibility of digital resources in Germany Working Group, Nov 2009.
Nielsen, M. A., & Chuang, I. L. (2000). *Quantum computation and quantum information*. Cambridge: Cambridge University Press.
Nolfi, S., & Floreano, D. (2000). *Evolutionary robotics. The biology, intelligence, and technology of self-organizing machines*. Cambridge, MA: MIT Press.
Nowak, M. A., & Ohtsuki, H. (2008). Prevolutionary dynamics and the origin of evolution. *Proceedings of the National Academy of Sciences USA, 105*(39), 14924–14927.
O'Brien, M. J., & Lee Lyman, R. (2005). Cultural phylogenetic hypotheses in archaeology: some fundamental issues. In R. Mace, C. G. Holden, & S. Shennan (Eds.), *The evolution of cultural diversity. A phylogenetic approach* (p. 85). London: UCL Press.
O'Hara, K. (2010, September 16–17). Narcissus to a man: lifelogging, technology and the normativity of truth. In *Keynote lecture on 16 September 2010 at the second annual SenseCam symposium*, Guinness Storehouse, Dublin, Ireland. http://eprints.soton.ac.uk/271904/1/ohara-sensecam-keynote.pdf
O'Hara, K., & Shadbolt, N. (2008). *The spy in the coffee machine. The end of privacy as we know it*. Oxford: Oneworld.
O'Hara, K., Morris, R., Shadbolt, N., Hitch, G. J., Hall, W., & Beagrie, N. (2006). Memories for life: A review of the science and technology. *Journal of the Royal Society Interface, 3*, 351–365.
O'Hara, K., Tuffield, M. M., & Shadbolt, N. (2008). Lifelogging: Privacy and empowerment with memories for life. *Identity in the Information Society, 1*(1), 155–172.
Olguin Olguin, D., Gloor, P. A., & Pentland, A. S. (2009). *Capturing individual and group behavior with wearable sensors*. Paper presented at the AAAI spring symposium on human behavior modeling, Association for the Advancement of Artificial Intelligence, Stanford.
O'Reilly, J. (2009). Your life as a number. All you do – From sexual activity to carb consumption – Can be reduced to data. And by publishing your data online, other people can help you spot patterns. Welcome to the life-tracking trend. *Wired, UK Edition,* pp. 144–149.
Orosz, C. G. (2001). An introduction to immuno-ecology and immuno-informatics. In L. A. Segel & I. R. Cohen (eds.), *Design principles for the immune system and other distributed autonomous systems*. Santa Fe Institute studies in the sciences of complexity. Oxford: Oxford University Press.
Ostrowski, E. A., Ofria, C., & Lenski, R. E. (2007). Ecological specialization and adaptive decay in digital organisms. *The American Naturalist, 169*(1), e1–e20.
Oyama, S. (2000). *The ontogeny of information. Developmental systems and evolution* (2nd Aufl.). Durham: Duke University Press.
Pagel, M. (2008). Rise of the digital machine. *Nature, 452*, 699. 10 Apr 2008.

Pagel, M. (2009a). Human language as a culturally transmitted replicator. *Nature Reviews Genetics, 10*, 405–415.
Pagel, M. (2009b). Natural selection 150 years on. *Nature, 457*, 808–811. 12 Feb 2009.
Pagel, M. (2013). *Wired for culture. The natural history of human communication*. London: Penguin.
Palmer, W., May, P., & Cliff, P. (2013). *An analysis of contemporary JPEG2000 codecs for image format migration*. Paper presented at the iPRES 2013 conference. The tenth international conference on preservation of digital objects, Lisbon, 3–5 Sept 2013, http://purl.pt/24107/1/iPres2013_PDF/iPres2013-Proceedings.pdf
Pirolli, P. (2007). *Information foraging theory. Adaptive interaction with information* (Oxford series in human-technology interaction). Oxford: Oxford University Press.
Pogue, D. (2012). The future is for fools. A few guidelines for anyone attempting to predict the future of technology. *Scientific American, 306*(2), 19. Feb 2012.
Pullin, A. S. (2002). *Conservation biology*. Cambridge: Cambridge University Press.
Reeve, H. K., & Sherman, P. W. (1993). Adaptation and the goals of evolutionary research. *Quarterly Review of Biology, 68*(1), 1–32.
Renfrew, C., & Scarre, C. (1999). *Cognition and material culture: The archaeology of symbolic storage*. Cambridge: McDonald Institute for Archaeological Research, University of Cambridge.
Reuters. (2012). *More mobile devices than people by 2016: Cisco*. 14 Feb 2012. Thomson Reuters, http://www.reutersreprints.com
Rodden, T. (2008). Living in a ubiquitous world. *Philosophical Transactions of the Royal Society of London Series A, 366*, 3837–3838.
Ross, S., & Gow, A. (1999). *Digital archaeology: Rescuing neglected and damaged data resources. Technical report. British library research and innovation report*. London: British Library.
Sakellariou, C., & Bentley, P. J. (2012). Describing the FPGA-based hardware architecture of systemic computation (HAOS). *Computing and Informatics, 31*, 1001–1021.
Salathé, M., Kouyos, R. D., & Bonhoeffer, S. (2008). The state of affairs in the kingdom of the Red Queen. *Trends in Ecology and Evolution, 23*(8), 439–445.
Schull, J. (2007). Predicting the evolution of digital rights, digital objects, and digital rights management languages. In D. Satish (Ed.), *Digital rights management: An introduction*. Andhra Pradesh: ICFAI Books.
Shennan, S. (2002). *Genes, memes and human history. Darwinian archaeology and cultural evolution*. London: Thames & Hudson.
Sipper, M. (1997). A phylogenetic, ontogenetic, and epigenetic view of bio-inspired hardware systems. *IEEE Transactions on Evolutionary Computation, 1*(1).
Smolin, L. (1992). Did the universe evolve? *Classical and Quantum Gravity, 9*, 173–191.
Smolin, L. (1997). *The life of the cosmos*. Oxford: Oxford University Press.
Smolin, L. (2006). The status of cosmological natural selection, arXiv:hep-th/0612185v1.
Spencer, M., Windram, H. F., Barbrook, A. C., Davidson, E. A., & Howe, C. J. (2006). Phylogenetic analysis of written traditions. In P. Forster & C. Renfrew (Eds.), *Phylogenetic methods and the prehistory of languages* (McDonald Institute monographs, pp. 67–74). Cambridge: McDonald Institute for Archaeological Research/University of Cambridge.
Stanley, K. O., & Miikkulainen, R. (2003). A taxonomy for artificial embryogeny. *Artificial Life, 9*(2), 93–130.
Steed, A., & Milton, R. (2008). Using tracked mobile sensors to make maps of environmental effects. *Personal Ubiquitous Computing, 12*, 331–342.
Sterritt, R., Hinchey, M., Rouff, C., Rash, J., & Truszkowski, W. (2006). *Sustainable and autonomic space exploration missions*. Paper presented at the second IEEE international conference on space mission challenges for information technology (SMC-IT 2006), IEEE Computer Society.
Stoica, A. (2002). *DARPA adaptive computing systems program. Evolvable hardware for adaptive computing, 28 March 2002. Pasadena*. California: Jet Propulsion Laboratory.
Stringer, C. B. (2012). *The origin of our species*. London: Penguin.

Sultan, C., Seereram, S., & Mehra, R. K. (2007). Deep space formation flying spacecraft path planning. *International Journal of Robotics Research, 26*(4), 405–430.

Szilard, L. (1929). On the decrease of entropy in a thermodynamic system by the intervention of intelligent beings. *Zeitschrift fuer Physik, 53*, 840–856.

Takken, S., & William Wong, B. L. (2013). *Tactile reasoning: Hands-on vs hands-off – What's the difference?* Paper presented at the 11th international conference on naturalistic decision making (NDM 2013), Paris, 21–24 May 2013.

Tehrani, J. (2013). The phylogeny of Little Red Riding Hood. *PLoS One, 8*(11), e78871.

Terrile, R. J., Adami, C., Aghazarian, H., Chau, S. N., Dang, V. T., Ferguson, M. I., Fink, W., et al. (2005). *Evolutionary computation technologies for space systems*. Paper presented at the IEEE aerospace conference (IEEE AC 2005), Big Sky, Montana, Mar 2005.

Thakoor, S., Javaan Chahl, M. V., Srinivasan, L., Young, F. W., Hine, B., & Zornetzer, S. (2002). Bioinspired engineering of exploration systems for NASA and DoD. *Artificial Life, 8*, 357–369.

Thompson, A. (1996). Silicon evolution. In J. R. Koza, D. E. Goldberg, D. B. Fogel, and R. L. Riolo (eds.), *Genetic programming 1996* (pp. 444–452). Cambridge, MA: MIT Press. Proceedings of the first annual conference, 28–31 July 1996, Stanford University.

Thompson, A. (1998). *Hardware evolution. Automatic design of electronic circuits in reconfigurable hardware by artificial evolution*. London: Springer.

Tinbergen, N. (1963). On aims and methods of ethology. *Zeitschrift für Tierpsychologie, 20*, 410–433.

Todd, M. (2009). *File formats for preservation*. DPC technology watch report series, 09–02. Digital Preservation Coalition, http://www.dpconline.org/advice/technology-watch-reports

Turing, A. M. (1936). On computable numbers, with an application to the *Entscheidungsproblem*. *Proceedings of the London Mathematical Society 42*, 230–265.

Valdya, N., Walker, S. I., & Lehman, N. (2013). Recycling of informational units leads to selection of replicators in a prebiotic soup. *Chemistry & Biology, 20*, 241–252.

van der Hoeven, J. R., van Diessen, R. J., & van der Meer, K. (2005). Development of a universal virtual computer for long-term preservation of digital objects. *Journal of Information Science, 31*(3), 196–208.

van der Knijff, J. (2009). Adobe portable document format. Inventory of long-term preservation risks, v0.2, 20 Oct 2009, 56 pp, 26 Oct 2009. Koninklijke Bibliotheek, National Library of the Netherlands.

West, S. A., & Gardner, A. (2013). Adaptation and inclusive fitness. *Current Biology, 23*, R577–R584.

West-Eberhard, M. J. (2003). *Developmental plasticity and evolution*. Oxford: Oxford University Press.

White, J. (2013). Design print. A test special: 3D printers – And the ingenious things they make. *Wired, UK Edition*, Oct 2013, pp. 135–141.

Whiten, A., Hinde, R. A., Laland, K. N., & Stringer, C. B. (2011). Culture evolves. *Philosophical Transactions of the Royal Society Series B, 366*, 938–948.

Wilke, C. O., & Adami, C. (2002). The biology of digital organisms. *Trends in Ecology and Evolution, 17*, 528–532.

Williams, G. C. (1996). *Plan and purpose in nature*. London: Weidenfeld & Nicolson.

Williams, P., John, J. L., & Rowlands, I. (2009). The personal curation of digital objects: a lifecycle approach. *Aslib Proceedings: New Information Perspectives, 61*(4), 340–363.

Windram, H. F., Shaw, P., Robinson, P., & Howe, C. J. (2008). Dante's Monarchia as a test case for the use of phylogenetic methods in stemmatic analysis. *Literary and Linguistic Computing, 23*(4), 443–463.

Wolf, G. (2009). Know thyself. The personal metrics movement goes way beyond diet and exercise. It's about tracking every fact of life, from sleep to mood to pain, 24/7/365. *Wired, American Edition*, 22 June 2009, pp. 92–95.

Wood, B. (1992). Yellow Wagtail Motacilla flava migration from West Africa to Europe: Pointers towards a conservation strategy for migrants on passage. *Ibis, 34*(Supplement), 66–76.

Woods, K., Lee, C. A., & Garfinkel, S. (2011). *Extending digital repository architectures to support disk image preservation and access*. Paper presented at the joint conference on digital libraries (JCDL 2011), Ottawa, Ontario, 13–17 June 2011.
Worden, R. P. (1995). A speed limit for evolution. *Journal of Theoretical Biology, 176*, 137–152.
Xu, K., Hui, P., Li, V. O. K., Crowcroft, J., Latora, V., & Liò, P. (2009). *Impact of altruism on opportunistic communications*. Paper presented at the first IEEE international conference on ubiquitous and future networks (ICUFN 2009), Hong Kong, 7–9 June 2009.
Yedid, G., Stredwick, J., Ofria, C., & Agapow, P.-M. (2012). A comparison of the effects of random and selective mass extinctions on erosion of evolutionary history in communities of digital organisms. *PLoS One, 7*(5), e37233.
Zurek, W. H. (1990). Complexity, entropy and the physics of information. In: *Santa Fe Institute studies in the sciences of complexity*. Cambridge, MA: Perseus Publishing.
Zurek, W. H. (2004). Quantum Darwinism and envariance. In J. D. Barrow, P. C. W. Davies, & C. L. Jr Harper (Eds.), *Science and ultimate reality* (pp. 121–137). Cambridge: Cambridge University Press.

Chapter 14
IQ: Purpose and Dimensions

Phyllis Illari

Abstract In this article I examine the problem of categorising dimensions of information quality (IQ), against the background of a serious engagement with the hypothesis that IQ is purpose-dependent. First, I examine some attempts to offer categories for IQ, and a specific problem that impedes convergence in such categorisations is diagnosed. Based on this new understanding, I suggest a new way of categorising both IQ dimensions and the metrics used in implementation of IQ improvement programmes according to what they are properties of. I conclude the paper by outlining an initial categorisation of some IQ dimensions and metrics in standard use to illustrate the value of the approach.

14.1 Introduction

Understanding information quality (IQ) is a pressing task. Undertaking it involves two related aspects, one conceptual and the other implementational. This is because what is needed is a settled analysis (or analyses) of IQ that matches definitions of IQ measures and improvement programs as well as ways to implement them. Unfortunately, current literature on IQ offers no settled agreement on answers to at least four closely related questions:

1. What is a good general definition of IQ?
2. How should we classify the multiple dimensions of IQ?
3. What dimensions of IQ are there, and what do key features such as 'timeliness', 'accuracy' and so on mean?

P. Illari (✉)
Department of Science and Technology Studies,
University College London, London, UK
e-mail: phyllis.illari@ucl.ac.uk

L. Floridi and P. Illari (eds.), *The Philosophy of Information Quality*,
Synthese Library 358, DOI 10.1007/978-3-319-07121-3_14,

4. What metrics might one use to measure the dimensions of IQ, bearing in mind that more than one metric may be required to yield an overall measure for a particular dimension?

These questions begin with the most clearly conceptual one, and descend to questions much more closely concerned with implementation. This dual nature of the problem of understanding IQ is recognised in the literature: 'Both data dimensions and schema dimensions are usually defined in a qualitative way, referring to general properties of data and schemas, and the related definitions do not provide any facility for assigning values to dimensions themselves. Specifically, definitions do not provide quantitative measures, and one or more *metrics* are to be associated with dimensions as separate, distinct properties' (Batini and Scannapieco 2006, p. 19). Qualitative descriptions of the meanings of words or phrases such as 'information quality', or 'timeliness' are not the same as formal metrics required to measure them, and which are needed for implementation.

In this paper, I intend to address only the conceptual aspect of the question, not the implementational one. However, this will involve touching upon all four questions, because these four questions ultimately need to be answered collectively. On the one hand, trying to answer the questions sequentially question 1 first, then moving forward to question 2 and so forth is tempting but unlikely to succeed because, without some understanding of sensible implementable metrics and measures, it seems impossible to give a really meaningful general definition of IQ. On the other hand, it is equally unlikely to be fruitful to try to answer question 4 first, and then attempt to move backward to the others, because designing effective metrics for measuring IQ requires grasping what IQ itself is. Since this set of questions needs to be answered collectively, anyone trying to answer any of these questions is in a way concerned with all four. This might sound paradoxical, but in fact it is simply realistic. The idea is that, just as it takes two to tango, it takes both conceptual understanding and implementation, in alliance, to succeed with regard to IQ. We need to improve our conceptual understanding, then implementation measures, then back to conceptual understanding, and so on, until we get it right.

This illustrates the challenge of understanding IQ: there's no one place to start in assessing, improving or understanding IQ: you can legitimately choose any one of these questions as the place to start. But the ensuing job is messy, because you can't answer any one of these questions adequately in complete isolation from answering all of the others, as an answer to any one of these questions constrains possible answers to all the rest. With this in mind, I shall proceed in this article by developing a conceptual framework for approaching these questions, and then seek to map available metrics on to the developing conceptual picture. In this way, I hope to show that much of the task of answering the question of what IQ is indeed requires conceptual effort, and indicate what can be achieved by mapping implementable metrics to the conceptual framework I develop. In the light of this, I will not attempt in this paper to make a novel study of IQ practice, nor to extend any formal IQ metrics, although those studies must ultimately complement the conceptual study I engage in here. The ultimate test of this conceptual work is forward-looking: it will succeed if it does prove useful in moving forward the overarching project of improving IQ.

Here is a quick outline of the article. In Sect. 14.2, I shall discuss question 1 above, explaining the first major challenge for IQ: being stuck between purpose-dependence and the need to re-purpose data. In Sect. 14.3 I shall jump to question 4 above, to explain the second major challenge for IQ: the domain-specificity of successful IQ metrics. There is less to be said about this question, conceptually, as it is the most clearly implementational of the four questions. However, understanding the implementation challenges is important to the work of understanding IQ conceptually. In Sect. 14.4, having examined both end-questions to set up the challenges of IQ, I then move to the middle-ground, looking at the issue of dimensions and their classification, to address questions 2 and 3 above. I shall discuss existing efforts to classify dimensions, and identify a problem that is impeding convergence of these efforts. I shall then offer my own classification, in terms of what IQ is a property of, and give an initial mapping of some IQ dimensions to that classification. It shall become clear that this intermediate theorising is important to IQ. To anticipate, I shall attempt to clear up some of the current confusion, but I shall not attempt to offer a single answer to questions 2 and 3. I will return to this point in Sect. 14.4. In the conclusion, I shall summarise the results obtained. A final terminological note: throughout this article I shall confine myself to considering 'information quality' or 'IQ'. Much of the literature also writes of 'data quality' or 'DQ'. Yet in the following pages nothing theoretically significant depends on the distinction between IQ and DQ because, given the level of abstraction at which I am working, conceptual issues about IQ and DQ do not need to be distinguished.

14.2 Purpose: The Rock-and-a-Hard-Place of IQ

To begin at question 1, a major conceptual problem in the literature is the *purpose-dependence* of good information. The general idea is simple. For example, information is timely if it gets to you before you need to use it, and that depends on the purpose for which you intend to use it. Information that gets to you soon after it is gathered is not timely if it is too late to use; while information that gets to you the day before you need it is timely even if that information has been held up by inefficient processing before it reaches you. Indeed, the obvious importance of purpose to IQ has gained so much currency that many working in, or influenced by, the MIT group accept 'fit for purpose' as a general definition of IQ. For example: 'Quality has been defined as fitness for use, or the extent to which a product successfully serves the purposes of consumers …' (Kahn et al. 2002, p. 185). More recently, definitions of quality dimensions in the ISO/IEC 25012:2008 all make reference to a 'specific context of use' (ISO 2008). One important feature, included in a specific context of use, is normal purposes in that context of use.

However, further and deeper analysis of the purpose-dependence of IQ and the effective connection of such analysis to implementation have proven to be serious challenges: 'While fitness for use captures the essence of quality, it is difficult to measure quality using this broad definition' (Kahn et al. 2002, p. 185). In particular, there is a need to understand how to lay out more specific IQ dimensions (questions 2 and 3) and specific metrics for these dimensions (question 4), against the background of a general definition of IQ (question 1) as broad as 'fit for purpose'. Further, there is a limit to how much information can reasonably be tailored for a particular purpose, as re-purposing good quality information is becoming increasingly important. This is the rock-and-a-hard-place of IQ, which I examine in this section.

14.2.1 The Rock of Purpose-Dependence

While the MIT group thinks IQ is best generally defined as information that is 'fit for purpose', both they and many others still think that at least some dimensions of IQ, and even some aspects of IQ itself, are purpose-independent. These might be called 'inherent' or 'intrinsic' dimensions of IQ. Consider for example: 'Inherent information quality is, simply stated, data accuracy. Inherent information quality is the degree to which data accurately reflects the real-world object that the data represents' (English 1999, p. 22). Even the MIT group, which of course has done an enormous amount to gain recognition for the purpose-relativity of IQ, think that some dimensions are independent of purpose. Describing one of their fourfold classifications of dimensions, which is one of the most widely used, Lee et al. write: 'Intrinsic IQ implies that information has quality in its own right' (Lee et al. 2002, p. 135).

However, take accuracy. Accuracy for one purpose is not sufficient for accuracy for another purpose. The accuracy required for address data to be usable for a marketing campaign might very well not do if the purpose is more urgent and significant, such as vital security decisions. A reasonable response is to say that purpose changes how accurate information has to be to count as accurate *enough* – and so for the information to be of high enough IQ for the task. But purpose doesn't change what accuracy itself means. This is understandably tempting, but is not wholly satisfactory for all cases. When gathering data to represent a worldly thing, only some aspects of that thing can be represented. To grasp the problem, consider recording heights of a population. The heights can be recorded to various decimal points, using various kinds of measuring devices. It might be natural to think that the more decimal points height is measured to, the more accurate that measurement is. But a moment's reflection on measuring the height of a person as 163.467732452524677 cm should undermine this. Most of the decimal points are positively a disadvantage for most purposes, if anything impeding the accuracy of the final result. The idea is that accuracy is affected by relevance. It is not merely that accurate *enough* is set by purpose, but that even accuracy itself is infected by relevance of this kind.

Ultimately, the only completely accurate model of the system is the system itself. But the system itself is no good to you – that is why you need to extract information about some aspects of the system, and store it in a database.[1] The aspects recorded are the relevant aspects, and accuracy in this context is determined also by relevance – relevance to the intended purpose. The general problem here is that all dimensions of IQ are infected with relevance – relevance for the purposes intended for the information. This is why I call this 'the relevance problem'. The best interpretation of all dimensions of IQ is affected by purpose. This is true even though some IQ metrics can be *defined* independently of purpose – such as tuple completeness, which measures whether there are missing values in tuples in the data. Metrics are indicators of the quality of a dimension; they are not the dimension itself. I will return to this point below.

The same view is shared by others: 'These considerations show that even a dimension such as accuracy, which is considered only from the inherent point of view in the ISO standard, is strongly influenced by the context in which information is perceived/consumed' (Batini et al. 2012). However, there is no need to conclude from the purpose-relativity of IQ, that IQ is *subjective*. Purpose is a *relational* rather than a relative concept: something has (or fails to have) a purpose for something else. Consider food, for example, it is a relation, but not a relative concept/phenomenon: something as a type (e.g., grass) is food for a specific type of eater (e.g., a cow) but not for another type (e.g., a human). Likewise, IQ does not depend merely on the opinion of the user. The purpose is chosen by the user, but how well different metrics and dimensions fit the same purpose is a matter of objective assessment; the user is constrained by the chosen purpose, and it is the purpose that determines IQ, not the user. What must be concluded instead is that what IQ means, and the best interpretations of the various IQ dimensions, are all dependent on the purpose of the information in question. I shall refer to this as the purpose problem.

14.2.2 The Hard Place of Re-purposable Data

Severe as it is, the purpose problem is only the beginning. There is an important response to what I have called the relevance problem, which deserves careful consideration. Consider the following: 'Quality is *not* fitness for purpose. The diagnosis code of "broken leg" was "fit for purpose" to pay a claim. But it was *not* fit to analyze risk. Quality is fitness for *all* purposes made of the data, including the *likely* future uses. Quality information will be used in many new ways in the intelligent learning organization. Information fit for one purpose but lacking inherent quality will stunt the intellectual growth of the learning organization' (English 1999, p. 16).

I call this the 'multiple purposes response'. It is important because it identifies a crucial worry: if you design a system to give you maximal IQ for one particular

[1] The only exception to this point is when data itself is a creation of a process, and so the data is all there is. There is no distinction between data about the system and the system itself, which is what generates the problem in other cases. Even so, in most cases, accuracy is infected by relevance in the ways I have argued.

purpose, you might very well design it so that the information is too fragile to be turned easily to another purpose. This is a familiar point in design – the more carefully a tool is honed for one purpose, the more limited it becomes in terms of reapplication. Consider trying to eat soup with a fork, or spaghetti with a spoon.

This problem is exacerbated by the fact that good data costs money, and is very valuable. If the government or a company or a research institution is to invest a substantial amount to improve the quality of its information, it is a reasonable requirement that the improved information still be usable at least for some time into the future. In all these organizations, repurposing of data is pretty important. In science, there are various movements afoot to maintain data in a reusable form, particularly data from medical trials, such as that led by the FDA in the US, or Health Level Seven in Europe.

The challenge now is to recognise the need to repurpose data, *without* ignoring the real depth of the purpose-dependence problem. This is where IQ is: stuck between the rock and the hard place. To address this, return to the idea that some metrics used to help assess IQ can be defined independently of the purpose to which the information is to be put. But, recall, these metrics can only be used as indicators of IQ once they are interpreted in the light of that purpose. Nevertheless, this shows the possibility of disentangling indicators that can be defined on your information – or more precisely, defined on your information system – from metrics that measure different aspects of the relation of your information to the purposes for which it is to be used. An overall assessment of IQ will always require metrics of the second type.

This offers a practical solution. There will always be deficiencies of some sort in information that is actually available, but deficiencies can be managed so long as you know what they are. One wishes to avoid being faced with information that looks good, but isn't, or information where one cannot tell whether it is any good. One also wants to avoid information that looks bad, but is good, as one risks throwing away a valuable resource. But ultimately, information that looks bad, and is bad, isn't as big a problem as information that looks good, but isn't.

The metric or measure we get when we succeed is merely an estimate or indicator of IQ: 'Although it is common in the IQ literature to talk of "measuring", "evaluating" or "assessing" the quality of information, in practice the best we can hope for is to compute a close *estimate* of quality. … At the end of all this, the best we can achieve is to combine the results from the various checks to make a defensible guess at the quality of the data, rather than a definitive, absolute measure of its quality' (Embury 2012). The result of making IQ indicators available to the user is to empower the user. This is in broad agreement with the following observation: 'unless systems explicitly track their information quality, consumers of the information they provide cannot make judgments and decisions with high confidence. Information providers don't have to provide perfect IQ, but they need to be explicit about what IQ they do provide' (Keeton et al. 2009 p. 28). This, then, is how IQ improvement or assessment is often done, although the strategy is not always clearly articulated. Clear articulation will help, alongside a clear understanding of the nature of the problem that requires such a strategy to be adopted.

Recognising this tension between the rock and the hard place should help to avoid misunderstanding, particularly the mistake of looking at metrics that have been designed to look purpose-independent, and taking them to be truly purpose independent, in spite of the fact that they have to be allied with purpose-dependent metrics to give an overall indication of IQ itself, and any IQ dimension.

14.3 Domain Specificity

Now I have discussed the first major challenge of IQ, which enters at question 1, the most obviously conceptual question. The integration of the questions is hopefully already very clear: purpose-independent metrics are going to be crucial to help address the purpose-dependence problem. To continue to lay out the major challenges of IQ, I jump to the other end of the list, question 4, the most clearly implementational question. I am not going to make any attempt at implementation, but question 4 is relevant to the conceptual project of understanding IQ, because the conceptual questions can't be answered without understanding the severity of the domain specificity problem for implementation.

The domain specificity problem can be stated fairly simply. Successful metrics to estimate IQ can be defined, but they tend to be very specific to the context for which they are designed. When the ISO standard talks about the importance of a 'specific context of use' (ISO 2008) for IQ, one other thing it means is that successful IQ metrics are designed for a specific domain of application. This takes two forms. First, metrics are designed to cope with the particular structure the data is maintained in. Most metrics are designed for highly structured data, such as that maintained in severely restricted databases. Such metrics do not transfer to data structured in a different way, or to unstructured data, such as data found sprawling on the internet. The second element is that successful metrics are frequently designed with domain knowledge in mind. For example, a metric for estimating how current address data is might use information about how often, on average, people move house in the population of interest. Such a metric would not transfer to other populations, without adjustment.

There is less for a philosopher to say about question 4, as of course much of the work on metrics is highly technical. But there are two points worth noting. First, the problem domain specificity creates for IQ is that it impedes the building up of a common resource for IQ academics and practitioners. It is hard to build a library of well-understood metrics that can be seized on and used in many different situations. As it is, practitioners have to do a great deal of their work designing metrics from scratch. They build up expertise in such design, of course, but not in the form of a library of metrics. Second, this is, like purpose-dependence, a relational problem. Domain specific metrics are dependent on a domain. This problem, however, seems to be dealt with much better by computer scientists. This is perhaps because domain specificity does not appear to create a subjectivity problem. However, the two problems are closer in nature than may appear.

14.4 Dimensions and Their Classification

Having laid out the major challenges of IQ, I move now into the middle-ground, to questions 2 and 3, i.e. the theorising between the more conceptual understanding of IQ and its implementation. This mid-ground theorising should, hopefully, more clearly connect the conceptual understanding of IQ and the design of metrics that allow implementation of IQ improvement measures. In particular, answering questions 2 and 3 should enhance understanding of how the metrics used to measure IQ meet the major challenges I have identified. I will follow the tradition current in the computer science literature of working top-down, trying to reach from the more conceptual questions such as question 1, down to the metrics of question 4. However, my most important aim is to work on the *connection* between the conceptual and the implementational. I do not mean to imply that I take question 1 to be in any way privileged. Working bottom-up from question 4, working out what successful metrics might imply about the nature of IQ, would be a perfectly acceptable project, although I do not pursue it here.

I shall now try to show what can be achieved by keeping in mind that the process of improving IQ, including defining it, defining and categorising its dimensions, and designing metrics to measure those dimensions, involves identifying metrics that can be defined on the data, and combining them with metrics that pay specific attention to purpose, and to the domain of interest.

In this section, I shall look at existing attempts to classify IQ dimensions, diagnose what may be wrong with them, and identify a fruitful approach. I shall then map some existing IQ metrics discussed by Batini and Scannapieco (2006) onto that approach. To anticipate, the main goal of this section is to show how important it is to understanding IQ that we can be precise about what IQ itself and what various IQ dimensions and metrics are actually properties of. For example, are they properties of the data held by a single information producer? Or are they properties of the dynamic relationship between a whole information system, which is changing through time, and long-term users of that system?

The importance of answering such questions is a direct result of the purpose-dependence of IQ, and of the fact that a great deal of work designing and improving IQ involves trying to find a purpose-independent, intrinsic feature of the data itself to measure and use as an indicator of what is in fact a complex purpose-dependent feature of a relationship between data and user. Increased precision on these matters will help us understand how to think in a usefully clearer way about categories, dimensions and metrics. At core, the aim is to allow greater precision and visibility about those features of the data that travel with it, as purposes change during repurposing, and which have to be reassessed. Ultimately I will argue for moving from a hierarchical organization of IQ dimensions and metrics to a relational model linking IQ dimensions and purpose.

Table 14.1 Wang's categorisation (Source: Wang 1998)

IQ category	IQ dimensions
Intrinsic IQ	Accuracy, objectivity, believability, reputation
Accessibility IQ	Access, security
Contextual IQ	Relevancy, value-added, timeliness, completeness, amount of data
Representational IQ	Interpretability, ease of understanding, concise representation, consistent representation

14.4.1 Why Existing Classifications of IQ Dimensions Won't Converge

An important feature of the literature on IQ is an attempt to classify IQ dimensions, to answer question 2. These attempts are proliferating, and there seems to be little convergence so far in the classifications produced. In this section, I shall examine some of the best known attempts at producing such categorisations of dimensions, and seek to diagnose the problem that is impeding a useful convergence in the debate on this issue.

I begin with the categorisation of Wang (1998), which is one of the earliest and most influential categorisations of IQ dimensions, and is still frequently cited. Table 14.1 is the table given in the original paper (Wang 1998, p. 60).

There are now quite a few dimension arrangements in this style. Indeed, Lee et al. (2002) even give us two comparison tables of classifications of IQ dimensions, one for academics and one for practitioners, reproduced in Table 14.2 (Lee et al. 2002, p. 136), laid out according to the Wang (1998) categories.

This is enough to illustrate a lack of convergence that should be cause for concern to those interested in the project of categorising dimensions. The problem is explicitly noted: 'In comparing these studies two differences are apparent. One is whether the viewpoint of information consumers is considered, which necessarily requires the inclusion of some subjective dimensions. The other is the difficulty in classifying dimensions, for example, completeness, and timeliness. In some cases, such as in the Ballou and Pazer study, the completeness and timeliness dimensions fall into the intrinsic IQ category, whereas in the Wang and Strong study, these dimensions fall into the contextual IQ category. As an intrinsic dimension, completeness is defined in terms of any missing value. As a contextual dimension, completeness is also defined in terms of missing values, but only for those values used or needed by information consumers' (Lee et al. 2002, pp. 135–136). Here, they are commenting only on part of the overall comparisons they make, but the concern is clear: there is no settled agreement even on the most deeply embedded dimensions. Now, lack of convergence, of itself, may not be a problem. However, the particular form of lack of convergence currently impedes the building of intermediate theory and so progress in IQ, in ways I shall describe.

Table 14.2 Classification for practitioners (Source: Lee et al. 2002)

	Intrinsic IQ	Contextual IQ	Representational IQ	Accessibility IQ
DoD [10]	Accuracy, completeness, consistency, validity	Timeliness	Uniqueness	
MITRE [25]	Same as (Wang and Strong 1996)	Same as (Wang and Strong 1996)	Same as (Wang and Strong 1996)	Same as (Wang and Strong 1996)
IRWE [20]	Accuracy	Timeliness		Reliability (of delivery)
Unitech [23]	Accuracy, consistency, reliability	Completeness, timeliness		Security, privacy
Diamond technology partners [24]	Accuracy			Accessibility
HSBC asset management [13]	Correctness	Completeness, currency	Consistency	Accessibility
AT&T and Redman [29]	Accuracy, consistency	Completeness, relevance, comprehensiveness, essentialness, attribute granularity, currency/cycle time	Clarity of definition, precision of domains, naturalness, homogeneity, identifiability, minimum unnecessary redundancy, semantic consistency, structural consistency, appropriate representation, interpretability, portability, format precision, format flexibility, ability to represent null values, efficient use of storage, representation consistency	Obtainability, flexibility, robustness
Vality [8]			Metadata characteristics	

The reason for this is that there is a particular source of this problem, holding up any successful mapping of IQ dimensions onto categories. Batini and Scannapieco (2006, p. 39) note: 'According to the definitions described in the previous section, there is no general agreement either on which set of dimensions defines data quality or on the exact meaning of each dimension. In fact, in the illustrated proposals, dimensions are not defined in a measurable and formal way. Instead, they are defined by means of descriptive sentences in which the semantics are consequently disputable.' The first important point is the descriptive, qualitative understanding of both categories such as 'intrinsic' and 'contextual', and dimensions such as 'timeliness' and 'accuracy', however disputable, are performing a useful role in our conceptualisation of IQ. Categories such as 'intrinsic' and 'representational' and so on have an intuitive meaning, easy to understand and use, that is helpful to IQ practitioners and academics alike. The concepts of these categories are performing some kind of useful function in the academic literature, and in practice. Similarly for the concepts of IQ dimensions themselves, such as 'accuracy', 'completeness' and 'timeliness'. They have intuitively understood meanings that are functioning usefully in the thinking of both practitioners and academics (see Batini and Scannapieco (2006, p. 19)).

This is problematic because the IQ dimensions, defined according to the intuitively meaningful words that are generally used for dimensions, do not map onto the IQ categories, defined in turn according to the intuitively meaningful words that are commonly used for categories. I will spell this out in much more detail in the next subsection, by trying to offer a mapping between IQ *metrics* and categories, showing how the dimensions are built up, that will work, which will require adapting both categories and dimensions. Before, let me indicate the problem as briefly as possible. The heart of it is that the current meaningful dimensions have to be *split*, and split into the metrics used as indicators, to map properly onto existing meaningful categories. 'Accuracy', 'timeliness', 'completeness' and so on do not fit onto categories like 'intrinsic' and 'contextual' – only parts of these dimensions fit into each of these categories.

This is difficult to get clear, and so I shall illustrate the problem here very crudely (see Table 14.3), using the intrinsic-accessibility-contextual-representational categories of Wang (1998), and the well-known dimensions of accuracy and completeness. The core idea is that accuracy has aspects that are intrinsic, but may also have aspects that fall under accessibility, contextual *and* representational features, as does completeness. Accuracy itself is not entirely intrinsic or representational, and so on, but shows aspects of all of the categories. Ultimately, as I have argued, all dimensions are purpose-dependent.

I hope the intended point is clear: aspects of *all four columns* in Table 14.3 feed into an overall measure of the accuracy, and the completeness, of the information, in so far as these are dimensions of IQ itself.

This means that, while useful, this fourfold categorisation does not categorise dimensions themselves, but something else. Dimensions do not map onto these categories, using intuitively understood words that do seem to have a function in the IQ literature and practice, 1-1: they do not map in such a way that each dimension can be allocated to one, and only one, category. This is what creates a problem. And

Table 14.3 Dimensions fall into multiple categories

Intrinsic	Accessibility	Contextual	Representational
Metrics that measure elements of accuracy, defined only on the data	Information about such 'intrinsic' metrics, concerning availability to user	Features of some or all of the 'intrinsic' metrics, relevant to the purpose for which information to be used	Features of presentation of the 'intrinsic' metrics and information that allow the user to use it effectively for his/her purpose
Metrics that measure elements of completeness, defined only on the data	Information about such 'intrinsic' metrics, concerning availability to user	Features of some or all of the 'intrinsic' metrics, relevant to the purpose for which information to be used	Features of the presentation of the 'intrinsic' metrics and information that allow the user to use it effectively for his/her purpose

although there may be other difficulties, this one by itself is already so significant as to be sufficient to explain the lack of convergence in the debate on categories of IQ dimensions. Different scholars, with different intuitions about the most important *aspect* of accuracy or completeness, or different metrics in mind, will naturally allocate these dimensions to different categories.

This at least drives lack of convergence, but note that the problem is more serious than this. There are not multiple competing, but coherent and sensible options for middle-ground theorising, but instead significant muddle. Meaningful terms like 'intrinsic' and 'contextual', which are highly relevant to the severe challenges of IQ that I have identified, cannot be used effectively. This is a severe impediment to developing this kind of badly needed middle-ground theorising.

The search for categories continues despite this problem, because there is a real need for something intervening between dimensions of IQ, and IQ itself, to give structure for thinking about IQ and its dimensions. Those engaged in this project are absolutely right that there is a need for something in this middle ground, given how far apart the two ends are: the conceptual understanding of IQ as fit for purpose but repurposable, and domain-specific task-specific metrics. At the moment, the major sustained attempt has been better to understand dimensions of IQ, question 3, and offer dimension categories, question 2. But such approaches are not likely to succeed, since they all attempt to map each dimension to a single category. The risk is that, in order to fit square pegs in round holes, the relations between the two are made increasingly loose, until fit is achieved only by means of irrecoverable vagueness.

I shall attempt to use the insights developed here to make a positive suggestion to move the debate forward by splitting the dimensions. Initially, this will make both categories and dimensions less intuitively meaningful, but I hope to show how the overall framework ultimately recovers the meaningful aspects of both category and dimension terms currently in use, while still clearing away some of the current confusion. It is worth noting two purposes here. Initially, my aim is to help the IQ field in computer science – primarily in the academic literature – move forwards in building intermediate theory, by making a suggestion to them for where to work. I hope that this might ultimately also be of use to help in the practice of IQ improvement programmes in terms of both theoretical knowledge and practical tools available as standard, but of course that is a much more distant goal. I shall comment further in Sect. 14.4.4 on what I take my theoretical contribution to be.

14.4.2 What Is IQ a Property of? Towards a Classification for IQ Dimensions

I shall now try to get more precise about the lesson learned from the discussion above, and begin the task of designing something like a classification of IQ dimensions that can generate settled agreement. I shall argue that what is vital to understanding IQ is the answer to the question what *exactly* IQ itself, its dimensions and its metrics are properties of. Note that what I offer is not a classification in the same spirit as existing ones, but more like a representation of elements worth representing in any particular classification, put together for a particular job. Further, while dimensions are represented, the *basic* elements in the classification are not dimensions, but metrics, for reasons I shall explain.

I first note the complexity of the problem. Batini and Scannapieco (2006) write: 'definitions do not provide quantitative measures, and one or more metrics are to be associated with dimensions as separate, distinct properties. For each metric, one or more measurement methods are to be provided regarding … (i) where the measurement is taken, (ii) what data are included, (iii) the measurement device, and (iv) the scale on which results are reported. According to the literature, at times we will distinguish between dimensions and metrics, while other times we will directly provide metrics' (Batini and Scannapieco 2006, p. 19). In order to answer the four questions I began with, and so lay out a framework for consistent settled thinking about IQ, it is not just dimensions that I need to map onto the categories I have in mind: ultimately I also need to lay out the relations between dimensions, their categories, and metrics and measures.

Consider what IQ could be a property of. Naturally, it is a property of information, but what information, exactly? There is a surprisingly large number of candidates:

- Single data item;
- Set of data about a particular worldly item;
- All data about a particular class of worldly items;
- All data in a database;
- Whole information system, even if it accesses multiple databases;
- Single data source;
- Whole information system, even if it accesses multiple databases, some or all of which use multiple sources;

- Whole dynamically evolving information system, so including IQ improvement measures which operate over time;
- Relation between entire (dynamically evolving) information system and a data consumer with a particular purpose (possibly a long-term one) in mind.

This list is probably not exhaustive. It may seem odd to count the later possibilities as possible bearers of IQ. But data is usually a collective. We do not usually worry about the quality of a datum, although we might, of course. However, clearly multiple data, or a collective of information, are legitimate bearers of information quality. As soon as that is noticed, the question of what collective we have in mind when assessing IQ is a natural one, and a question that is important for understanding IQ. It matters for what we count as, most obviously, completeness, but it also matters for other dimensions. If we think of the collective as the whole functioning information system, then *dynamic* properties of that system, such as correction mechanisms, also become legitimate parts of the bearer of IQ.

Recall what I have indicated as the fundamental problem: that defining, modelling, and implementing good IQ requires transforming purpose-dependent features of a whole information system into, as far as is possible, proxy indicators of IQ. These proxy indicators are, as far as is possible, intrinsic features qualifying only parts of the system itself, rather than properties of the relationship between the system and its context. This means that they are features that can be defined on, and are properties of, the system itself, isolated from the world and from the purposes of any user. Now, a settled classification of standard IQ dimensions and metrics along the lines of what they are properties of would seem likely to help in the enterprise that engages with the fundamental problem.

This idea offers a way of categorising IQ dimensions that might lead to considerably more agreement and so convergence. I also hope to show that it will maintain some of the intuitive notions already in use, such as 'intrinsic' and 'contextual', which are already functioning usefully in the debate, as these notions will be recoverable from the end result.

14.4.3 A New Classification

The idea of the new classification is to look carefully at the information system, and identify parts of it that are different bearers of properties relevant to IQ, creating a diagram with spaces for each. Then start identifying the elements of the IQ improvement program: IQ itself, dimensions and metrics that you want to map. Then map the elements of the IQ improvement program onto the spaces representing the bearers of the property. Note that the mapping from dimension to category is not 1:1 but 1:N. Note also that there are two *kinds* of things that might be bearers of properties relevant to IQ, and the two must be distinguished:

1. Parts of the information system before you:
   (a) in which case the important thing is to get clear on which parts, as there may be several that are useful to distinguish.
2. Relations between the information system and something external to it, its 'context'. This most notably includes:
   (a) the relation (deployment) between the information system and the purpose of the user, and,
   (b) the relation (reference) between the information system and the external world, particularly aspects of the world represented somewhere in your information system.

The difference between these two can no doubt be represented successfully in a myriad of ways. In our example below:

1. Properties of parts of the information system itself fall into columns, headed 'Data, or the data in a particular population', 'a particular source of information' 'information in the single information system in front of you', and 'information across several information systems' to discriminate different parts of an information system that may well be worth distinguishing.
2. Relations between the information itself and the two crucial features of its context are represented by the 'open' columns on either side of the columns for the information system:
   (a) The left hand one 'relation between the information system itself and the world' allows representation of relations between the proxy indicators that can be defined on the information system, and features of the external world that are *not* the user or the purpose of use.
   (b) The right hand one 'relation between information system and the purpose of the user' allows representation of the other relational features of IQ.

I have made an initial mapping of some existing dimensions and metrics into this space, beginning with timeliness and associated metrics. CAPITALISED words represent IQ dimensions, while words in lower case represent metrics or measures. A single row of the table contains metrics and measures that are related to the dimension also contained in that row – specifically, they are used as proxy indicators of the quality of the dimension. But they are metrics defined on the data, so they can also be used as proxy indicators of the dimension – suitably reinterpreted – if the purpose shifts.

This kind of mapping could usefully be done with any kind of element of IQ, including entirely new metrics, which may require more elements of the information system and its context than I illustrate below to be identified as bearers of the properties measured. However, I will illustrate the idea of the mapping rather crudely and briefly using dimensions and metrics discussed by Batini and Scannapieco (2006), and using abstract descriptions of some of the kinds of things

Table 14.4 Timeliness and associated metrics

What is IQ a property *of*?					
The **relation between information system and world**	**Data**, or the **data in a particular population**	A particular **source** of information e.g. database or informant	**Information** in the **single information system in front of you**	**Information** across **several information systems**	The **relation** between information system and the purpose of a user
Rapidity of change in the target population	Volatility		Currency	Currency	TIMELINESS

that I might want to identify as the bearers of the properties we are interested in when defining and constructing measures for IQ improvement. I begin with the dimension of timeliness in Table 14.4.

The idea is that timeliness is the dimension of IQ, which is relative to the purpose of use as already explained above. Currency is a metric which can be defined on the information itself, using something as simple as an update date, and it can be defined on information in one system or several, so that it falls into multiple columns. Currency does not yield timeliness, though, because whether an update date of 2 months ago is 'recent' depends on the volatility of the data in question – how rapidly the values of the data change. If your information is a house address, then 2 months ago is recent. If your information is levels of glucose within a metabolising cell, it is thoroughly obsolete. Volatility measures change in data, and of course this depends on the rapidity of change in the real-world target population.

With this simpler example in mind, I add other dimensions of usable accuracy and completeness in Table 14.5 below. The mapping is very far from complete or exhaustive. It is meant merely to illustrate. I suspect that this kind of mapping may be useful in many attempts to improve and better understand IQ, but that different aspects of the information system, on which different more specific metrics may be defined, will be more or less useful to identify in different cases.

As for timeliness, usable accuracy, and completeness with respect to purpose are the true dimensions of IQ, and, as I have argued above, they are dependent on the purpose of the user. Well-known metrics that are used as indicators of these dimensions can be defined on a single information system, and on multiple information systems. Some can be defined on a single attribute, such as attribute completeness. In both cases, again, there is also an important relation to the world. Semantic accuracy concerns whether the information in your system matches worldly values, while choosing between closed or open world assumptions involves making a big assumption – which should be marked – about the relation between the information in the system and the world. Again, useful relations between metrics as indicators of quality dimensions, the purpose of the user, and the nature of the world can be seen laid out in this manner.

Table 14.5 Other dimensions and their associated metrics

What is IQ a property *of*?					
The **relation between information system and world**	**Data**, or the **data in a particular population**	A particular **source** of information e.g. database or informant	**Information** in the **single information system in front of you**	**Information** across **several information systems**	The **relation** between information system and the purpose of a user
Rapidity of change in the target population	Volatility	Sources may be characterised by usual quality	Currency	Currency	TIMELINESS
Semantic accuracy	Semantic accuracy	Sources may be characterised by usual quality	Syntactic accuracy Comparison functions Edit distance	Syntactic accuracy Comparison functions Edit distance	USABLE ACCURACY
Open world assumption versus closed world assumption	Population completeness	Sources may be characterised by usual quality	Attribute completeness Entity completeness Column completeness	Attribute completeness Entity completeness Column completeness	COMPLETENESS

The simplified mapping above was achieved conceptually, by examining the definitions and measures to pick out precisely what aspects of the information system they are defined on. Nevertheless, some quite interesting conclusions can be drawn. First, it is worth putting quite a few different elements of the information system into the columns for this mapping, and it is not difficult to think of more things that could usefully be represented. Second, many of the elements of IQ are properties of relations. Even some, such as semantic rules and integrity constraints, which can be defined on the information system itself, are properties of quite complex relationships. They remain properties of the information system itself, because those complex relationships are themselves internal to the information system. But note that semantic rules are often, if not always, constructed *successfully* using world-knowledge, and they will not transfer to data structured differently. Third, as expected, even though the dimensions of IQ themselves are properties of the relation between the whole information system and the user, some elements of all of them, particularly metrics used to measure them, can sensibly be defined just on the information system itself, so allowing such metrics to be properties of that system. This allows them to be used when data is transferred for a different purpose, as indicators that can be used to construct new estimates of the IQ of the data when used for that purpose.

Finally, the domain-specificity of metrics is also made clear. If metrics depend on domain-knowledge, it is worth representing that explicitly, so that it not be forgotten in the case of worldly change – perhaps trying to transfer a metric for currency of address data to a more volatile population.

14.4.4 Discussion of the Classification

The idea has been to move from a hierarchical organization of IQ dimensions and metrics to a relational model linking IQ dimensions and purpose. To this end, the previous mapping offers several advantages, including the possibility of convergence of a single classification of IQ metrics and dimensions, or multiple non-competing classifications, classifications sensitive to what IQ improvement programs are really trying to do, a clear indication of potential pitfalls, and finally a valuable recovery of important concepts like 'intrinsic' and 'contextual'. I shall briefly comment on each of them in turn.

First, convergence should be encouraged by this mapping, because it should be possible to map metrics and dimensions onto this kind of space, and useful in sharpening up their definition, and their interrelations. Deciding what such things are properties of – what they can be defined on – is a matter of considerably more objective assessment and should be much easier to agree on than whether entire IQ dimensions are, for example, 'intrinsic'. This mapping also completely avoids the muddle at the heart of current attempts to map dimensions themselves onto categories.

Second, this kind of mapping lays out the tools of IQ improvement in a way that is sensitive to what IQ improvement programmes try to do. It lays out the relationship between metrics that are genuinely objective measures of the data itself, domain-specific metrics, and highly purpose-dependent features of the whole system. The place of such metrics as mere indicators of the relational IQ dimensions is clear. The tables give a representation of the scale of the challenge of IQ, and what is being done to meet it.

Third, as a complement to the table laying out useful features of tools, it also represents the gaps. These mappings visually represent where the enterprise of finding intrinsic features of the information to act as proxy indicators of properties of relational features is forced, where the metric or dimension is a property of a relation. The forced nature of proxy indicators of the quality of the information for the purposes of the user will not be blurred or easily forgotten with such maps in mind.

Finally, this mapping allows the recovery of some important intuitive terms in the literature, but in more precise form. I suggest that intrinsic IQ metrics are those that can be defined solely on the information system itself, such as some specific completeness metrics. These are properties of the information stored, and our mapping still has the advantage of encouraging continuous attention to exactly what feature of the information stored they are properties of. Note, though, that it tends to be only metrics, and only some of them, which are intrinsic in this sense. And in so far as such metrics relate to IQ, they are always proxy indicators of a more complex relational property. Contextual features of IQ are those which attempt to measure something about the relationship between the information system and its context. I have now identified the two crucial features of that context: (a) the relation between the information system and the purpose of the user, (b) the relation between the information system and the world, including of course features of the world explicitly represented, such as birth dates, but also features of the world used to construct appropriate semantic rules for checking consistency. Ideas of 'representational' and 'accessibility' relations are less easy to define precisely. But I suggest they are thought of explicitly as themselves features of the relationship between the information and the user, which is an idea that requires future work. Further, here it is particular characteristics of the users that are relevant, such as the language they speak, and what technical skills and theoretical understanding they have, rather than merely their purpose.

Ultimately, this mapping has many advantages, and recovers the intuitive usability of terms that are performing a useful role in both the literature and practice.

14.5 Conclusion

I have briefly summarised my reasons for thinking that the purpose problem for IQ is serious, and that much of the work on IQ responds by looking for proxy indicators of IQ that can be defined on features of the information system itself. I have offered my approach to mapping elements of all major concepts engineered for IQ improvement onto a space designed to represent what they are properties of. This is my first attempt to address the four interrelated questions with which I began:

1. What is a good general definition of IQ?
2. How should we classify the multiple dimensions of IQ?
3. What dimensions of IQ are there, and what do key features such as 'timeliness', 'accuracy' and so on mean?
4. What metrics might one use to measure the dimensions of IQ, bearing in mind that more than one metric may be required to yield an overall measure for a particular dimension?

My mapping offers a way of seeing the problems laid out collectively, showing how much in common they have. Fitness for purpose is vital to IQ, and should inform understanding of the purpose of a classification, and also identification of dimensions and the design of metrics. It is due to the difficulty of addressing the fitness for purpose problem that metrics are used, as they are, as proxy indicators of purpose-dependent dimensions. This research can continue by examining further metrics and adding to the mapping above, and expanding understanding of how they are designed to meet the purpose problem.

I finish by commenting on the challenges I began by identifying. They are indeed serious. But properties of relations are not in themselves intractable. Relational properties internal to the information system itself are frequently defined very well, such as integrity constraints. The purpose problem is just that the bearer of some features of IQ is the relation between system and purpose of user. But there is nothing here that can't be measured in principle. The relation might be imperfectly measured, perhaps, but no more imperfectly than some relational features internal to the information system itself are measured. If the purpose requires speed more than accuracy, this trade-off can be assessed, proxy measures found and implemented. If the purpose requires completeness, this too can be assessed, measures created and implemented, then tested and adjusted, and so on. From another point of view, we could track user choices, given stated purpose, and learn how to improve measures of the relation between the system and purpose that way. This is not very different from the domain-specificity of many metrics, which require the relation between the domain and the information system to remain unaltered.

To summarise, there are two major challenges of IQ. The first is that IQ itself is purpose-dependent, while we need to be able to repurpose data. The second is the domain-specificity of successful metrics. To succeed in IQ improvement and assessment, one side of the problem is just that we have to relate the information system to the world. This is probably going to mean that some measures will remain ineliminably domain-specific. The other side is that we have to relate the information system to the purpose of the user. So some measures will remain ineliminably purpose-specific. These two are both ineliminably contextual – but tractable – features of IQ.

Acknowledgements Research for this article was supported by a 2-year project, entitled "Understanding Information Quality Standards and Their Challenges", currently funded (2011–2013) by the British Arts and Humanities Research Council (AHRC). I owe thanks to Suzanne Embury and Carlo Batini for giving me very useful comments on earlier versions of this article, and to Luciano Floridi both for specific comments on this piece, and for the opportunity to collaborate with him on the wider project. Remaining errors are, of course, my own.

References

Batini, C., & Scannapieco, M. (2006). *Data quality: Concepts, methodologies and techniques*. New York: Springer.

Batini, C., Palmonari, M., & Viscusi, G. (2012). *The many faces of information and their impact on information quality*. Paper presented at the AISB/IACAP world congress, University of Birmingham. http://philosophyofinformation.net/IQ/AHRC_Information_Quality_Project/Proceedings.html

Embury, S. (2012). *Forget dimensions. Define your information quality using quality view patterns*. Paper presented at the AISB/IACAP world congress, University of Birmingham. http://philosophyofinformation.net/IQ/AHRC_Information_Quality_Project/Proceedings.html

English, L. (1999). *Improving data warehouse and business information quality*. New York: Wiley.

ISO. (2008). IEC FDIS *software engineering – software product quality requirements and evaluation – data quality model* (Vol. 25012).

Kahn, B. K., Strong, D. M., & Wang, R. Y. (2002). Information quality benchmarks: Product and service performance. *Communications of the ACM, 45*(4), 184–192.

Keeton, K., Mehra, P., & Wilkes, J. (2009). Do you know your IQ? A research agenda for information quality in systems. *ACM SIGMETRICS Performance Evaluation Review, 37*(3), 26–31.

Lee, Y. W., Strong, D. M., Kahn, B. K., & Wang, R. Y. (2002). AIMQ: A methodology for information quality assessment. *Information & Management, 40*(2), 133–146. doi:10.1016/s0378-7206(02)00043-5.

Wang, R. Y. (1998). A product perspective on total data quality management. *Communications of the ACM, 41*(2), 58–65. doi:10.1145/269012.269022.

Wang, R. Y., & Strong, D. M. (1996). Beyond accuracy: What data quality means to data consumers. *Journal of Management Information Systems, 12*(4), 5–34.

Chapter 15
Big Data and Information Quality

Luciano Floridi

Abstract This paper is divided into two parts. In the first, I shall briefly analyse the phenomenon of "big data", and argue that the real epistemological challenge posed by the zettabyte era is *small patterns*. The valuable undercurrents in the ocean of data that we are accumulating are invisible to the computationally-naked eye, so more and better technology will help. However, because the problem with big data is small patterns, ultimately, the game will be won by those who "know how to ask and answer questions" (Plato, *Cratylus*, 390c). This introduces the second part, concerning information quality (IQ): which data may be useful and relevant, and so worth collecting, curating, and *querying*, in order to exploit their valuable (small) patterns? I shall argue that the standard way of seeing IQ in terms of fit-for-purpose is correct but needs to be complemented by a methodology of abstraction, which allows IQ to be indexed to different purposes. This fundamental step can be taken by adopting a bi-categorical approach. This means distinguishing between purpose/s for which some information is *produced* (P-purpose) and purpose/s for which the same information is *consumed* (C-purpose). Such a bi-categorical approach in turn allows one to analyse a variety of so-called IQ dimensions, such as accuracy, completeness, consistency, and timeliness. I shall show that the bi-categorical approach lends itself to simple visualisations in terms of radar charts.

15.1 Big Data

Just a few year ago, researchers at Berkeley's School of Information estimated that humanity had accumulated approximately 12 exabytes of data in the course of its entire history (1 exabyte corresponds to 10^{18} bytes or a 50,000 year-long video of

L. Floridi (✉)
Oxford Internet Institute, University of Oxford,
1 St Giles, Oxford OX1 3JS, UK
e-mail: luciano.floridi@oii.ox.ac.uk

L. Floridi and P. Illari (eds.), *The Philosophy of Information Quality*,
Synthese Library 358, DOI 10.1007/978-3-319-07121-3_15,

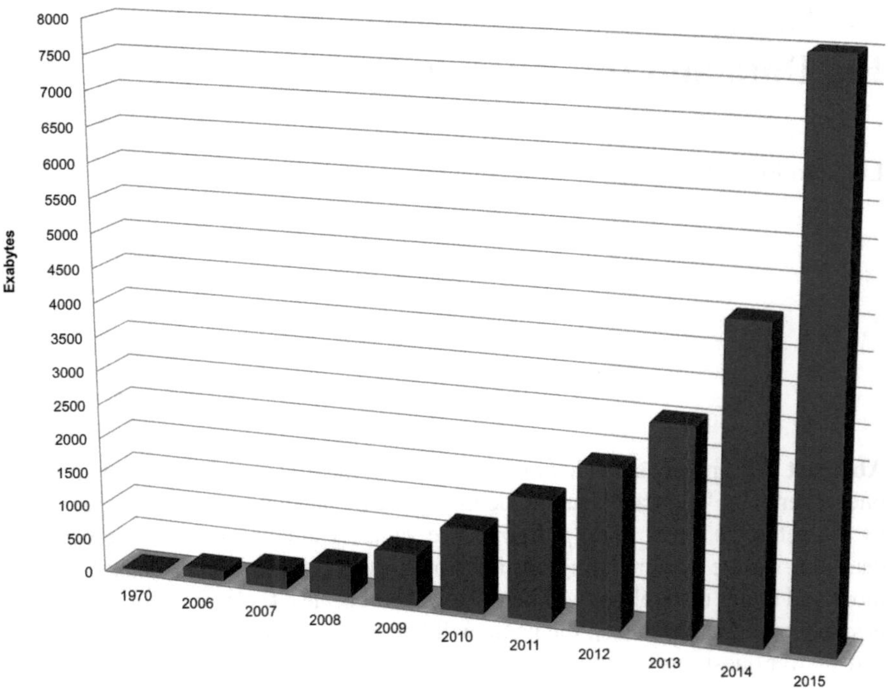

Fig. 15.1 The growth of big data. Based on IDC white paper, "The diverse and exploding digital universe", March 2008 and IDC white paper "Worldwide big data technology and service 2012–2015 forecast", March 2012

DVD quality), until the commodification of computers, but it had already reached 180 exabytes by 2006. According to a more recent study, the total grew to over 1,600 exabytes, between 2006 and 2011, thus passing the zettabyte (1,000 exabytes) barrier. This figure is now expected to grow fourfold approximately every 3 years, so that we shall have 8 zettabyes of data by 2015 (Fig. 15.1). Every day, enough new data is being generated to fill all U.S. libraries eight times over. Of course, trillions of Information and Communications Technology (ICT) systems are constantly working to keep us afloat and navigate through such an ocean of data. These are all numbers that will keep growing quickly and steadily for the foreseeable future, especially because those very systems are among the greatest sources of further data, which in turn require or simply make possible more ICTs. It is a self-reinforcing cycle and it would be unnatural not to feel overwhelmed. It is, or at least should be, a mixed feeling of apprehension for the risks, excitement for the opportunities, and astonishment for the achievements.

Thanks to ICTs, we have entered *the age of the zettabyte*. Our generation is the first to experience a Zettaflood, to introduce a neologism to qualify this tsunami of bytes that is submerging our environments. In other contexts, this is also known as "big data".

Despite the importance of the phenomenon, it is unclear what exactly the term "big data" means and hence refers to. The temptation, in similar cases, is to adopt the approach pioneered by United States Supreme Court Justice Potter Stewart to describe pornography: difficult to define, but "I know when I see it". Other strategies have been much less successful. For example, in the United States, the National Institutes of Health (NIH) and the National Science Foundation (NSF) have identified big data as a program focus. One of the main NSF-NIH interagency initiatives addresses the need for core techniques and technologies for advancing big data science and engineering. However, the two agencies specify that:

> The phrase 'big data' in this solicitation refers to large, diverse, complex, longitudinal, and/or distributed data sets generated from instruments, sensors, Internet transactions, email, video, click streams, and/or all other digital sources available today and in the future. (see NSF-12-499)

You do not need to be a logician to find this both obscure and vague. Wikipedia, for once, is also unhelpful. Not because the relevant entry is unreliable, but because it reports the common definition, which is unsatisfactory:

> data sets so large and complex that it becomes difficult to process using on-hand database management tools or traditional data processing applications. (16 August version)

Apart from the circular problem of defining "big" with "large" (the NSF and NHI seem to be happy with it), the aforementioned text suggests that data are too big or large only in relation to our current computational power. This is misleading. Of course, "big", as many other terms, is a relational predicate: a pair of shoes may be too big for you, but fine for me. It is also trivial to acknowledge that we tend to evaluate things non-relationally, in this case as absolutely big, whenever the frame of reference is obvious enough to be left implicit. A horse is a big animal, no matter what whales may think. Yet these two simple points may give the impression that there is no real trouble with "big data" being a loosely defined term referring to the fact that our current computers cannot handle so many gazillions of data efficiently. And this is where two confusions seem to creep in. First, that the *epistemological problem* with big data is that there is too much of it (the *ethical problem* concerns how we use them, see below). And, second, that the *solution* to the epistemological problem is *technological*: more and better techniques and technologies, which will "shrink" big data back to a manageable size. The epistemological problem is different, and it requires an equally epistemological solution, not a technological one.

15.2 The Epistemological Problem with Big Data

Consider the problem first. "Big data" came to be formulated after other buzz expressions, such as "infoglut" or "information overload", began to fade away, yet the idea remains the same. It refers to an overwhelming sense that we have bitten off more than we can chew, that we are being force-fed like geese, that our intellectual livers are exploding. This is a mistake. Yes, we have seen that there is an obvious exponential growth of data on an ever-larger number of topics, but complaining about such over-abundance would be like complaining about a banquet that offers more than we can ever eat. Data remain an asset, a resource to exploit. Nobody is forcing us to digest every available byte. We are becoming data-richer by the day; this cannot be the fundamental problem.

Since the problem is not the increasing wealth of data that is becoming available, clearly the solution needs to be reconsidered: it cannot be merely how many data we can technologically process. We saw that, if anything, more and better techniques and technologies are only going to generate more data. If the problem were too many data, more ICTs would only exacerbate it. Growing bigger digestive systems, as it were, is not the way forward.

The real, epistemological problem with big data is *small patterns*. Precisely because so many data can now be generated and processed so quickly, so cheaply, and on virtually anything, the pressure both on the data *nouveau riche*, such as Facebook or Walmart, Amazon or Google, and on the data *old money*, such as genetics or medicine, experimental physics or neuroscience, is to be able to spot where the new patterns with real added-value lie in their immense databases, and how they can best be exploited for the creation of wealth, the improvement of human lives, and the advancement of knowledge. An analogy with energy resources may help: we have now entered the stage of data fracking.[1]

Small patterns matter because today they represent the new frontier of competition, from science to business, from governance to social policies, from security to business. In a Baconian open market of ideas, if someone else can exploit them earlier and more successfully than you do, you might be out of business soon, like Kodak, or miss a fundamental discovery, or put your country in danger.

Small patterns may also be risky, because they push the limit of what is predictable, and therefore may be anticipated, about not only nature's, but also people's, behaviour. This is an ethical problem. Target, an American retailing company, relies on the analysis of the purchasing patterns of 25 products in order to assign each shopper a "pregnancy prediction" score, estimate her due date, and send coupons timed to specific stages of her pregnancy. In a notorious case, it caused some serious problems when it sent coupons to a family in which the teenager daughter had not informed her parents about her new status.

15.3 From Big Data to Small Patterns

Unfortunately, small patterns may be significant only if properly aggregated, e.g. in terms of loyalty cards and shopping suggestions, compared, as when a bank can use big data to fight fraudsters, and timely processed, as in financial markets. And

[1] Fracking (hydraulic fracturing) is a technique in which a liquid (usually water), mixed with sand and chemicals, is injected underground at high pressure in order to cause small fractures (typically less than 1 mm), along which fluids such as gas (especially shale gas), petroleum and brine water can surface.

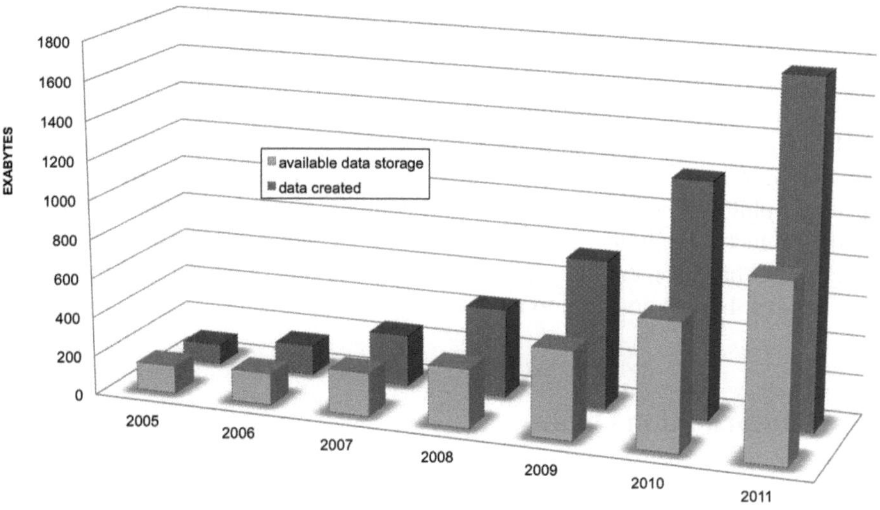

Fig. 15.2 Global information (data) created vs. memory (storage) available. Based on IDC white paper, "The diverse and exploding digital universe", March 2008; IDC white paper "Worldwide big data technology and service 2012–2015 forecast", March 2012; and "Data, data everywhere"; and *The Economist*, 25 February, 2010

because information is indicative also when it is not there, small patterns can also be significant if they are absent. Sherlock Holmes solves one of its famous cases because of the silence of the dog, which should have barked. If big data are not "barking" when they should, something is going on, as the Financial watchdogs (should) know.

The increasingly valuable undercurrents in the ever-expanding oceans of data are invisible to the computationally-naked eye, so more and better techniques and technologies will help significantly. Yet, by themselves, they will be insufficient. And mere data hoarding, while waiting for more powerful computers and software, will not work either. Since 2007, the world has been producing more data than available storage (see Fig. 15.2). We have shifted from the problem of what to save to the problem of what to erase. Something must be deleted or never be recorded in the first place. Think of your smart phone becoming too full because you took too many pictures, and make it a global problem.

The infosphere run out of memory space to dump its data years ago. This is not as bad as it looks. Rephrasing a common saying in advertisement, half of our data is junk, we just do not know which half. So what we need is a better understanding of which data are worth preserving. And this is a matter of grasping both what information quality is, as we shall see in the second half of this chapter, and which questions are or will be interesting. Which is just another way of saying that, because the problem with big data is small patterns, ultimately, the game will be won by those who "know how to ask and answer questions" (Plato, *Cratylus*, 390c), and therefore know which data may be useful and relevant, and hence worth collecting, curating, and *querying*, in order to exploit their valuable patterns. We need more and better techniques and technologies to see the small data patterns, but we need more and better epistemology to sift the valuable ones.

Big data is here to grow. The only way of tackling it is to know what we are or may be looking for. At the moment, such epistemological skills are taught and applied by a black art called *analytics*. Not exactly your standard degree at the University. Yet so much of our well-being depends on it that it might be time to develop a philosophical investigation of its methods. The epistemology of analytics is not just uncharted, it is still a virgin territory at the moment. Who knows, philosophers might have something to learn, but also a couple of lessons to teach. Plato would agree. Let me now turn to the quality problem.

15.4 Information Quality

The most developed post-industrial societies live by information, and Information and Communication Technologies (ICTs) keep them oxygenated (English (2009)). So the more (big data) and better (information quality) the information exchanged is, the more likely such societies and their members may prosper. But what is information quality (IQ) exactly? The question has become increasingly pressing in recent years.[2] Yet, in this case too, our answers have been less than satisfactory so far.

In the US, the *Information Quality Act*, also known as the *Data Quality Act*,[3] enacted in 2000, left undefined virtually every key concept in the text. So it required the Office of Management and Budget

> to promulgate guidance to agencies ensuring the quality, objectivity, utility, and integrity of information (including statistical information) disseminated by Federal agencies.

Unsurprisingly, the guidelines have received much criticism and have been under review ever since.[4]

In the UK, some of the most sustained efforts in dealing with IQ issues have concerned the National Health Service (NHS). Already in 2001, the Kennedy Report[5] acknowledged that: "All health care is information driven, so the threat associated with poor information is a direct risk to the quality of healthcare service and governance in the NHS". However, in 2004, the NHS Information Quality Assurance Consultation[6] still stressed that

---

[2] The body of literature on IQ is growing, see for example (Olson (2003), Wang et al. (2005), Batini and Scannapieco (2006), Lee et al. (2006), Al-Hakim (2007), Herzog et al. (2007), Maydanchik (2007), McGilvray (2008), Theys (2011)).

[3] http://www.whitehouse.gov/omb/fedreg_reproducible

[4] See more recently United States. Congress. House. Committee on Government Reform. Subcommittee on Regulatory Affairs (2006).

[5] http://webarchive.nationalarchives.gov.uk/20090811143745/http://www.bristol-inquiry.org.uk

[6] http://webarchive.nationalarchives.gov.uk/+/www.dh.gov.uk/en/Publicationsandstatistics/Publications/PublicationsPolicyAndGuidance/DH_4125508

> Consideration of information and data quality are made more complex by the general agreement that there are a number of different aspects to information/data quality but no clear agreement as to what these are.

We know that lacking a clear and precise understanding of IQ properties causes costly errors, confusion, impasse, dangerous risks, and missed opportunities. Part of the difficulty lies in constructing the right conceptual and technical framework necessary to analyse and evaluate them. As the reader knows, some steps have been taken to rectify the situation. The first *International Conference on Information Quality* was organised in 1996.[7] In 2006, the Association of Computing Machinery (ACM) launched this journal.[8] The Data Quality Summit[9] now provides an international forum for the study of information quality strategies. Pioneering investigations in the 1990s – including Wang and Kon (1992), Tozer (1994), Redman (1996), and Wang (1998) – and research programmes such as the Information Quality Program[10] at MIT, have addressed applied issues, plausible scenarios, and the codification of best practices. So there is already a wealth of available results that could make a difference. However, such results have had limited impact also because research concerning IQ has failed to combine and cross-fertilise theory and practice. Furthermore, insufficient work has been done to promote the value-adding synthesis of academic findings and technological know-how. There is a proliferation of taxonomies (Batini and Scannapieco (2006) offer an excellent introduction), which highlights one of the main epistemological difficulties in dealing with IQ, the one with which I shall be concerned in the rest of this chapter.

15.5 The Epistemological Problem with Information Quality

There is a lot of convergence in the literature on understanding IQ by starting from an analysis of the fit-for-purpose value of the data in question:

> There is no doubt that a database can be of high quality for a given application, while being of low quality for a different one. Hence the common definition of data quality as "fitness for use". However, such consideration often leads to the wrong assumption that it is not possible to have an objective assessment of quality of data. We claim that for most data quality dimensions (including accuracy, completeness and consistency at least) it makes sense to have objective measures on the basis of which the perceived quality can be evaluated in relation to a given user application requirements. Batini and Scannapieco (2006), p. 221

Once IQ is analysed teleologically, in terms of "fit for purpose", IQ properties, known in the literature as *dimensions* – such as accessibility, accuracy, availability, completeness, currency, integrity, redundancy, reliability, timeliness, trustworthiness, usability, and so forth – are clustered in IQ groups, known as *categories*, such as

[7] http://mitiq.mit.edu/ICIQ/2013/

[8] http://jdiq.acm.org/

[9] http://www.dataqualitysummit.com/

[10] http://mitiq.mit.edu/

IQ CATEGORIES	IQ DIMENSIONS
Intrinsic IQ	**Accuracy**, Objectivity, Believability
Accessibility IQ	**Access**, Security
Contextual IQ	Relevancy, Value-Added, **Timeliness**, **Completeness**, Amount of data
Representational IQ	Interpretability, Ease of understanding, Concise representation, **Consistent representation**

Fig. 15.3 Example of IQ categories and dimensions (Adapted from Wang (1998), in bold, dimensions from Batini and Scannapieco (2006))

intrinsic, extrinsic, contextual, representational and so forth (Fig. 15.3 provides an illustration).

All this is well known and does not need to be discussed in this context. However, since there are many ways of identifying and specifying dimensions and categories, the result is that the issuing maps do not overlap, and some of them resemble Borges' *Celestial Emporium of Benevolent Knowledge's Taxonomy*[11]:

> The list divides all animals into one of 14 categories: (1) Those that belong to the emperor; (2) Embalmed ones; (3) Those that are trained; (4) Suckling pigs; (5) Mermaids (or Sirens); (6) Fabulous ones; (7) Stray dogs; (8) Those that are included in this classification; (9) Those that tremble as if they were mad (10) Innumerable ones; (11) Those drawn with a very fine camel hair brush; (12) Et cetera; (13) Those that have just broken the flower vase; (14) Those that, at a distance, resemble flies.

A further consequence is that the all-important, practical issue of how to operationalize IQ evaluation processes is disregarded. This is not just a matter of lack of logical rigour and methodological negligence, although they too play a role. The main trouble seems to be caused by

(1) a failure to identify the potentially multipurpose and boundlessly repurposable nature of information as the source of significant complications. This is particularly significant when dealing with "big data"; because of
(2) a disregard for the fact that any quality evaluation can only happen at a given *level of abstraction*.[12] To simplify (see Fig. 15.4): the quality of a system fit for a particular purpose is analysed within a context, at a LoA, whose selection is determined by the choice of the *purpose* in the first place. If one wants to evaluate a bayonet for the purpose of holding some paper in place on the desk, then that purpose determines the LoA within that context, which will include, for example, how clean the bayonet is, but not whether it is sharp; leading to
(3) a missed opportunity to address the development of a satisfactory approach to IQ in terms of LoA and purpose-orientation.

---

[11] Borges, "The Analytical Language of John Wilkins", originally published in 1952, English translation in Borges (1964).

[12] On the method of abstraction and LoA see Floridi (2008) and Floridi (2011).

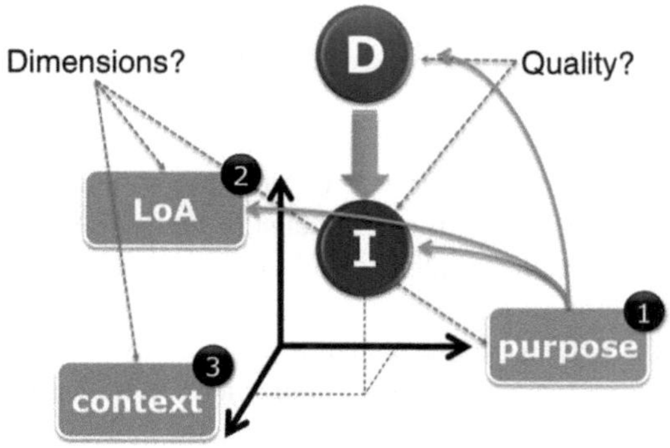

Fig. 15.4 Data become information within a context, at a Loa, chosen for a purpose

Admittedly, all this may be a bit hard to digest, so here are four examples that should clarify the point.

Some data are supposed to be re-purposable since their collection. In the UK, the 2011 Census population estimates were examined through a quality assurance (QA) process "to ensure that users of census data have confidence in the *quality* and *accuracy* of the information" (my italics).[13] The Census Data Quality Assurance Strategy stated that

> The proposed strategy reflects a considered balance between data relevance, accuracy, timeliness and coherence. The data accuracy that can be achieved reflects the methods and resources in place to identify and control data error and is therefore constrained by the imperative for timely outputs. 'Timeliness' refers to user requirements and the guiding imperative for the 2011 Census is to provide census population estimates for rebased 2011 mid-year population estimates in June 2012. 'Coherence' refers to the internal integrity of the data, including consistency through the geographic hierarchy, as well as comparability with external (non-census ONS) and other data sources. This includes conformity to standard concepts, classifications and statistical classifications. The 2011 Data Quality Assurance Strategy will consider and use the best available administrative data sources for validation purposes, as well as census time series data and other ONS sources. A review of these sources will identify their relative strengths and weaknesses. The relevance of 2011 Census data refers to the extent to which they meet user expectations. A key objective of the Data Quality Assurance Strategy is to anticipate and meet user expectations and to be able to justify, empirically, 2011 Census outcomes. To deliver coherent data at acceptable levels of accuracy that meet user requirements and are on time, will demand QA input that is carefully planned and targeted. Census (2011), pp. 8–9

Apart from a questionable distinction between information *quality* and *accuracy* (as if accuracy were something else from IQ), overall the position expressed in the document (and in the citation above) is largely reasonable. However, I specify "largely" because the statement about the "key objective" of anticipating and meeting user expectations remains quite problematic. It shows a lack of appreciation for the complexity of the "fit for purpose" requirement. The objective is problematic because it is unrealistic: such expectations are unpredictable, that is, the purpose for

[13] http://www.ons.gov.uk/ons/guide-method/census/2011/how-our-census-works/how-we-took-the-2011-census/how-we-processed-the-information/data-quality-assurance/index.html

which the information collected in the census is supposed to be fit may change quite radically, thus affecting the fitness itself. To understand why, consider a second example.

Some data are not supposed to be re-purposed, but they are, and sometimes for evil goals, which were not anticipated. This is our second example. There is a puzzling fact about the Holocaust in the Netherlands: 74 % of the "full" Jews (according to the Nazi definition) living in the Netherlands died. In relative terms, this was the highest death toll in any West European Jewish community, including Germany itself. One of the plausible explanations (Luebke and Milton (1994)) is that the Netherlands had an excellent census, which provided plenty of accurate and reliable information about people's religious beliefs and home addresses.

Some data are re-purposed more or less successfully, to purse goals that could not have been envisaged when the data were first produced. This is our third example. In the UK, postcodes for domestic properties refer to up to 100 properties in contiguous proximity. Their original purpose was to aid the automated sorting of the mail. That was what the postcode information was fit for (Raper et al. (1992)). Today, they are used to calculate insurance premiums, designate destinations in route planning software, and allocate different levels of public services, depending on one's location (postcode) in such crucial areas such as health and social services and education (the so-called postcode lottery). In short, the information provided by postcodes has been radically repurposed, and keeps being repurposed, leading to a possible decline in fitness. For instance, the IQ of postcodes is very high when it comes to delivering mail, but rather poorer if route planning is in question, as many drivers have experienced who expect, mistakenly, a one-to-one relation between postcodes and addresses.

Finally, some data are re-purposed despite the fact that we know that the new usage is utterly improper and could be risky. This is our last example. Originally, and still officially, Social Security Numbers (SSNs) in the US were intended for only one purpose: tracking a worker's lifetime earnings in order to calculate retirement benefits. So much so that, between 1946 and 1972, SSNs carried the following disclaimer: "For social security purposes not for identification". However, SSNs are the closest thing to a national ID number in the US, and this is the way they are regularly used today, despite being very "unfit" for such a purpose, especially in terms of safety (United States Federal Trade Commission (2010)).

15.6 A Bi-categorical Approach to Information Quality

The previous examples illustrate the fact that one of the fundamental problems with IQ is the tension between, on the one hand, *purpose–depth* and, on the other hand, *purpose–scope*. This point is also stressed by Illari in her chapter. Ideally, high quality information is information that is fit for both: it is optimally fit for the specific purpose/s for which it is elaborated (purpose–depth), and is also easily re-usable for new purpose/s (purpose–scope). However, as in the case of a tool, sometimes the

IQ DIMENSIONS	IQ CATEGORIES: Mail delivery	IQ CATEGORIES: Navigation
Accuracy	1	0.8
Objectivity	1	1
Accessibility	0.9	0.9
Security	1	1
Relevancy	1	0.9
Timeliness	1	1
Interpretability	0.8	0.7
Understandability	1	0.9

Fig. 15.5 Example of bi-categorical IQ analysis

better some information fits its original purpose, the less likely it seems to be re-purposable, and *vice versa*. The problem is that not only may these two requirements be more or less compatible, but that we often forget this (that is, that they may be), and speak of purpose-fitness as if it were a single feature, synonymous for information quality, to be analysed according to a variety of taxonomies. Recall the statement from the Census Data Quality Assurance Strategy. This is a mistake. Can it be avoided? A detailed answer would require more space than is available here, so let me offer an outline of a promising strategy in terms of a bi-categorical approach, which could be implemented through some user-friendly interfaces.

The idea is simple. First, one must distinguish between the purpose/s for which some information is originally *produced* (P-purpose) and the (potentially unlimited) purpose/s for which the same information may be *consumed* (C-purpose). These two categories somewhat resemble but should not be confused with what in the literature on IQ are known as the "intrinsic" vs. "extrinsic" categories. In our previous example, one would distinguish between postcodes as information fit for the purpose of mail delivery – the P-purpose – and postcodes as information fit for other uses, say driving navigation – the C-purpose. This bi-categorical approach could be introduced in terms of a simple Cartesian space, represented by P-purpose $=x$ and C-purpose $=y$, in such a way that, for any information I, I must have two values in order to be placed in that space. This in turn allows one to analyse a variety of dimensions, such as accuracy, objectivity, accessibility, etc. in a purpose-oriented way (see Fig. 15.5 for an illustration).

Second, one could then compare the quality of some information with respect to purpose P and with respect to purpose C, thus identifying potential discrepancies. The approach lends itself to simple visualisations in terms of radar charts (see Fig. 15.6, for an illustration based on the data provided in Fig. 15.5).

The result would be that one would link IQ to a specific purpose, instead of talking of IQ as fit-for-purpose in absolute terms.

There are many senses in which we speak of "fit for purpose". A pre-Copernican, astronomical book would be of very bad IQ, if its purpose were to instruct us on the nature of our galaxy, but it may be of very high IQ if its purpose is to offer evidence

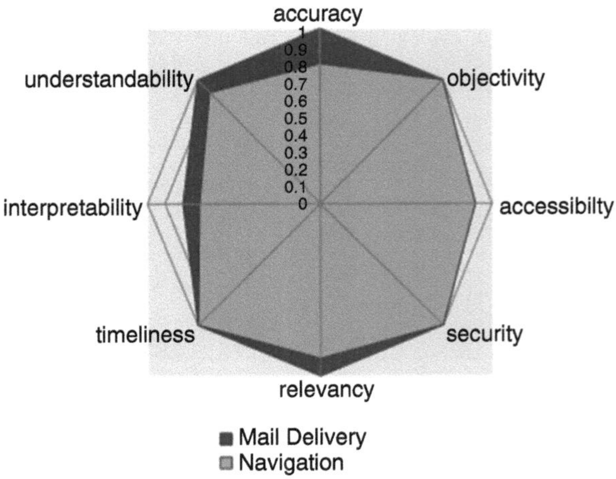

Fig. 15.6 Graph of a bi-categorical IQ analysis

about the historical development of Ptolemaic astronomy. This is not relativism; it is a matter of explicit choice of the purpose against which the value of some information is to be examined. Re-purposing is largely a matter of intelligence, not of mechanised procedures. You need to have a bright idea to re-purpose some data. Here is an elegant example. In the first study reliably showing that a lunar rhythm can modulate sleep structure in humans, Cajochen et al. (2013) were able to

> exclude confounders such as increased light at night or the potential bias in perception regarding a lunar influence on sleep

because they used data that had been collected for different purposes, and

> retrospectively analyzed sleep structure, electroencephalographic activity during non-rapid-eye-movement (NREM) sleep, and secretion of the hormones melatonin and cortisol found under stringently controlled laboratory conditions in a cross-sectional setting. At no point during and after the study were volunteers or investigators aware of the a posteriori analysis relative to lunar phase.

It was a clear case of successful repurposing. Once the repurposing step is carefully taken, then a bi-categorical approach is compatible with, and can be supported by quantitative metrics, which can (let users) associate values to dimensions depending on the categories in question, by relying on solutions previously identified in the literature on IQ: metadata, tagging, crowd sourcing, peer-review, expert interventions, reputation networks, automatic refinement, and so forth. The main advantage of a bi-categorical approach is that it clarifies that the values need not be the same for different purposes. It should be rather easy to design interfaces that enable and facilitate such interactive selection of purposes for which IQ is evaluated. After all, we know that we have plenty of information systems that are syntactically smart and users who are semantically intelligent, and a bi-categorical approach may be a good way to make them work together successfully.

References

Al-Hakim, L. (2007). *Information quality management: Theory and applications*. Hershey: Idea Group Pub.

Batini, C., & Scannapieco, M. (2006). *Data quality – Concepts, methodologies and techniques*. Berlin/New York: Springer.

Borges, J. L. (1964). *Other inquisitions, 1937–1952*. Austin: University of Texas Press.

Cajochen, C., Altanay-Ekici, S., Münch, M., Frey, S., Knoblauch, V., & Wirz-Justice, A. (2013). Evidence that the lunar cycle influences human sleep. *Current Biology: CB, 23*(15), 1485–1488.

Census. (2011). *Census data quality assurance strategy*. http://www.ons.gov.uk/ons/guide-method/census/2011/the-2011-census/processing-the-information/data-quality-assurance/2011-census---data-quality-assurance-strategy.pdf

English, L. (2009). *Information quality applied: Best practices for improving business information, processes, and systems*. Indianapolis: Wiley.

Floridi, L. (2008). The method of levels of abstraction. *Minds and Machines, 18*(3), 303–329.

Floridi, L. (2011). *The philosophy of information*. Oxford: Oxford University Press.

Herzog, T. N., Scheuren, F., & Winkler, W. E. (2007). *Data quality and record linkage techniques*. New York: Springer.

Lee, Y. W., Pipino, L. L., Funk, J. D., & Wang, R. Y. (2006). *Journey to data quality*. Cambridge, MA: MIT Press.

Luebke, D. M., & Milton, S. (1994). Locating the victim: An overview of census-taking, tabulation technology and persecution in Nazi Germany. *Annals of the History of Computing, IEEE, 16*(3), 25–39.

Maydanchik, A. (2007). *Data quality assessment*. Bradley Beach: Technics Publications.

McGilvray, D. (2008). *Executing data quality projects ten steps to quality data and trusted information*. Amsterdam/Boston: Morgan Kaufmann/Elsevier.

Olson, J. E. (2003). *Data quality the accuracy dimension*. San Francisco: Morgan Kaufmann Publishers.

Raper, J. F., Rhind, D., & Shepherd, J. F. (1992). *Postcodes: The new geography*. Harlow: Longman.

Redman, T. C. (1996). *Data quality for the information age*. Boston: Artech House.

Theys, P. P. (2011). *Quest for quality data*. Paris: Editions TECHNIP.

Tozer, G. V. (1994). *Information quality management*. Oxford: Blackwell.

United States Federal Trade Commission. (2010). *Social security numbers and id theft*. New York: Nova.

United States. Congress. House. Committee on Government Reform. Subcommittee on Regulatory Affairs. (2006). *Improving information quality in the federal government: Hearing before the subcommittee on regulatory affairs of the committee on government reform, house of representatives, one hundred ninth congress, first session, july 20, 2005*. Washington, DC: U.S. G.P.O.

Wang, R. Y. (1998). A product perspective on total data quality management. *Communication of the ACM, 41*(2), 58–65.

Wang, Y. R., & Kon, H. B. (1992). *Toward quality data: An attributes-based approach to data quality*. Cambridge, MA: MIT Press.

Wang, R. Y., Pierce, E. M., Madnik, S. E., Zwass, V., & Fisher, C. W. (Eds.). (2005). *Information quality*. Armonk/London: M.E. Sharpe.

References

MIX
Papier aus verantwortungsvollen Quellen
Paper from responsible sources
FSC® C105338

If you have any concerns about our products,
you can contact us on
ProductSafety@springernature.com

In case Publisher is established outside the EU,
the EU authorized representative is:
Springer Nature Customer Service Center GmbH
Europaplatz 3, 69115 Heidelberg, Germany

Printed by Libri Plureos GmbH
in Hamburg, Germany